Unconventional Water Resources

Manzoor Qadir · Vladimir Smakhtin ·
Sasha Koo-Oshima · Edeltraud Guenther
Editors

Unconventional Water Resources

Springer

Editors
Manzoor Qadir
United Nations University Institute for Water, Environment and Health (UNU-INWEH)
Hamilton, ON, Canada

Sasha Koo-Oshima
Land and Water Division, Food and Agriculture Organization of the United Nations (FAO)
Rome, Italy

Vladimir Smakhtin
United Nations University Institute for Water, Environment and Health (UNU-INWEH)
Hamilton, ON, Canada

Edeltraud Guenther
United Nations University Institute for Integrated Management of Material Fluxes and of Resources (UNU-FLORES)
Dresden, Germany

ISBN 978-3-030-90145-5 ISBN 978-3-030-90146-2 (eBook)
https://doi.org/10.1007/978-3-030-90146-2

This Springer imprint is published by the registered company Springer Nature Switzerland AG
The registered company address is: Gewerbestrasse 11, 6330 Cham, Switzerland

Foreword

Water is a keystone element in achieving the 2030 Sustainable Development Agenda with two unique Sustainable Development Goals (SDGs) addressing water in oceans, seas, and marine resources (SDG 14), and inland on water and sanitation (SDG 6). In addition, several other SDG targets explicitly or implicitly related to water, including those in SDG 2 (zero hunger), SDG 3 (healthy lives), SDG 11 (sustainable cities), and SDG 13 (climate action)—to mention a few.

Inland freshwater resources are threatened by growing water scarcity, which is recognized as a key challenge to sustainable development and a potential cause of social unrest within and between countries. Water scarcity, together with other water insecurity factors, could reshape the world, not the least by intensifying the already significant involuntary human migration flows. As water scarcity intensifies in dry and overpopulated areas, we are in danger of leaving this challenge to future generations who will have to deal with the consequences of today's practices or lack of action.

Conventional water sources like snowfall, rainfall, river runoff, and shallow groundwater are being affected by climate change, and supplies are shrinking as demand in water-scarce areas intensifies. We therefore need not just to improve water use efficiency, but also—to look beyond conventional water resources, if we are to avoid water scarcity becoming a chronic global challenge.

Water-scarce countries need a fundamental change in planning and management through the creative exploitation of alternate, unconventional water resources for food production, livelihoods, ecosystems, climate change adaption, and overall sustainable development. From Earth's seabed to its upper atmosphere, the world is blessed with a variety of unconventional water resources that can be tapped. Their potential remains vastly under-explored, although recent years have seen some sporadic examples of exploiting such unconventional options to augment water supplies for different uses.

The time is coming to harness the full potential of unconventional water resources with continued applied research and technological advances, effective policy messages, private sector involvement, and capacity development. With this, the prospects of addressing global, regional, and local water scarcity can be greatly enhanced for the benefit of billions of people.

Based on the most up to date information and data and with the contributions from world renowned scientists, experts, and practitioners, this book showcases the potential of different types of unconventional water resources, such as harvesting water from air and on the ground; tapping offshore and onshore deep groundwater; reusing used water; moving water physically to water-scarce areas; and developing new water.

This is the first ever book that offers a pertinent and credible analysis of all major types of unconventional water resources comprehensively in terms of their biophysical aspects and extent along with related policy, institutional, economics, gender, and environmental tradeoffs.

David Malone
Rector
United Nations University
Tokyo, Japan

Contents

Editors and Contributors

About the Editors

Manzoor Qadir is an environmental scientist working on water-related sustainable development through contribution to policy, institutional and biophysical aspects of unconventional water resources, water recycling and safe reuse, water quality and environmental health, and water and food security under changing climate. Manzoor has led teams of eminent professionals as Coordinating Lead Author to contribute to global assessments, such as the Comprehensive Assessment of Water Management in Agriculture, and the International Assessment of Agricultural Science and Technology for Development. He has implemented multidisciplinary projects on water and land management with major work in the Middle East and North Africa, Central Asia, South Asia, and Sub-Saharan Africa regions. In addition to supervising post-doctoral fellows, postgraduate students, and interns, he has undertaken several international and regional capacity development initiatives such as organizing knowledge bridging workshops and training courses for young researchers. Before joining United Nations University Institute for Water, Environment and Health (UNU-INWEH) in Canada, Manzoor previously held professional positions at the International Center for Agricultural Research in the Dry Areas (ICARDA) and International Water Management Institute (IWMI); Alexander-von-Humboldt Fellow and Visiting Professor at Justus-Liebig University, Germany; and Associate Professor at the University of Agriculture, Pakistan.

Vladimir Smakhtin has over 35 years of experience as a Researcher and Manager in the broad area of water resources. He holds a Ph.D. in hydrology and water resources from the Russian Academy of Sciences. He worked at Rhodes University and Council for Scientific and Industrial Research (CSIR) in South Africa, and as a research program director at IWMI headquartered in Sri Lanka. His experience spreads across agricultural and environmental water management, low-flow and drought analyses, assessment of basin development and climate change impacts on water availability, provision of hydrological information for data-poor regions, water-related disaster risk management. He initiated, managed, or contributed to numerous research initiatives in over 20 countries worldwide, including the state program for mitigating the consequences of Chernobyl Accident in Russia and Ukraine, development of Ecological Reserve methods in South Africa, a first global analysis of ecosystem water requirements, and several projects focusing on managing water resources variability through enhanced surface and groundwater storage. Vladimir has authored over 200 publications and has consulted for several national governments and international organizations including the Department of Water Affairs of South Africa, Ontario Ministry of Natural Resources, World Commission on Dams, WB, ADB, IUCN, and UNEP.

Sasha Koo-Oshima has nearly 35 years of experience in international assistance and policy development in agriculture water and environment/natural resource management. Currently, she is the Deputy Director and Head of Water at the UN Food & Agriculture Organization (FAO), leading programs on sustainable land and water management and governance, geospatial data, integrated water resources management with linkages to climate, energy, health, and food and nutrition security. She formerly served as Senior Advisor to the leadership at the U.S. EPA, and as Secretariat of the Organization for Economic Co-operation and Development (OECD), where she directed and managed international environment and water programs, strategized in the development of various sustainable investment mechanisms such as wastewater financing in the GEF Caribbean Wastewater Revolving Fund and the Millennium Challenge Corporation Cabo Verde Compact on sustainable water infrastructure financing. She is now on the Governing Boards of the World Water Council and the CGIAR's Water Land Ecosystems, and UNEP Global Partnerships on nutrient and wastewater management. She published, sponsored, and peer-reviewed extensively on international water issues, such as the UN World Water Development Reports, FAO-WHO Wastewater Reuse Guidelines for Agriculture, FAO reports on Agriculture-Nature Based Solutions, Wealth of Waste: The Economics of Wastewater Reuse, Desalination and Agriculture, Agriculture Water Quality Guidelines for China, and the OECD country Water Governance reviews.

Edeltraud Guenther is an internationally recognized economist and professor of business management. She is currently serving as the Director of the United Nations University Institute for Integrated Management of Material Fluxes and of Resources (UNU-FLORES), based in Dresden, Germany, and has held the position of Chair of Business Management, Sustainability Management, and Environmental Accounting, at the Faculty of Business and Economics at Technische Universität Dresden (TU Dresden) since 1996. Additionally, Edeltraud is the founding member and current Chair of the Center for Performance and Policy Research in Sustainability Measurement and Assessment (PRISMA) and has undertaken visiting professorships at Namibia University of Science and Technology (NUST), Kobe University, and University of Virginia. Edeltraud was also one of the establishing Directors and the first Chair for the UNU Water Network, which was initiated in 2019. In 2020, she was appointed UNU Senior Official for the Environmental Management Group (EMG).

Edeltraud's extensive research covers sustainability management, environmental accounting, and management control systems, with an emphasis on corporate responsibility, life cycle assessment, resilience, and sustainability assessment—particularly the question of 'how does it pay to be sustainable'?

Contributors

Ali M. Abshaev Hail Suppression Research Center "ANTIGRAD", Nalchik, Russia

Kim Andersson Stockholm Environment Institute, Stockholm, Sweden

Águeda Bellver-Domingo Institute of Local Development (ILD-WATER), Water Economics Group, University of Valencia, Valencia, Spain

Serena Caucci United Nations University Institute for Integrated Management of Material Fluxes and of Resources (UNU-FLORES), Dresden, Germany

Gueladio Cisse Department of Epidemiology and Public Health (EPH), Swiss Tropical and Public Health Institute (Swiss TPH), University of Basel, Basel, Switzerland

Aaron L. M. Daigh Soil Science Department, North Dakota State University, Fargo, ND, USA

Marlos De Souza Land and Water Division, Food and Agriculture Organization (FAO), Rome, Italy

Rebecca L. Farnum Syracuse University, London, UK

Andrea Flossmann Université Clermont-Auvergne Laboratoire de Météorologie Physique, Aubière Cedex, France

Edeltraud Guenther United Nations University Institute for Integrated Management of Material Fluxes and of Resources (UNU-FLORES), Dresden, Germany

Francesc Hernández-Sancho Institute of Local Development (ILD-WATER), Water Economics Group, University of Valencia, Valencia, Spain

Sasha Koo-Oshima Land and Water Division, Food and Agriculture Organization (FAO), Rome, Italy

Birguy Lamizana Marine and Freshwater Branch, Ecosystem Division, UN Environment Programme, Nairobi, Kenya

Christina Leb World Bank, Washington, D.C., WA, USA

Clever Mafuta GRID-Arendal, Arendal, Norway

Olfa Mahjoub National Research Institute for Rural Engineering, Water, and Forestry (INRGREF), Laboratory of Agricultural Sciences and Techniques (LR16INRAT05), National Institute of Agronomic Research of Tunisia (INRAT), University of Carthage, Ariana, Tunisia

Renée Martin-Nagle Environmental Law Institute, A Ripple Effect pllc, Ebensburg, PA, USA

A. Micallef Marine Geology and Seafloor Surveying, Department of Geosciences, University of Malta, Msida, Malta

J. D. Oster Environmental Sciences Department, University of California, Riverside, CA, USA

Theib Y. Oweis International Center for Agricultural Research in the Dry Areas (ICARDA), Amman, Jordan

Mark A. Person New Mexico Tech, Hydrology Program, Socorro, NM, USA

Thara Prabhakaran Indian Institute of Tropical Meteorology, Pune, Maharashtra, India

Manzoor Qadir United Nations University Institute for Water, Environment and Health (UNU-INWEH), Hamilton, ON, Canada

Nigel W. T. Quinn Climate & Ecosystem Sciences Division, Berkeley National Laboratory, Berkeley, CA, USA

Francesc-Hernandez Sancho Institute of Local Development (ILD-WATER). Water Economics Group, University of Valencia, Valencia, Spain

Nafis Sazeed New Mexico Tech, Hydrology Program, Socorro, NM, USA

Elia Scudiero U.S. Salinity Laboratory, United States Department of Agriculture—Agricultural Research Service, Riverside, CA, USA;
Environmental Sciences Department, University of California, Riverside, CA, USA

Steven T. Siems School of Mathematics, Monash University, Melbourne, VIC, Australia

Nisal Siriwardana United Nations University Institute for Water, Environment and Health (UNU-INWEH), Hamilton, ON, Canada

Vladimir Smakhtin United Nations University Institute for Water, Environment and Health (UNU-INWEH), Hamilton, ON, Canada

Sarah Tessendorf Research Applications Laboratory, National Center for Atmospheric Research, Boulder, CO, USA

Nikolay Voutchkov Water Globe Consultants, LLC, Florida, USA

Zhanyu Yao Chinese Academy of Meteorological Sciences, Beijing, People's Republic of China

Part I
Setting the Scene

Chapter 1
Global Water Scarcity and Unconventional Water Resources

Manzoor Qadir, Vladimir Smakhtin, Sasha Koo-Oshima, and Edeltraud Guenther

Abstract Freshwater scarcity is a global systemic risk. Its impacts reach far beyond socio-economic and environmental challenges and influence people's livelihoods and wellbeing. As water scarcity deepens in arid and overpopulated regions, there is a need to explore water supply options beyond conventional water resources—snowfall, rainfall, river runoff, and easily accessible groundwater—since they already often fall short of meeting the growing water demands. Water-scarce countries need a radical re-thinking of water resource planning and management and, among other options, turn to unconventional water sources for food production, livelihood, ecosystems, and overall requirements—for sustainable development. Such water resources exist ranging from the Earth's seabed to its upper atmosphere. Securing access to them requires specific technologies and innovations. This introductory chapter takes stock of water scarcity trends, puts forward unconventional water resources as a critical response to global water scarcity, provides insights into linkages with water-related sustainable development, and introduces the book's parts and chapters.

Keywords Water scarcity · Water stress · Arid areas · Water resources · Sustainable development

M. Qadir (✉) · V. Smakhtin
United Nations University Institute for Water, Environment and Health (UNU-INWEH), 204-175 Longwood Road South, Hamilton L8P 0A1, ON, Canada
e-mail: Manzoor.Qadir@unu.edu

V. Smakhtin
e-mail: vladimir.smakhtin@unu.edu

S. Koo-Oshima
Land and Water Division, Food and Agriculture Organization (FAO), Viale delle Terme di Caracalla, 00153 Rome, Italy
e-mail: Sasha.Koo@fao.org

E. Guenther
United Nations University Institute for Integrated Management of Material Fluxes and of Resources (UNU-FLORES), Ammonstrasse 74, Dresden 01067, Germany
e-mail: Guenther@unu.edu

M. Qadir et al. (eds.), *Unconventional Water Resources*,
https://doi.org/10.1007/978-3-030-90146-2_1

1.1 Water Scarcity: Evolution of Concepts

Water has been consistently ranked among the top five global risks for several years in terms of its impacts because it sits at the heart of the growing number of complex and interconnected challenges that the world at large is facing (World Economic Forum 2020). The water crisis, incorporated into the revised-risk category, "Natural Resource Crises" in the Global Risks Report 2021, remains among the top five risks. This is true despite infectious diseases taking over the first place as a global risk in 2021 (World Economic Forum 2021). The water crisis, translated as water scarcity, is also recognized as a potential cause of social unrest and conflict within and between countries (UN-Water 2020).

There is a growing disparity between water resources availability and human population because freshwater resources and population are unevenly distributed worldwide. Increasing competitions among agricultural, domestic, industrial, and energy sectors make water scarcity prominent in areas characterized as water-stressed or expected to become so in the future (Wada et al. 2011; Liu et al. 2017). Thus, understanding the magnitude of water scarcity and its potential impacts is essential for formulating responsive policies and practices at different scales.

The evolution of the water scarcity concept started in the 1980s with a population-driven water scarcity indicator and emerged into assessments that also included water quality and water allocation for food and agriculture and ecosystems (Falkenmark 1986; Shiklomanov 1991; Seckler et al. 1998; Smakhtin et al. 2004; Rijsberman 2006; Hanasaki et al. 2008; Rockström et al. 2009; Zeng et al. 2013; Mekonnen and Hoekstra 2016; Liu et al. 2017; van Vliet et al. 2021; FAO 2020). Publications on different aspects of water scarcity have increased gradually but spiked since the beginning of the 21st century, when water scarcity intensified and started to noticeably affect people's livelihoods and well-being, particularly in water-scarce areas.

Formal quantification of water resources per capita or the status of water scarcity at the national level began with developing the water stress index (WSI) by linking food security and freshwater availability (Falkenmark 1986). Commonly known as the *Falkenmark Indicator*, the WSI was initially defined in terms of the number of people who compete for a single-flow unit of water (Falkenmark et al. 1989), i.e., population-driven water availability described as annual renewable water resources (ARWR) per capita. Falkenmark et al. (1989) recommended 1,700 m^3 of ARWR per capita as the threshold below which water scarcity starts to manifest itself, based on the estimates of water requirements in the household, agricultural, industrial, and energy sectors. Countries with ARWR per capita exceeding 1,700 m^3 were considered water sufficient (no scarcity), and those with less than 1,700 m^3 were termed water stressed. Countries with ARWR per capita falling below 1,000 m^3 were categorized as extremely water scarce, and those with ARWR per capita below 500 m^3 in the grip of absolute water scarcity.

Due to its simplicity, the *Falkenmark Indicator* has been the currency for numerous water scarcity-related debates and publications. The advantage of using WSI is that the data to estimate ARWR per capita is readily available. It requires only knowing an average annual total renewable water resources (TRWR) at the national level and the population in a specific year in the same country. Despite such an advantage, WSI provides only coarse national-level estimates that mask water scarcity variability at finer scales. Thus, areas within a country may fall into different categories of WSI, perhaps ranging from no water scarcity to absolute water scarcity.

While WSI is based on water supply, other studies have focused on water demand. Shiklomanov (1991) compared nationwide annual water availability with national water demand in the agricultural, industrial, and domestic sectors. Using data from Shiklomanov (1991), Raskin et al. (1997) considered annual water withdrawals from rivers, streams, and groundwater as a percentage of total available water resources. Based on the resulting *Water Exploitation Index* (WEI), it was proposed that a country is "water scarce" if the annual water withdrawals are in the range 20–40% of the total water resources available, and "severely water-scarce" if this number exceeds 40% (Raskin et al. 1997; European Environment Agency 2005).

In the late 1990s, the International Water Management Institute (IWMI) developed a different concept for water scarcity assessment based on access and water infrastructure development (Seckler et al. 1998). The *IWMI approach* differentiated water-scarce areas into four categories: (1) Little or no water scarcity—abundant water resources relative to use, with less than 25% of the water from rivers withdrawn for human purposes; (2) Physical water scarcity—water resources development is approaching or has exceeded sustainable limits. More than 75% of river flows are used for agriculture, industry, and domestic purposes; (3) Approaching physical water scarcity—more than 60% of river flows withdrawn. These basins will soon experience physical water scarcity; and (4) Economic water scarcity—human, institutional, and financial capital limit access to water even though the water in nature is available locally to meet human demands. Water resources are abundant relative to water use, with less than 25% of the water from rivers withdrawn for human purposes, but malnutrition exists (Fig. 1.1).

Sullivan et al. (2003) assessed whether individuals are water secure at the household and community levels by developing the *Water Poverty Index* (WPI) based on (1) water resources explained by total amount of surface and groundwater; (2) access to water for domestic, agricultural, and industrial sectors; (3) water use by these sectors; (4) effectiveness of people's ability to manage water; and (5) water-related environmental aspects. They used the weighted averages of these components by first standardizing each component so that it falls within the range 0–100; thus, the resulting WPI value is also between 0 and 100. The highest value, 100, is the optimal situation (or the lowest level of water poverty), while 0 is the worst. The index has been applied at the community level in selected areas in Sri Lanka, Tanzania, and South Africa (Rijsberman 2006).

Smakhtin et al. (2004) redefined the water stress indicator, explicitly including environmental flow requirements (EFR) into it. WSI was defined as the ratio of total water withdrawals to "utilizable water"; later, the definition was changed to mean the difference between the total water resources and estimated EFR that was "set aside"

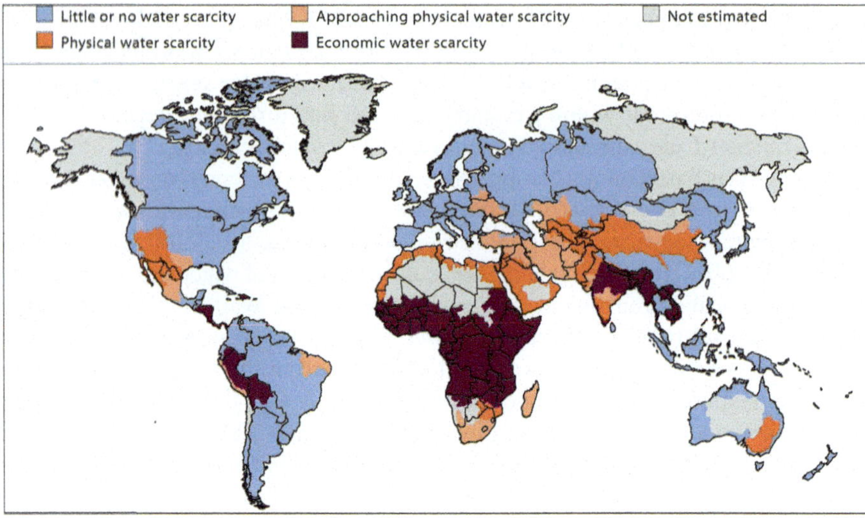

Fig. 1.1 Global map of physical and economic water scarcity (Comprehensive Assessment of Water Management in Agriculture 2007)

to protect freshwater ecosystems. The estimated EFR needed to maintain freshwater-dependent ecosystems in fair conditions ranged between 20 and 50% of the mean annual river flow. Moreover, the estimated WSI in some river basins or parts thereof was found to be close to or even exceeding unity, pointing to the fact that water withdrawals were already tapping into EFR (Fig. 1.2). WSI in this form has become the prototype of the SDG indicator 6.4.2 on water stress (Sood et al. 2017; FAO 2019).

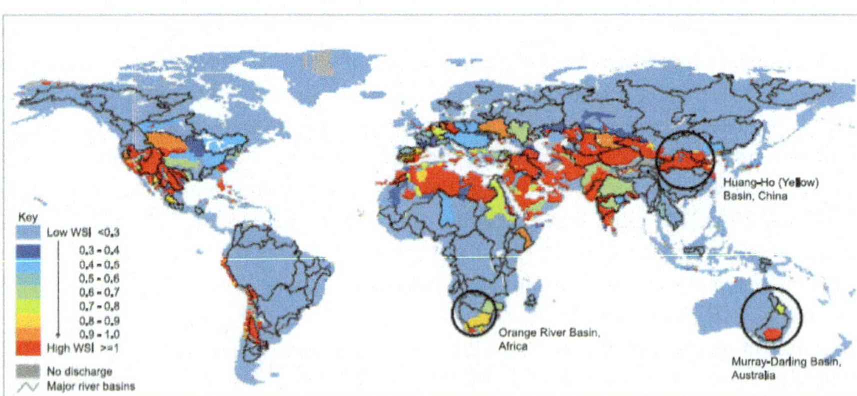

Fig. 1.2 Global map of environmental water stress index—total water withdrawals as a proportion of water available once the environmental flow requirements (EFR) are satisfied (Smakhtin et al. 2004)

Considering temporal variations in water scarcity, Hanasaki et al. (2008) introduced the *Cumulative Abstraction to Demand* (CAD) index as the ratio of cumulative daily freshwater abstraction from a water body to cumulative daily potential water demand for agricultural, industrial, and domestic uses. It was assumed that if CAD falls below unity, water scarcity can occur. CAD reflects increases in water-scarce conditions during dry months. However, it is data-intensive and so far has found to be of limited use.

Mekonnen and Hoekstra (2016) assessed *Blue Water Scarcity* on monthly intervals at the 30 × 30 arc min spatial resolution. It was defined as the ratio of blue water consumption (net water withdrawal) to the total blue water availability in a grid cell. The later was calculated as the sum of runoff generated within a cell plus runoff generated in all upstream grid cells and minus—the EFR and blue water consumption in upstream cells. The approach holds good potential; however, a rather simplistic assumption of very high EFR (80% of the total water resources) across river basins, may result in inflated water-scarcity estimates.

While most water-scarcity assessments have focused on water quantity, recent estimates have also included water quality and water allocation for EFR. Zeng et al. (2013) developed an integrated indicator "*QQE*" based on water quantity, water quality, and EFR and expressed it as the sum of a quantity-induced indicator and a quality-induced indicator. Considering the development of new water resources such as desalinated water and treated wastewater, van Vliet et al. (2021) built on *QQE* by subtracting these water resources from the water demand. They found that water scarcity levels and the percentage of people affected by severe water scarcity were substantially higher when one considers both water quantity and quality (on average 40%) rather than solely water quantity (30%).

1.2 Water-Scarcity Trends

Since freshwater resources and population densities are unevenly distributed across the world, there are significant variations in ARWR per capita ranging from more than 30,000 m^3 in countries, such as Bhutan, Guyana, and Iceland, to less than 100 m^3 in countries, such as Yemen, Algeria, and Saudi Arabia. Based on a recent assessment of water scarcity stemming from population growth across regions and countries (Baggio et al. 2021), almost half of the countries (87 out of 180) are projected to become water-scarce by 2050, i.e., ARWR per capita will drop below 1,700 m^3. The number of countries with absolute water scarcity is projected to increase from 25 in 2015 to 45 by 2050. Degraded water quality may result in unsuitability for sectoral water uses, thus exacerbating water-scarcity levels (van Vliet et al. 2021), and may contribute to the severity of water-scarce hotspots globally (Zeng et al. 2013; Liu et al. 2016).

After the Middle East and North Africa region, sub-Saharan Africa may become the next hotspot of water scarcity (Baggio et al. 2021). The highest average rate of decline in ARWR per capita (2.4%) is expected in sub-Saharan Africa, followed by the Middle East and North Africa (1.1%), Asia (0.8%), and Latin America and the Caribbean (0.5%). Europe is the only global region where the average rate of change in ARWR per capita may be positive in the next 30 years, due to the expected decrease in population in the region. Regarding the total change in ARWR for 2020–2050, the water-resources allocation per capita is expected to reduce to half in sub-Saharan Africa as compared with the current levels. In the Middle East and North Africa, the total reduction in ARWR will be 33.2% followed by 24.3% in Asia and 17.8% in Latin America and the Caribbean (Fig. 1.3a).

Water scarcity has been a chronic challenge and a pressing issue in recent decades in the Middle East and North African region. The region contains only 1% of global water resources and 6% of the global population (World Bank 2018). However, the countries in the region are expected to further approach absolute water scarcity. The worsening water scarcity in the region also indicates that the countries already struggling with scarce freshwater resources for several decades will face additional water challenges. In Asia, Afghanistan, Tajikistan, Timor–Leste, and Pakistan are the top four countries so challenged, with an annual decline in ARWR per capita of

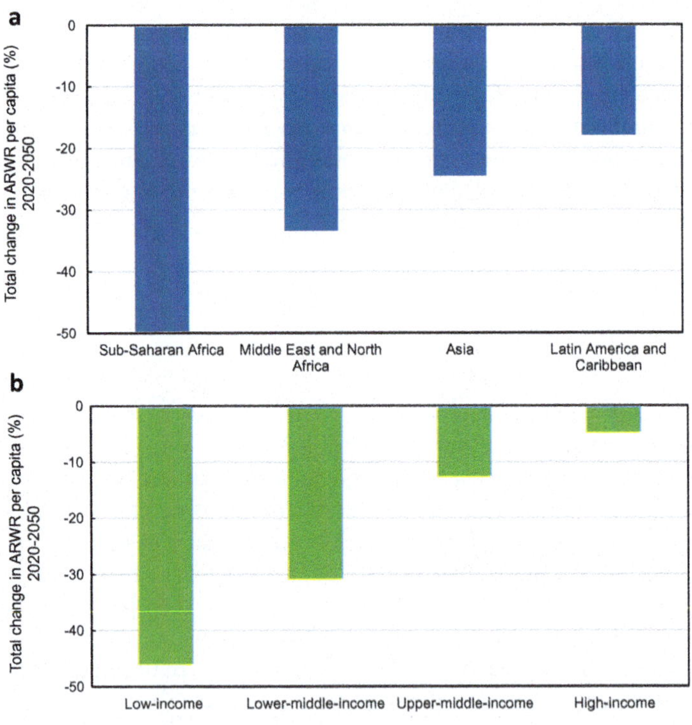

Fig. 1.3 Total change in annual renewable water resources (ARWR) per capita across countries within global regions **a** and in countries grouped into low-income, lower-middle-income, upper-middle-income, and high-income economies **b** during 2020–2050 (Based on the data from Baggio et al. 2021)

1.9–2.5%. The total decrease in the ARWR per capita in these countries by 2030 and 2050 will be 26–32% and 48–59%, respectively (Baggio et al., 2021).

The world's top 15 countries with the highest rate of decline in ARWR per capita in the next 30 years are in sub-Saharan Africa. Water-scarce countries such as Niger, Angola, Somalia, Uganda, Chad, Burundi, Tanzania, and Burkina Faso, for instance, are projected to face intensified water scarcity, with many of them reaching absolute water scarcity. Today's water-rich countries, such as the Democratic Republic of Congo, Mali, Mozambique, Zambia, and the Gambia, are projected to have an ARWR <1,700 m^3 per capita by 2050. Although Equatorial Guinea and Guinea are projected to remain water-rich until 2050, they will experience a significant decline in ARWR per capita (Baggio et al. 2021).

Population growth is expected to lead to an unprecedented drop in water resources in countries with insufficient financial resources (Fig. 1.3b), with low-income countries projected to have a total average drop in ARWR per capita of 45.8%, followed by lower-middle-income countries (30.6%), upper-middle-income countries (12.4%), and high-income countries (4.6%). Such trends vividly suggest that rapid population growth in low-income and lower-middle-income countries will make water problems more complicated to address and may hinder the efforts to mitigate water scarcity (Baggio et al. 2021). In addition, water scarcity combined with water–quality deterioration in low-income and lower-middle-income countries may limit possible solutions because additional investments will be needed to improve the quality of the scarce water resources in such groups of countries (Liu et al. 2016). A recent reassessment of water quality for irrigation suggests that the agricultural use of poor-quality water will increase, and the sustainability of irrigated agriculture will become an even more challenging issue (Qadir et al. 2021).

Assessments of the water-scarcity trends has relevance for the 2030 Agenda for Sustainable Development because there is a dedicated water goal—Sustainable Development Goal (SDG) 6 on ensuring water and sanitation for all—including a specific water stress indicator (SDG 6.4.2), which tracks how much freshwater is being withdrawn by all economic activities, compared to the total renewable freshwater resources available. The indicator also considers EFR (UN-Water 2021). Through the UN-Water Integrated Monitoring Initiative for SDG 6 (IMI-SDG6), the United Nations seeks to support countries in monitoring water- and sanitation-related issues and in compiling national data to report on global progress towards SDG 6. Data on water withdrawals and total water resources available are commonly collected by national line ministries and institutions and national statistical offices (UN-Water 2021).

1.3 Responding to Water Scarcity with Unconventional Water Resources

As water scarcity is expected to continue and intensify in arid and overpopulated areas, there is an urgency to utilize every available water-resource management option that can help minimize water scarcity. The response options vary and depend on a range of settings, such as the extent of water scarcity, availability of financial resources, technical and institutional capacity to manage water effectively, and climatic conditions, among others. Water-scarce countries need to promote sustainable water-resources augmentation based on one or more unconventional water resources, which may stem from by-products of specialized processes, need suitable pre-use treatment, require proper on-farm management when used for irrigation, or result from specific techniques to collect/access water (Smakhtin et al. 2001; Qadir et al. 2007).

Given the limited consolidated information and data, quantifying how far unconventional water resources can bridge water demand-supply gap at different scales remains largely unclear. However, recent studies have quantified the potential of specific unconventional water resources, such as desalinated water (Jones et al. 2019) and municipal wastewater (Qadir et al. 2020; Jones et al. 2021). Although the potential of most unconventional water resources remains far from achievement, recent years have witnessed some emerging examples of using unconventional water resources effectively across the world (UN-Water 2020). Such momentum needs to continue and intensify because:

- Water scarcity concerns are moving up on the global political agenda as reflected by SDG 6 and water-related targets embedded in other SDGs.
- Economic development, higher living standards, and population growth lead to increasing demands for water and substantially higher pressure on the water supply.
- Water-quality deterioration is increasing globally, particularly in arid and semi-arid areas where environmental degradation intensifies the water crisis. Water quality below standards required by specific sector use has limited financial and ecological value.
- Increasing the supply and efficiency of conventional water sources has limits because there are just no more conventional resources to be developed in most water-scarce zones of the world.

Unconventional water resources are essential in building a water future in arid areas where water is recognized as a precious resource and a cornerstone of the circular economy. Such water resources range from Earth's seabed to its upper atmosphere. However, securing their access in useable forms needs distinct but specific technologies and innovations. Harvesting water from the air consists of rain enhancement through cloud seeding and collection of water from fog, while capturing water on the ground addresses microscale capture of rainwater where it would otherwise evaporate; these techniques are pertinent to address local water scarcities. There are water-augmentation opportunities below the ground, where tapping offshore and deep onshore groundwater and extending sustainable extraction of undeveloped groundwater are essential options in water-scarce areas with the potential to develop additional groundwater resources. Water recycling and reuse is the key to support water conservation and enhancement opportunities that lead to fit-for-purpose use of treated municipal wastewater and agricultural drainage water. Other opportunities to develop water resources exist in the form of desalinated potable water. Physical transport of water, such as through towed icebergs and ballast water held in tanks and cargo holds of ships, receives attention, but corresponding practices remain in infancy (UN-Water 2020).

The volumes of some unconventional water resources, such as municipal wastewater (380 km^3) and desalinated water (35 km^3), are known (Qadir et al. 2020; Jones et al. 2019), and there are broader estimates available for deep groundwater volume of 16–30 million km^3 (Gleeson et al. 2016). Of this volume, 0.1–5.0 million km^3 is less than 50 years old and renewable, while the remaining nonrenewable but larger volume is embedded in deep geological settings found offshore and onshore (Ferguson et al. 2018). The Earth's atmosphere contains about 13,000 km^3 of water in the vapor phase, the source of which is evaporation from the surface of the oceans, seas, moist soil, and plants (Bengtsson 2010). Antarctic ice contains 27 million km^3 of water, of which 2,000 km^3 breaks off annually as icebergs (Lewis 2015). Seawater stands at 1.335 billion km^3 (UN-Water 2020). Accessing even a tiny fraction of such gigantic volumes of deep groundwater, atmospheric water, Antarctic ice, and seawater could significantly reduce water scarcity in arid and semi-arid areas.

1.4 Unconventional Water Resources and Water-Related SDGs

Achieving SDG 6 and water-related targets in other SDGs is a grand challenge for the world due to increasing water scarcity. Given that most countries are not on track to achieve SDG 6 by the deadline set for 2030 (United Nations 2018), a new water paradigm for water-scarce countries and river basins based on a range of unconventional water resources is crucially important for achieving water-related sustainable development.

The term "unconventional water resources" has not been mentioned explicitly in any SDG. Still, specific terms related to unconventional water resources, such as water harvesting, desalination, water efficiency, and wastewater treatment, recycling, and reuse technologies (SDG 6.a), have been explicitly mentioned. Wastewater treatment and safe reuse have also been part of SDG 6.3. Diving deep into the 2030 Sustainable Development Agenda reveals that there are close linkages between unconventional water resources and SDG 6 and its targets and water-related targets in other SDGs:

- SDG 6.1: *achieving universal and equitable access to safe and affordable drinking water for all.* Unconventional water resources, such as fog-water collection, desalinated water, transportation of water through icebergs, and groundwater confined in deep land-based geological formations or offshore aquifers, can provide enough potable water in areas where water is scarce, or the quality of available water resources does not meet drinking water-quality standards.

- SDG 6.3: *improving water quality by reducing pollution, eliminating dumping, minimizing the release of hazardous chemicals and materials, halving the proportion of untreated wastewater, and substantially increasing recycling and safe reuse globally*. Disposal of large volumes of untreated or inadequately treated wastewater to freshwater bodies has deteriorated the quality of water resources in arid regions, where achieving SDG 6.3 via halving the volumes of untreated wastewater by 2030 would help improve water quality and provide a water resource in the form of treated wastewater that could be used in various sectors, such as agriculture, aquaculture, agroforestry, aquifer recharge, and for environmental flows.
- SDG 6.4: *substantially increasing water-use efficiency across all sectors, ensuring sustainable withdrawals and supply of freshwater to address water scarcity, and substantially reducing the number of people suffering from water scarcity*. The increase in water-use efficiency and water productivity supported by some unconventional water resources in arid regions would help reduce overall water scarcity and the number of people suffering from it. Examples include, but are not limited to, micro-scale capture of rainwater where it otherwise evaporates, use of drainage water from irrigated areas, and water supply increase by rainfall enhancement via cloud seeding.
- SDG 6.5: *implementing integrated water resources management at all levels, including through transboundary cooperation as appropriate*. Efficient use of unconventional water resources can support implementing integrated water-resource management in water-scarce countries and transboundary basins and ensure transboundary planning and actions to develop a supportive environment for new approaches harnessing the potential of unconventional water resources.
- SDG target 6.a: *promoting international cooperation and capacity-building support to developing countries in water- and sanitation-related activities and programs, including water harvesting, desalination, water efficiency, and wastewater treatment, recycling, and reuse technologies*. As explicitly mentioned in this SDG target, there is a need for international cooperation to support capacity-building activities for developing-country professionals in harnessing the potential of unconventional water resources.
- Other water-related SDGs where water plays its role in ensuring their achievement include: SDG 1 on *ending poverty*; SDG 2 on *achieving food security;* SDG 3 on *ensuring healthy lives and well-being*; SDG 7 *on accessing affordable and sustainable energy for all*; SDG 9 on *building resilient infrastructure, promoting sustainable industrialization, and fostering innovation*; SDG 10 on *reducing inequalities within and among countries*; SDG 11 *on making cities inclusive, safe, resilient, and sustainable*; and SDG 17 on *strengthening the means of implementation and revitalizing the global partnership for sustainable development*.
- SDG 13: *taking urgent actions to combat climate change and its impacts*. Because lack of water is the key factor in triggering the impacts of frequent drought events, unconventional water resources could partially offset increased water

needs caused by increased temperature and extended periods of drought and the increased frequency of extreme-weather events.

1.5 Key Features of the Book

In undertaking the critical task of providing insights into the various types of unconventional water resources, this book is structured into seven parts: (I) setting the scene; (II) harvesting water from the air and on the ground; (III) tapping offshore and onshore deep groundwater; (IV) reusing used water; (V) moving water physically to water-scarce areas; (VI) developing new water; and (VII) promoting an enabling environment.

The introductory part (Part I) sets the scene with Chap. 1 (this chapter). The part on harvesting water from the air and on the ground (Part II) consists of three chapters addressing localized water scarcity, which is worsening in many regions and so affecting associated communities because conventional water resources are far from meeting their basic water needs. Under pertinent conditions, cloud seeding may enhance rainfall by dispersing specific substances into the air for cloud condensation, resulting in an increase in rainfall in the target area (Chap. 2). Similarly, water in fog may be transformed into a source of potable water in arid areas where fog intensity and events are common, by using a vertical mesh to intercept the air, producing the collision and coalescence of suspended water droplets (Chap. 3). Micro-catchment rainwater harvesting provides a unique opportunity in areas where rainfall is limited and subject to high intra- and inter-seasonal variability, and much of the rainwater that does fall on the ground is lost through surface runoff and evaporation. Such water loss is further exacerbated due to poor vegetative covering of soils, particularly those that are shallow and form crusts at the surface. These conditions provide strong motivation to develop interventions that ensure making the best use of even the small amounts of rainfall and the resulting runoff water by establishing micro-catchment rainwater harvesting systems to supply water for crop production and to address the local needs of the associated communities (Chap. 4).

Part III focuses on tapping offshore and onshore deep groundwater. Offshore freshwater refers to water hosted in aquifers that are beneath the seafloor. Such water is found at water depths of less than 50 m and distances of less than 100 km. Offshore freshwater is anticipated to be emplaced during sea-level low stands during the last 2.5 million years in periods when continental ice sheets extended across the continents. During the last glacial maximum, for example, the sea level was 120-m lower than present-day conditions, and on average, the sea level has been 40-m lower than at the present (Chap. 5). While located inland, the onshore water reserves are also available in deep geological settings, such as groundwater from confined aquifers and fossil aquifers. Onshore fossil aquifers can be found between low-permeable confining layers and in unconfined formations far below the surface where no recharge is possible (Chap. 6).

Reusing water not only helps in water-resources conservation in water-scarce areas but also in the sustainability of environmental health and protection of the quality of conventional water resources. Part IV consists of two chapters that address reusing municipal wastewater (Chap. 7) and agricultural drainage water (Chap. 8). By recycling and reusing water until it becomes useless for any economic activity, a significant contribution to food, feed, and renewable energy production could be achieved.

Innovative ways of water augmentation via physically moving water to water-scarce areas are treated in Part V. There is a growing interest in recent years to move water from Antarctica in the form of icebergs to be towed and delivered to water-scarce countries. Although iceberg-towing technology is available—the Canadian oil and gas industry regularly tows icebergs away from offshore platforms when there is a risk of collision—the challenge is the scale of transportation, both in terms of iceberg mass and towing distance (Chap. 9). Ballast water is another option because such water is held in the tanks and cargo holds of seagoing ships to increase stability and maneuverability during transit. The UN-led International Convention on the Control and Management of Ships' Ballast Water and Sediments has established a regulatory framework to which the shipping industry and countries must comply concerning ballast-water quality by providing onboard ballast water-treatment options. The regulations essentially created a new unconventional water source, i.e., treated ballast water for potential reuse (Chap. 10).

Part VI concentrates on developing new methods to provide a stabilized supply of potable water, a game-changing opportunity in water-scarce areas. In this regard, desalination of seawater or highly brackish water is an important water-augmentation opportunity, extending water supplies beyond what is available from the hydrological cycle and providing a climate-independent and steady supply of high-quality water. Chapter 11 provides an overview of the status of desalination. It discusses critical barriers and solutions associated with its broader adoption as an unconventional water supply, including technological advances, desalinated water production costs, energy use, environmental impacts in the form of brine generation and its disposal, and institutional challenges. Ocean-brine mining has been gaining momentum over the last few years and is expected to yield commercially viable products that are likely to offset the cost of desalinated water production in the following decades.

Although there are growing examples of using unconventional water resources across the world to boost water supplies, their potential is masked by a lack of supportive national water policies and action plans, low institutional and human capacity, environmental tradeoffs, and the perceived high costs of such water resources without undertaking comprehensive economic analyses and innovative financing mechanisms. These challenges are addressed in Part VII , where Chap. 12 aims at the governance and policy aspects of unconventional water resources. The chapter attempts to predict how rights and obligations for various forms of unconventional water resources will emerge under current principles and practices. Chapter 13 covers the economics and innovative financing mechanisms of unconventional water resources in a circular economy via economic incentives to promote and ensure the use of pertinent types of unconventional water resources. Chapter 14 is concluding chapter of the book, providing insights into the way forward to harness the potential of unconventional water resources. The chapter stresses the urgent need to develop a diversified portfolio of water management to enter a new era of water management by addressing bottlenecks to efficient water management and ensuring that water in all its forms is monitored and accounted for, rather than being considered simply as a supply source, while caring about its value in ecosystems and health, and its role in supporting the basic needs and well-being of humanity.

References

Baggio G, Qadir M, Smakhtin V (2021) Population-driven water scarcity trends across countries and national income levels: challenges and response options. Sci Total Environ 792:148230

Bengtsson L (2010) The global atmospheric water cycle. Environmental Research Letters 5, 025202. https://doi.org/10.1088/1748-9326/5/2/025202

Comprehensive Assessment of Water Management in Agriculture (2007) Water for food, water for life: a comprehensive assessment of water management in agriculture. London: Earthscan, and Colombo: International Water Management Institute

European Environment Agency (EEA) (2005) The European environment—State and outlook 2005. Available at: https://www.eea.europa.eu/publications/state_of_environment_report_2005_1

Falkenmark M (1986) Freshwater: Time for a modified approach. Ambio 15:192–200. Available at: www.jstor.org/stable/4313251

Falkenmark M, Lundqvist J, Widstrand C (1989) Macro-scale water scarcity requires micro-scale approaches. Nat Resour Forum 13(4):258–267. https://doi.org/10.1111/j.1477-8947.1989.tb00348.x

FAO (2019) Incorporating environmental flows into "water stress" indicator 6.4.2—Guidelines for a minimum standard method for global reporting. Rome. 32 pp. Licence: CC BY-NC-SA 3.0 IGO

FAO (2020) The State of Food and Agriculture 2020: Overcoming water challenges in agriculture. Rome. 210 pp. http://www.fao.org/documents/card/en/c/cb1447en/

Ferguson G, McIntosh JC, Perrone D, Jasechko S (2018) Competition for shrinking window of low salinity groundwater. Environ Res Lett 13(11):114013. https://doi.org/10.1088/1748-9326/aae6d8

Gleeson T, Befus KM, Jasechko S, Luijendijk E, Cardenas MB (2016) The global volume and distribution of modern groundwater. Nat Geosci 9(2):161–167. https://doi.org/10.1038/ngeo2590

Hanasaki N, Kanae S, Oki T, Masuda K, Motoya K, Shirakawa N, Shen Y, Tanaka K (2008) An integrated model for the assessment of global water resources—Part 2: applications and assessments. Hydrol Earth Syst Sci 12:1027–1037. https://doi.org/10.5194/hess-12-1027-2008

Jones E, Qadir M, van Vliet MTH, Smakhtin V, Kang S (2019) The state of desalination and brine production: a global outlook. Sci Total Environ 657:1343–1356. https://doi.org/10.1016/j.scitotenv.2018.12.076

Jones ER, van Vliet MTH, Qadir M, Bierkens MFP (2021) Country-level and gridded estimates of global wastewater production, collection, treatment, and reuse. Earth Syst Sci Data 13:237–254

Lewis CJ (2015) Iceberg harvesting: suggesting a federal regulatory regime for a new freshwater source. Boston College Environ Affairs Law Rev 42:439–472. http://lawdigitalcommons.bc.edu/ealr/vol42/iss2/6

Liu J, Liu Q, Yang H (2016) Assessing water scarcity by simultaneously considering environmental flow requirements, water quantity, and water quality. Ecol Indic 60:434–441. https://doi.org/10.1016/j.ecolind.2015.07.019

Liu J, Yang H, Gosling SN, Kummu M, Flörke M, Pfister S, Hanasaki N, Wada Y, Zhang X, Zheng C, Alcamo J, Oki T (2017) Water scarcity assessments in the past, present, and future. Earth's Future (5):545–559. https://doi.org/10.1002/2016EF000518

Mekonnen MM, Hoekstra A (2016) Four billion people facing severe water scarcity. Sci Adv 2(2):e1500323. https://doi.org/10.1126/sciadv.1500323

Qadir M, Sharma BR, Bruggeman A, Choukr-Allah R, Karajeh F (2007) Non-conventional water resources and opportunities for water augmentation to achieve food security in water-scarce countries. Agric Water Manag 87(1):2–22. https://doi.org/10.1016/j.agwat.2006.03.018

Qadir M, Drechsel P, Cisneros BJ, Kim Y, Pramanik A, Mehta P, Olaniyan O (2020) The global and regional potential of wastewater as a water, nutrient, and energy source. Nat Resour Forum 44(1):40–51. https://doi.org/10.1111/1477-8947.12187

Qadir M, Sposito G, Smith CJ, Oster JD (2021) Reassessing irrigation water quality guidelines for sodicity hazard. Agric Water Manag 255:107054. https://doi.org/10.1016/j.agwat.2021.107054

Raskin P, Gleick P, Kirshen P, Pontius G, Strzepek K (1997) water futures: assessment of long-range patterns and prospects. Stockholm Environment Institute, Stockholm, Sweden

Rijsberman FR (2006) Water scarcity: fact or fiction? agricultural water management, 80 (1–3):5–22. https://doi.org/10.1016/j.agwat.2005.07.001

Rockström J, Falkenmark M, Karlberg L, Hoff H, Rost S, Gerten D (2009) Future water availability for global food production: the potential of green water for increasing resilience to global change. Water Resour Res 45:W00A12. https://doi.org/10.1029/2007wr006767

Seckler D, Amarasinghe U, Molden DJ, de Silva R, Barker R (1998) World water demand and supply, 1990 to 2025: scenarios and issues. IWMI Research Report 19. IWMI, Colombo, Sri Lanka

Shiklomanov IA (1991) The world's water resources. Proceedings of the International Symposium to Commemorate 25 Years of the IHP, UNESCO/IHP, Paris, France, pp 93–126

Smakhtin V, Ashton P, Batchelor A, Meyer R, Murray E, Barta B, Bauer N, Naidoo D, Olivier J, Terblanche D (2001) Unconventional water supply options in South Africa. Water Int 26(3):314–334. https://doi.org/10.1080/02508060108686924

Smakhtin V, Revenga C, Döll P (2004) Taking into account environmental water requirements in global-scale water resources assessments. Comprehensive Assessment Research Report 2. Colombo, Sri Lanka: Comprehensive Assessment Secretariat

Sood A, Smakhtin V, Eriyagama N, Villholth KG, Liyanage N, Wada Y, Ebrahim G, Dickens C (2017) Global environmental flow information for the sustainable development goals. International Water Management Institute (IWMI). https://doi.org/10.5337/2017.201

Sullivan CA, Meigh JR, Giacomello AM, Fediw T, Lawrence P, Samad M, Mlote S, Hutton C, Allan JA, Schulze RE, Dlamini DJM, Cosgrove WJ, Delli Priscoli J, Gleick P, Smout I, Cobbing J, Calow R, Hunt C, Hussain A, Acreman MC, King J, Malomo S, Tate EL, O'Regan D, Milner S, Steyl I (2003) The water poverty index: development and application at the community scale. Nat Resour Forum 27:189–199

United Nations (2018) Sustainable Development Goal 6 Synthesis Report 2018 on Water and Sanitation. New York: UN. Available at: https://www.unwater.org/publications/sdg-6-synthesis-report-2018-on-water-and-sanitation/

UN-Water (2020) UN-Water Analytical Brief on Unconventional Water Resources. Geneva, Switzerland. Available at: https://www.unwater.org/publications/un-water-analytical-brief-on-unconventional-water-resources/#:~:text=At%20the%20same%20time%20water,change%20adaption%2C%20and%20sustainable%20development

UN-Water (2021) UN-Water Integrated Monitoring Initiative for SDG 6 (IMI-SDG6). Available at: https://www.sdg6monitoring.org/about/integrated-monitoring-initiative/
van Vliet MTH, Jones E, Franssen WHP, Flörke M, Hanasaki N, Wada Y, Yearsley JR (2021) Global water scarcity including surface water quality and expansions of clean water technologies. Environ Res Lett 16:024020. https://doi.org/10.1088/1748-9326/abbfc3
Wada Y, van Beek LPH, Viviroli D, Dürr HH, Weingartner R, Bierkens MFP (2011) Global monthly water stress: 2. Water demand and severity of water stress. Water Resour Res 47:W07518
World Bank (2018) Beyond Scarcity: Water Security in the Middle East and North Africa. MENA Development Report. The World Bank: Washington, DC. Available at: https://openknowledge.worldbank.org/handle/10986/27659
World Economic Forum (2020) The Global Risks Report 2020. 14th Edition. Geneva: World Economic Forum. Available at: http://www3.weforum.org/docs/WEF_Global_Risk_Report_2020.pdf
World Economic Forum (2021) The Global Risks Report 2021. 15th Edition. Geneva: World Economic Forum. Available at: http://www3.weforum.org/docs/WEF_The_Global_Risks_Report_2021.pdf
Zeng Z, Liu J, Savenije HHG (2013), A simple approach to assess water scarcity integrating water quantity and quality. Ecol Indic 34:441–449. https://doi.org/10.1016/j.ecolind.2013.06.012

Part II
Harvesting Water from Air and on the Ground

Chapter 2
Rain Enhancement Through Cloud Seeding

Ali M. Abshaev, Andrea Flossmann, Steven T. Siems, Thara Prabhakaran, Zhanyu Yao, and Sarah Tessendorf

Abstract An increasing number of countries are planning to carry out rain-enhancement activities in response to water shortages and other societal needs. Rain enhancement can work with reasonable cost–benefit ratios under the right conditions. However, many components of the technology need improvement and testing, and many physical processes are not yet fully understood due to their complexity. Global research on cloud-seeding technology indicates that precipitation can be increased up to 15% of the annual norm, depending on the available cloud resources and technical systems used. However, there is still ambiguity in the results of the studies conducted and the effects and scale of the rain enhancement. When evaluating the results of rain enhancement projects, it is necessary to adhere to rigorous scientific approaches and proven methods.

A. M. Abshaev (✉)
Hail Suppression Research Center "ANTIGRAD", Nalchik, Russia
e-mail: abshaev.ali@mail.ru

A. Flossmann
Université Clermont-Auvergne Laboratoire de Météorologie Physique 4, Avenue Blaise Pascal TSA 60026 CS 60026, 63178 Aubière Cedex, France
e-mail: Andrea.flossmann@uca.fr

S. T. Siems
School of Mathematics, 9 Rainforest Walk, Monash University, Melbourne, VIC 3800, Australia
e-mail: steven.siems@monash.edu

T. Prabhakaran
Indian Institute of Tropical Meteorology, Dr. Homi Bhabha Road, Pashan, Pune 411 008, Maharashtra, India
e-mail: thara@tropmet.res.in

Z. Yao
Chinese Academy of Meteorological Sciences, 46 Zhong Guan Cun South Street, Beijing 100081, People's Republic of China
e-mail: yaozy@cma.gov.cn

S. Tessendorf
National Center for Atmospheric Research, Research Applications Laboratory, 3090 Center Green Drive, Boulder, CO 80301, USA
e-mail: saraht@ucar.edu

M. Qadir et al. (eds.), *Unconventional Water Resources*,
https://doi.org/10.1007/978-3-030-90146-2_2

Keywords Atmosphere · Cloud seeding · Rain enhancement · Weather modification · Arid areas

2.1 Introduction

The Earth's atmosphere contains about 13,000 km^3 of water in the vapor phase, the source of which is evaporation from the surface of the oceans, seas, soil moisture, and transpiration from plants. These water-vapor reserves are continuously renewed due to the circulation of evaporation–condensation and precipitation, making eight to nine hydrological cycles annually (Bengtsson 2010). Thus, the water vapor present in the atmosphere is an infinite freshwater source and an opportunity for rain enhancement (RE).

Cloud seeding (CS) has been used for more than 75 years for the purpose of RE, hail suppression (HS), fog dispersion, and the improvement of weather conditions. Under pertinent conditions, CS involves the application of extensive technology for the modification of the precipitation regime in convective clouds, large-scale stratus clouds, and ground fogs, by dispersing special glaciogenic or hygroscopic substances within clouds or in their vicinity. These particles enable water droplets or ice crystals to activate on heterogeneous nuclei through water-vapor condensation–freezing processes (Flossmann et al. 2019). Subsequent collision–coalescence growth of water droplets and ice crystals leads to the formation of large rain-sized hydrometeors (drops, graupels, hailstones, snowflakes, etc.) that fall as precipitation.

Some estimates show that only up to 10–15% of the total cloud water content of typical cumulonimbus convective clouds is released to the ground as precipitation, while the rate of precipitation from these clouds varies in the range of 10,000 to 50,000 t/min (Abshaev et al. 2009), a number that exceeds the capacity of all currently operational desalination plants (Eke et al. 2020) and suggests the huge potential of RE technologies.

This chapter focuses on rain enhancement through CS and provides concise information on: the basics; a variety of technical options for its execution; and on unresolved scientific aspects. The main objective is to briefly convey to the reader the potential and the limitations of CS methods for addressing global water scarcity. A major part of the chapter stems from a report of the World Meteorological Organization (Flossmann et al. 2018), which was funded by the United Arab Emirates.

2.2 History

For thousands of years, people have sought to modify weather and climate to increase water resources and mitigate severe weather. Aristotle was already able to formalize the observations of weather accumulated by the 3rd century BC, which gave meteorology the status of a science (Burtsev et al. 2018). From his work, even in those days, he was studying the mechanisms of precipitation, particularly that of hail.

The modern technology of weather modification (WM) was launched by the discovery in the late 1940s that supercooled cloud droplets could be converted to ice crystals by the insertion of a cooling agent such as dry ice or an artificial ice nucleus such as silver iodide (AgI). About 80 years of subsequent research has greatly enhanced our knowledge about the microphysics, dynamics, and precipitation processes of natural clouds (rain, hail, snow) and the impacts of human interventions on those processes (Rauber et al. 2019).

The main obstacle has always been the need for enormous amounts of energy to manage meteorological processes. For example, the energy associated with the formation of convective clouds is equivalent to several of the largest hydroelectric power plants. If we want to change the direction of the wind within a 100-km region, then we would need to use energy generated by all the power plants in the world. Moreover, if we decided to change the weather of a territory or a small country, the current total global energy generation would not be sufficient.

However, in addition to the enormous energy needs, the unstable state is necessary to govern different evolutionary processes that are favorable through small impulses in the meteorological processes. In other words, the only credible approach to modify weather is to take advantage of microphysical and dynamic sensitivities through human interventions. It is assumed that a relatively small human-induced disturbance in the system can substantially alter the natural evolution of atmospheric processes. Currently, there are three instability states in clouds that are being utilized by human interventions; (1) Colloidal instability, which is a condensation–coagulation growth of droplets in clouds and rainfall from warm clouds; (2) thermodynamic (liquid–ice phase) instability of colloidal systems in clouds and fogs containing supercooled water; and (3) convective instability of the atmosphere.

The scientific basis of WM depends on the understanding of how clouds composed of tiny droplets evolve into precipitation. One principal path is provided by the fact that the ice particles in the presence of supercooled cloud drops can constitute a nonequilibrium state that results in the growth of the ice and evaporation of the drops. The German meteorologist Alfred Wegener was the first researcher who addressed WM (Wegener 1910, 1912). Later in 1933, the Bergen school meteorologist Tor Bergeron at the Lisbon meeting of the International Union of Geodesy and Geophysics developed the argument that relatively few ice particles in a supercooled cloud could individually grow large enough to provide a release as precipitable particles. Bergeron's hypothesis that rain can have its origin in snowflakes was a cornerstone for cloud physics. The acceptance and rapid further development of this hypothesis was greatly advanced by the work of another German scientist Walter Findeisen (1938) and others contributing to the theoretical development of cloud physics and its application to WM in the 1940s and 1950s (Al Mandous et al. 2006).

In the former USSR, the basis of condensation growth of particles was developed through experimental/theoretical studies of precipitation-formation processes in clouds and coagulation phenomena and studies in the microstructure of clouds and precipitation, obtained in the early 1930s at the Leningrad Institute of Experimental Meteorology under the supervision of V.N. Obolensky. Later, the first data on the water content and the size of droplets in clouds were obtained, and empirical and numerical models of clouds were created at the Voeikov Main Geophysical Observatory, using aircraft meteorological laboratories. This new knowledge about the processes taking place in cloud systems made it possible to construct a concept of the increase in liquid precipitation based on an artificial increase in the concentration of crystallization nuclei in a cloud. To check and clarify the provisions of the concept, a technically equipped experimental test site was created. As a result of research and development work of scientific teams of the former-USSR National Hydrometeorological Service, reagents that can be dispersed have been developed and used as artificial crystallization nuclei and, in some cases, as condensation nuclei. Further research was conducted on methodological techniques and technical means for bringing them into clouds, as well as the development of criteria to assess the degree of readiness of the cloud to produce additional precipitation (Burtsev et al. 2018).

In 1946, Vincent Schaefer demonstrated that dry ice dropped into supercooled clouds rapidly transformed the droplets into ice crystals. The same effect at cloud temperatures below −10 °C was demonstrated by his colleague Bernard Vonnegut at the General Electric Laboratory using AgI particles in 1947. These experiments were carried out under the direction of Nobel Laureate Irving Langmuir, who was also instrumental in promoting a five-year series of field experiments (Langmuir 1950). Those experiments, plus similar experiments carried out during the same year, yielded convincing visual proof that cloud-seeding induced changes in cloud composition depth and other characteristics can be readily achievable. Experiments and CS operations commenced all over the world and many continue to this day. The precipitation-forming processes in clouds from which substantial amounts of precipitation might be expected are much more complex than the simple ones where visible evidence of seeding might be provoked. Thus, it has proven to be a much greater challenge to quantify microphysical seeding signatures and to obtain statistical documentation of added precipitation on the ground. Each of the two has been achieved separately, but their combination, which is necessary for maximum scientific credibility, has not yet been fully achieved. It was found, however, that the energy involved in weather systems is so large that it is impossible to create entire artificial rainstorms or to alter wind patterns to transport water vapor into a region.

In the mid-latitudes, hail damages exceed three billion dollars per year worldwide, and for many countries HS is an attempt to reduce economic damages in agriculture. Reduction of hail damage by CS became a major part of WM activities after the introduction of the concept of *"beneficial competition"* (Sulakvelidze et al. 1965) at the High Mountain Geophysical Institute (Nalchik, former USSR) in the early 1960s. The idea was to introduce large numbers of artificial embryos (using rockets and/or artillery cannons), which compete for the limited water content in clouds and, as a result, reduce the size of the growing hailstones. As revealed by later research, the complexities of hailstorms and the details of hailstone growth turned out to be much more complex than it was assumed in the simple notions underlying the idea of beneficial competition. Hereafter, in the 1970s, Magomet Abshaev proposed a new concept of *"premature precipitation"* (Abshaev 1966, 1994), artificially induced in feeder clouds of mature hailstorms earlier than it would occur naturally; this would lead to the depletion of a cloud's liquid-water content required for hail growth. Another concept of *"trajectory lowering"* was proposed by Brant G. Foote from the National Center for Atmospheric Research in the USA, which anticipated limiting the hail growth level by the enlargement of artificial ice particles in hailstorms (Borland et al. 1977). Implementation of CS for hail-damage mitigation has evolved considerably, though modification of mature hailstorms is still controversial.

Another area of application of RE has been for extinguishing forest fires. Meteorological observations on the state of the atmosphere in forest-fire zones made by the Research Institute of Forestry (Russia) showed that powerful convective clouds with volumes of tens of cubic kilometers that do not produce precipitation often appear above forest fires. Each km^3 of clouds contains an average of 1,000 t of water, so clouds above regions of active forest fires are natural reserves of water. In the former USSR, back in the late 1960s, a method for extinguishing forest fires using artificially induced precipitation was proposed (Burtsev et al. 2018). Experimental and practical work in various regions of Siberia indicated prospects for its application.

Over the 80-year history of WM experimentation, the interest in WM has varied significantly. There was a major increase in commercial CS activities in the early 1950s. Since numerous attempts to modify weather systems did not produce verifiable positive outcomes, it became obvious that some basic questions had yet to be addressed. Consequently, intense research activities were undertaken in the universities and government agencies during the 1960s and the early 1970s (Al Mandous et al. 2006).

One major international effort toward deeper understanding of the effects of CS on cloud and precipitation development was the World Meteorological Organization (WMO)-initiated major international RE Project in Spain in the late 1970s and early 1980s (WMO 1985). Through more than 30 relevant reports, the RE Project established important scientific principles for the planning and execution of such experiments. Currently, according to the WMO Registers, there are dozens of nations operating hundreds of WM projects, particularly in arid and semi-arid regions. More than 70 countries have expressed their interest in how to use RE as part of their water resource-management strategy.

2.3 Status

The development of new equipment—such as weather radars, satellites, microwave radiometers, wind profilers, automated rain gauges, mesoscale network stations, and aircraft platforms with microphysical and air-motion measuring systems—has introduced new dimensions into WM operations and research over the last three decades. Equally important are the advances in computer systems and algorithms for cloud processes that permit higher resolution and more physically based simulations to be run in relatively short times. New field observations used in conjunction with increasingly sophisticated numerical cloud models have helped in testing various WM hypotheses. Through some of these innovative facilities, a better understanding of clouds and precipitation climatology can be achieved to test seeding hypotheses prior to the commencement of WM projects.

The complexity and variability of clouds cause certain difficulties in understanding and detecting the effects of artificially modifying clouds. The ability to influence cloud microstructures has been demonstrated in the laboratory, simulated by numerical models, and verified through physical measurements in some natural systems such as fogs, stratus, and cumulus clouds. The confidence level in being able to generate predictable results is quite high for the dissipation of supercooled fog and moderate to high for increasing snowfall from orographic clouds. However, statistical evidence for the degree of artificial modification on precipitation, hail, lightning, or winds is limited. Experiments have suggested a positive effect on individual convective cells, but conclusive evidence that such seeding can increase rainfall over large areas has yet to be established.

The expected effects of seeding are often within the range of natural variability (low signal-to-noise ratio), and our ability to predict the natural behavior is still limited. Randomization methods, augmented by physical predictors, are considered to be the most desirable for detecting cloud-seeding effects. Coupling of physical experiments with ongoing operational projects would be a productive and cost-effective approach to collect information for the clarification of questions related to WM. RE projects are generally expected to yield increases of 10-20%, however that level of success is difficult to achieve in measurements and in the simulation of precipitation.

The success of WM depends on the understanding of the related disciplines (cloud physics and dynamics, mesoscale weather forecasting, numerical modelling of cloud processes, and measuring technology). This explains why progress has been slow in establishing the validity of WM results. There is growing evidence that the basic concepts are correct and that successful implementation is feasible.

2.4 Technological Interventions

The purpose of CS is to change the microphysical processes in clouds to increase the efficiency of precipitation formation. This can be obtained by accelerating the condensation–coalescence–collision–freezing processes that promote early development of precipitation and thus harvest more of the available water from clouds.

The CS operation can be divided into six basic elements (Fig. 2.1). As a first step, suitable clouds or cloud clusters need to be identified. Cloud characteristics such as type, vertical depth of the cold and warm parts, liquid-water content, spatial dimensions, tendencies of development, etc. must be observed and evaluated to select a suitable cloud. The second step is to select proper seeding materials (SM) depending on the cloud type (warm/cold or mixed phase). Glaciogenic agents are used for clouds with high super cooled liquid-water content (SLW) in the cold part, while

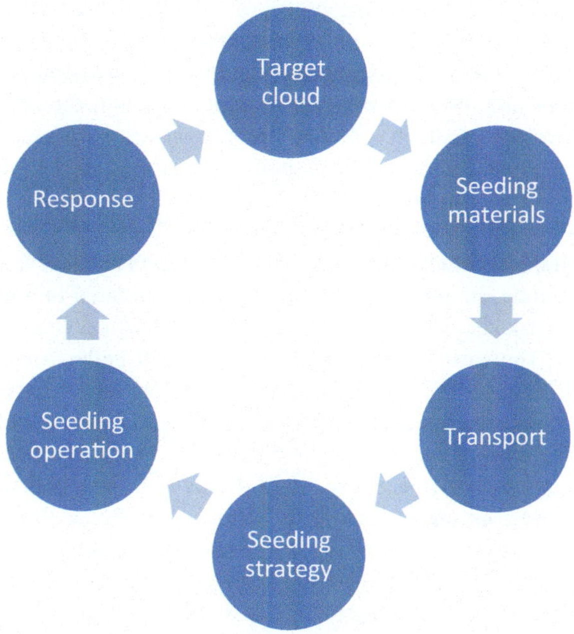

Fig. 2.1 Simplified components of cloud seeding through six basic elements

hygroscopic agents are the only option for warm clouds. An optimal delivery system should then be selected (e.g., aircraft, artillery shells, rockets, ground generators, etc.). A seeding strategy is implemented to determine the optimal location, time, and dosage of seeding, as well as its frequency, until a decision is made to stop the seeding. The next element in the chain is the seeding operation itself, and the last is the measurement of the seeding response based on the analysis of radar-satellite-ground data.

2.4.1 Seeding Materials

Most WM methods are based on the introduction of a large amount of special artificial aerosol particles into the cloud environment. According to the principle of action, such particles called seeding materials can be divided into two large classes: *glaciogenic* and *hygroscopic.* Glaciogenic seeding introduces ice-nucleating particles (INP) into the cloud to enhance the ice/mixed phase, while hygroscopic seeding introduces cloud-condensation nuclei (CCN) to enhance the formation of larger drops and activate the coalescence process. Depending on the type of cloud and the purpose of the WM, one or the other is selected.

For *glaciogenic seeding,* AgI and dry ice are still the most widely used SM. Both materials enhance ice-crystal concentrations in clouds by either nucleating new crystals or by freezing cloud droplets. AgI nucleates ice particles at temperatures below −2 °C to a minimum of −10 °C. Based on past experiments, two seeding concepts have been proposed, namely, the "static" and "dynamic" (Braham 1986). While the first attempts to increase precipitation embryos, the latter attempts to increase the buoyancy in the cloud through the release of latent heat due to the freezing of supercooled liquid drops (National Research Council 2003).

Silver iodide can be dispersed either by pyrotechnic flares from generators at the surface or in the air. For the ejectable flare, ignition occurs as it leaves the aircraft. Pyrotechnic flares typically produce 10–100 g of active seeding agents per minute of burn, whereas aerial acetone generators produce 2–3 g of active seeding agent per minute. In anti-hail rockets, AgI is sublimated in a special chamber in the head of the rocket or in its engine combustor (Abshaev et al. 2014).

The earliest experiments on CS used pellets of dry ice dropped from aircrafts (Dennis 1980). Dry ice is dispensed through openings located in the floor of baggage compartments of CS aircraft. Dispensers disperse pelletized (0.6–2.5 cm) or small particles of dry ice. Dry-ice pellets have a surface temperature of around −78 °C, and so they freeze any cloud droplets in their paths, and they also activate cloud-condensation nuclei to form droplets that freeze through homogeneous nucleation.

Hygroscopic seeding is potentially applicable to all clouds that have a liquid region. Typically, hygroscopic seeding particles are larger (a few microns) and more hygroscopic than the natural aerosol particles. The resulting droplets grow to larger than normal sizes through condensation, and then they rapidly grow further through collisions with other droplets (Cooper et al. 1997), initiating the rain process within the convective cell. There are two main concepts under consideration regarding hygroscopic seeding: the competition effect and the tail effect (Segal et al. 2007). The tail effect envisages the droplets' spectrum broadening by large seed particles, while competition is based on more efficient droplet formation compared to natural CCN.

Hygroscopic seeding in convective clouds is carried out with the help of aircraft-based flares through a burning process or the dispersion of prepared micro-powders of either pulverized salt; alternatively, or pyrotechnic flares are used. The principle of hygroscopic flare seeding is based on the flares producing effective CCN particles in larger sizes (large or giant nuclei) than occur in the natural environment (Bruintjes et al. 2012). Hygroscopic flares contain sodium chloride or calcium chloride, which produce small salt particles in the size range of 0.1–10 μm in diameter.

Cooper et al. (1997) found that an optimum particle size of 1 μm is required to form drizzle drops and to enhance collision-coalescence processes. The optimum size of soluble particles was found to be 1–5 μm (Segal et al. 2004; Rosenfeld et al. 2010). It is also important to have information on the particle-size distribution of naturally present aerosols, recognizing that, if the procedure occurs close to a coastline, it can be dominated by the already large sea-salt aerosols.

Drofa et al. (2013) studied the effect of seeding of a cloudy environment with salt powder in a large cloud chamber (3,200 m^3) in conditions corresponding to the formation of convective clouds and observed that the introduction of salt powder before a cloud is formed in the chamber results in the formation of a "tail" of additional large drops. In this case, seeding with the salt powder also leads to an increase in size of the entire population of cloud drops and to a decrease of their total concentration as compared to a cloud that is formed of background aerosols, showing that a salt powder milled to a size of several μm is more effective in initiating warm rain than hygroscopic flares. While the chamber experiments and numerical simulations provide some evidence of the effects of salt-powder seeding, their validation in the real atmosphere is still needed. Belyaeva et al. (2013) showed with numerical simulations that the use of polydispersed salt powders has an advantage over hygroscopic agents from pyrotechnic flares and that precipitation could be induced from warm convective clouds of moderate thickness that do not precipitate naturally. Zhekamukhov and Abshaev (2009a and b) showed that anti-hail rockets equipped with hygroscopic micro-powders could be effectively used for seeding the cores of developed cumulus-congestus clouds for the purpose of RE. The optimum suggested size of NaCl crystals is 7.5–10 μm because these "salty" droplets can rapidly grow to raindrops size through condensation–coalescence mechanisms.

In recent years, new formulations of SM are being developed for release from pyrotechnic flares (National Research Council 2003). These materials require less AgI than previous formulations, and they are much more active in ice nucleation at temperatures colder than about −5 °C. Considerable work to improve the efficiency of SM is being carried out by numerous groups using complex chemical compositions, nano-technologies, various types of cloud chambers, and full-size testing stands of seeding devices (Drofa et al. 2013; Liang et al. 2019; Tai et al. 2017).

2.4.2 Transport and Dispersion

Seeding material can be dispersed into clouds and their surroundings by aircraft, artillery shells, small-sized rockets, high-altitude fireworks, unmanned aircraft, balloons, and ground-based generators (Figs. 2.2 and 2.3).

Any CS program should first determine the transport, dispersion, and dilution of SM in the clouds to supply the right quantities at the right time to the right place in the clouds. The temperature range, cloud type, delivery mechanism, and seeding targets are all crucial factors to be considered. Seeding materials act locally and dissipate with time due to thermodynamics, transport by advection–convection forces, and turbulent mixing.

There is no universally ideal delivery system, and each method has advantages and disadvantages. Aircrafts are used to disperse SM from the sub cloud and cloud-top levels and directly in super cooled regions of winter orographic clouds. However, penetrations of convective clouds are rarely practiced for safety reasons, as well as nocturnal flights, because of the limited visual contact. Due to the risks associated with carrying flammable liquids on aircraft, pyrotechnic flares have been developed for aircraft-based seeding (Dennis 1980). For seeding from cloud base or top, a certain period of time is required (5–10 min) until the SM reaches the level of maximum water content. Aircrafts enable coverage of large areas and the measurement of meteorological characteristics along the flight path. Hygroscopic seeding is mainly

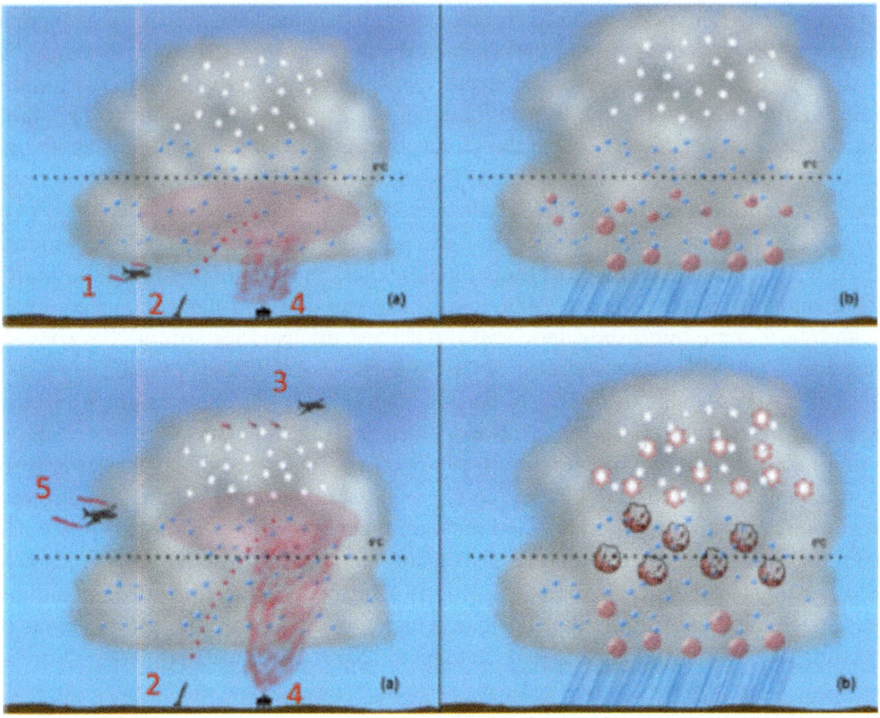

Fig. 2.2 Rain enhancement through could seeding using aviation, rocket, or ground-based generators of nuclei for crystallization or condensation: **a** refers to seeding options and **b** indicates the intended outcome of the seeding. The seeding options include: (1) cloud-base aircraft; (2) rockets; (3) cloud-top aircraft; (4) ground generators; and (5) direct cloud penetration by aircraft (Flossmann et al. 2019; © American Meteorological Society. Used with permission)

dispensed from aircraft at the cloud base through flares or salt powders, while dry ice and ejectable flares are used for cloud-top applications.

Rockets and artillery shells are used for direct and almost simultaneous seeding of the required cloud part in the required dosage in 24/7 mode irrespective of cloud conditions (turbulence, lightning activity, heavy solid/liquid precipitation). One rocket site serves a ground circle of 100–300 km^2. But this requires well-developed infrastructure and logistics on the ground maintained by personnel. State-of-the-art progress now makes possible robotizing ground sites using automated rocket launchers without personnel (Abshaev et al. 2011a). Launching is prohibited when ground-level winds exceed 20–25 m/s. Rocket launch permission should first be obtained from the regional aviation authorities.

Ground-based generators mainly apply AgI in pyrotechnic flares or in acetone. The main problem here is the large temporal and spatial distance from the aerosol generator to the cloud. Only a fraction of the SM, if at all, reaches the cloud base when airflows are favorable. Deactivation of SM due to the impact of high humidity,

a

b

c

Fig. 2.3 Technical systems for delivering seeding materials into clouds and their environments: **a** airborne pyroflares; **b** small-sized anti-hail rocket "Alazan-6" and automated anti-hail launcher "ELIA-2"; **c** liquid acetone-based ground generator. *Sources* **a** National Center of Meteorology of the United Arab Emirates; **b** Magomet Abshaev; **c** Viktor Korneev; all used with permission ambient aerosol and ultraviolet is another challenge when employing this method (Shilin et al. 2015).

A tracer (chaff and/or SF6) can be used as a tag for a seeded region to understand the dispersion and transport of the SM. Chaffs can be monitored by radar (Reinking and Martne 1995), while detection of the SF6 is done by aircraft (Rosenfeld et al. 2010). SF6 tracers have been used by Rosenfeld et al. (2010) for identifying seeding signature in convective clouds over Texas in the USA.

Studies of the transport and dispersion of SM in convective clouds were conducted in Moldova for 20 consecutive summers. Special tracers based on deuterium $210P_o$ and D_2O (Dinevich and Shalaveyus 2010) were introduced into the clouds by rockets and aircrafts. A dual-wavelength radar MRL-5 was used to measure cloud characteristics, while rain gauges and laboratory equipment were used to detect tracers in precipitation on the ground.

2.4.3 Seeding Strategies

The selection of a seeding strategy predominantly depends on the type and characteristics of the clouds, SM, and delivery methods. Hygroscopic seeding involves seeding summer time convective clouds below the cloud base with pyrotechnic flares that produce salt particles (~0.5 μm) that are larger than naturally available CCN. The particles are supposed to activate at lower supersaturations and condense water more readily, as well as limit the total number of droplets activated. The degree of concentration of the background aerosol population needs to be considered (Semeniuk et al. 2014). Cloud droplets should nucleate preferentially on the seeding particles, and this inhibits smaller natural CCN from activation, resulting in a broader-than-natural droplet spectrum near the cloud base, triggering collision coalescence within 15 min (Cooper et al. 1997), and initiating the rain process earlier within the 30-min lifetime of a typical cumulus cloud. This is expected to increase the potential for precipitation to develop earlier and more efficiently in the lifetime of the cloud.

For glaciogenic seeding, introducing INP close to cloud base will yield an effect like hygroscopic seeding because the AgI particles are large enough (~0.1 μm; Dessens et al. 2016) to serve also as CCN. Reaching higher altitudes, the INP freeze the SLW drops and trigger precipitation via the formation of graupel particles. Depending on the height of the freezing level, the particles will melt before reaching the ground. Because the direct penetration of the SLW is dangerous for aircraft, the release of SM is realized from sub cloud or cloud-top levels for the majority of aircraft-based seedings of convective clouds. Applying this seeding strategy, one must account for the time required for the SM to attain the necessary levels of SLW in the cloud, which can take several minutes. Artillery shells (Zhekamukhov and Abshaev 2012) and anti-hail rockets (Abshaev et al. 2014) are widely used to deliver glaciogenic SM directly to a specific altitude for the SLW of convective clouds in the required dosage. Dispersed from sub-cloud and cloud-top levels or directly into regions of SLW, deposition of water vapor onto the introduced INP is supposed to be the main mechanism for the growth of ice particles.

Given the variability of operating conditions, it is important to ensure that seeding generators produce a steady flow of SM and that the particle-size distribution and number and mass concentrations of the SM are documented (Huggins et al. 2008). These precautions are needed to ensure that any seeding effects can be related accurately to the source characteristics. It is also vital to ensure that the plumes from generators are dispersed into regions of the cloud where they can interact with available SLW.

The targeting and dispersion of SM remains an important issue in seeding experiments and needs to be validated by observations and numerical model simulations (Xue et al. 2013a; French et al. 2018; Abshaev et al. 2004, 2011b). Routine targeting is achieved using a suitable model, which can vary from rather simple dispersion-microphysical models to full three-dimensional numerical models. A CS parameterization in the weather research and forecasting model used by Xue et al. (2013b) suggests that the effects of aircraft-based and ground-based seedings are different. For aircraft-based seeding, where the SM is dispersed directly into regions of SLW, deposition of water vapor onto the introduced ice nuclei is the main mechanism for the growth of ice particles. However, for ground-based seeding, where the SM must be mixed vertically into the SLW region, the dominant effect is probably from AgI acting as CCN when temperatures are warmer and subsequent immersion freezing because the introduced ice nuclei are incorporated into the SLW droplets before freezing occurs. Three-dimensional modelling by Xue et al. (2013b) confirms that in general direct (by aircraft, rockets, and artillery shells) and near cloud (by aircraft) methods of CS are more efficient.

Recent advances in technology provide a new dimension to the targeting and evaluation of CS experiments. Radar polarimetric parameters can be used to identify the zones of hydrometeor classes that may be targeted with more precision and may help with the selection of the areas for seeding. High-quality real-time radar observations illustrating various types of hydrometers and other analytical products from dual polarization radar networks in the world seem to have a large potential for the targeting and evaluation of CS experiments.

The new generation of geostationary satellites enables tracking of clouds at higher resolutions, and so cloud tops and other spatial structures will be discernible. Information on cloud types and microphysics during potential seeding days, especially the relationship between cloud-top temperature and effective radius (Rosenfeld and Lensky 1998), together with estimates of vertical velocity and CCN information, can be derived. This information, together with sophisticated numerical modelling, gives guidance for seeding decisions.

Optimal seeding conditions may be determined based on guidance from high-resolution weather-forecast radar visualization using improved analysis tools (such as TITAN, ASU-MRL, etc.) that can handle the selection of variable target/control areas or individual radar cells (Woodley et al. 2003; Abshaev et al. 2010).

2.4.4 Seeding Operation

Cloud Seeding is usually done at the cloud base and top and by direct introduction of the reagent at the desired height inside the cloud. The reagent delivery is carried out by light and medium-sized aircrafts, small-sized rockets, artillery shells, or ground generators (Fig. 2.2).

2.4.5 Response and Impacts of Cloud Seeding

Quantifying the impact of CS has been a longstanding challenge and has been attempted via many methods (Rauber et al. 2019). The main problem is to detect the signal caused by CS against the background of natural variability in the development of cloud processes (noise). For this purpose, various statistical methods have often been used (Woodley et al. 2003; Brier et al. 1973; Yao 2006; Guo 2015; Manton et al. 2011; Breed et al. 2014; Rasmussen et al. 2018; and others). However, relatively small sample sizes have limited the conclusiveness of these statistical approaches (Rauber et al. 2019).

The seeding response of clouds is often understood as a change in their micro- and macro-physical parameters (Abshaev et al. 2014, 2009; Al Mandous et al. 2006; Bruintjes 1999; Flossmann et al. 2019; Koloskov et al. 2011), and attempts have been made to measure these parameters to physically detect the response. Specific parameters expected to be impacted by CS vary depending on the type and purpose of seeding, but include the cloud's liquid and ice-water content, the vertical extent of the cloud, the volume of the cloud, the intensity and amount of precipitation, the area covered by precipitation, the precipitation path, radar reflectivity, and other parameters measured in situ (Fig. 2.4), remote-sensing (weather radars, microwave satellite and ground radiometers, lidars, etc.) and ground-based instruments (rain gauges, river runoff, snow depth, etc.). For example, for the purpose of detecting physical efficacy of cloud seeding for hail mitigation, Abshaev et al. (2003) suggest applying map of hail kinetic energy (E, J/m^2) integrated over relatively long periods in terms of month and years calculated based on radar data (Fig. 2.5). Comparison of integrated E multiplied by square of protected and adjacent (control) areas can be used for estimation of physical effect of HS.

A novel approach is to use numerical cloud models to augment statistical and experimental cloud-seeding programs (Geresdi et al. 2017; Segal et al. 2004; Xue et al. 2013a; Zhekamukhov and Abshaev 2009a). Recent improvements in the sophistication of numerical models, aided by the advances in supercomputing, make possible very high-resolution three-dimensional simulations that have the capability to simulate cloud seeding in a physically meaningful manner and account for model uncertainty using ensemble modeling methods (Rasmussen et al. 2018). Newly obtained observations of the seeding impact in winter orographic clouds (French et al. 2018, Tessendorf et al. 2019) have provided unprecedented datasets

a

b

Fig. 2.4 Highly equipped middle-sized research aircraft "ROSHYDROMET YAK-42D" of the Russian Hydrometeorological Service for measurement of atmospheric and cloud characteristics and cloud seeding, using various types of seeding materials: **a** external view; **b** sensors for aerosol, cloud droplets, and ice crystals spectrum, temperature, humidity, water content. *Source* Viktor Petrov; used with permission

to thoroughly quantify the physical response of precipitation production to seeding (Friedrich et al. 2020; Fig. 2.6) and to validate numerical models (Rauber et al. 2019).

Recently, French et al. (2018) carried out complex measurements of orographic clouds seeded from aircraft using ground-based X-band radars, an airborne W-band cloud radar, and instrumented aircraft for employing in-situ cloud-physics probes. They found the strongest evidence for the initiation and growth of ice crystals as a result of glaciogenic seeding with AgI, leading to precipitation (snow) on a mountain surface within a specific target region. These observations, in two separate cases, showed the initial appearance of cloud-seeding signatures within 30 min following the release of AgI in the cloud. Seeding lines were tracked, and the evolution of ice crystals to precipitating snow was documented (Fig. 2.7). These comprehensive

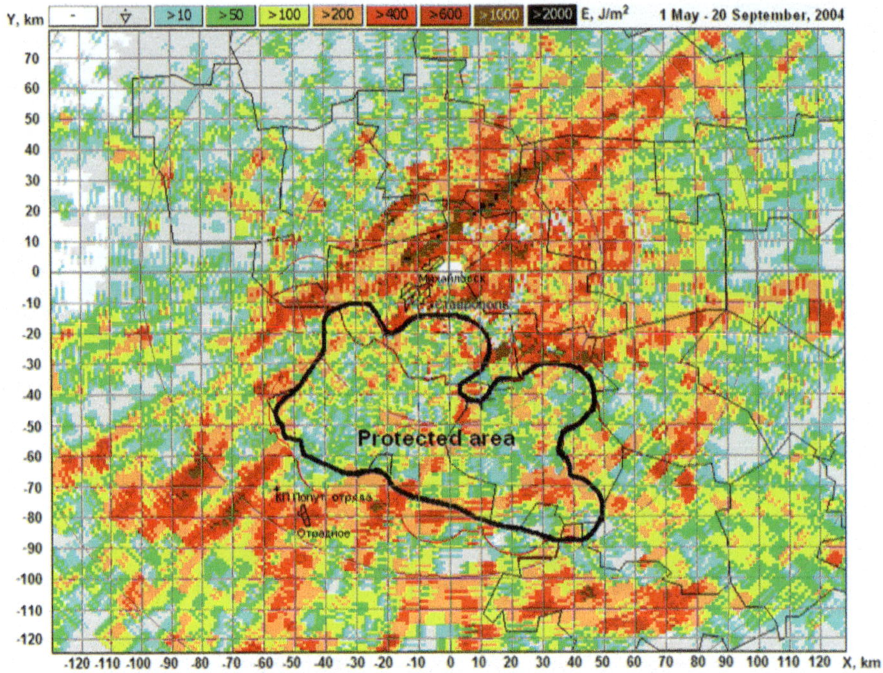

Fig. 2.5 Example of radar-measured kinetic energy (E, J/m^2) of hail in the Stavropol district of Russia, collected from 1 May to 20 September 2004. The black contour denotes the area where hail clouds were seeded (*Source* Abshaev et al. 2006)

observations provide unambiguous evidence that glaciogenic seeding of a supercooled liquid cloud can enhance natural precipitation growth in a seeded cloud, leading to precipitation that would otherwise not fall within the targeted region.

2.5 Could-Seeding Conditions

It is important to specify the environmental conditions that must be satisfied before the seeding is commenced. For example, these seedability criteria need to ensure that there is SLW that can be converted into ice crystals by the glaciogenic SM in the cloud. In turn, the ice crystals can grow and ultimately fall into the target area. To ensure that the SM properly interacts with the SLW, there are also conditions on the dispersion and targeting of the SM, with the specific conditions dependent upon the seeding strategy. The targeting conditions require a range of wind speeds and directions to allow for the mixing of SM from ground-based generators to a level where the SM activates ice nuclei. Manton et al. (2011) found that seeding from ground-based generators was not effective at low wind speeds.

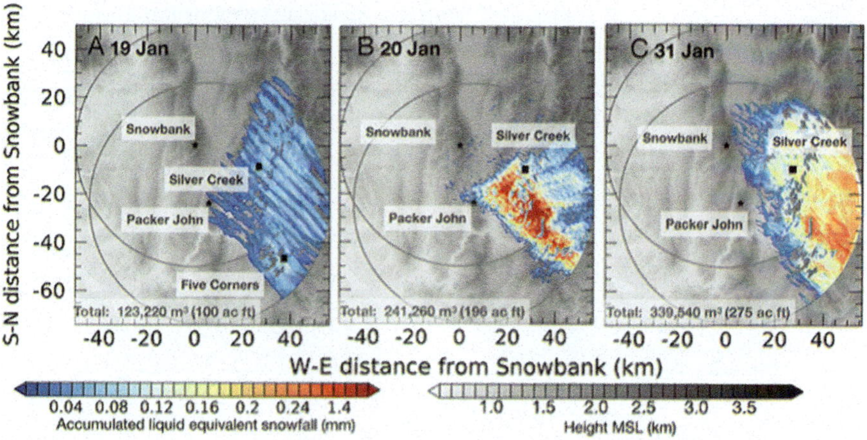

Fig. 2.6 Distribution of accumulated liquid-equivalent snowfall (S) attributed to parallel cloud seeding lines over the observational period between **a** 1705 and 1806 UTC on 19 January; **b** 0042 and 0315 UTC on 20 January; and **c** 2117 and 2151 UTC on 31 January 2017, using the best-match Ze–S relationship for that day. Data are shown on a 100 × 100 m grid. Total accumulations over the entire domain and observational period are highlighted (*Source* Friedrich et al. (2020), used with permission of PNAS)

Also, the aerosol particle population of the ambient air needs to correspond to the selected seeding method, potentially excluding certain population scenarios for mixed-phase clouds. The dispersion of SM to interact with available SLW is the first step to ensuring that any enhanced precipitation falls into the target area. SM such as AgI leads to ice nucleation at temperatures colder than about −5 °C, and so it is usual to require cloud-top temperatures to be less than about −8 °C (Breed et al. 2014). By considering the ratio of seeded to unseeded precipitation in the target area, it is often implicitly assumed that the seeding impact is multiplicative. This assumption implies that there needs to be some natural precipitation during seeding that is enhanced by seeding. Manton et al. (2017) demonstrated that there was a negligible impact of seeding when the natural precipitation was low, and so it was preferable that some precipitation was falling at the time of seeding commencement. Related to this condition is the need for a forecast of seedable conditions to persist over the duration of seeding. Such forecasts are usually developed through analysis of numerical-model results.

2.6 Economics

The economic benefits of RE arise from the value of the increased water reaching the ground. That water will either directly feed agricultural crops or more likely lead to increased runoff into regional hydrological systems. However, the major challenge

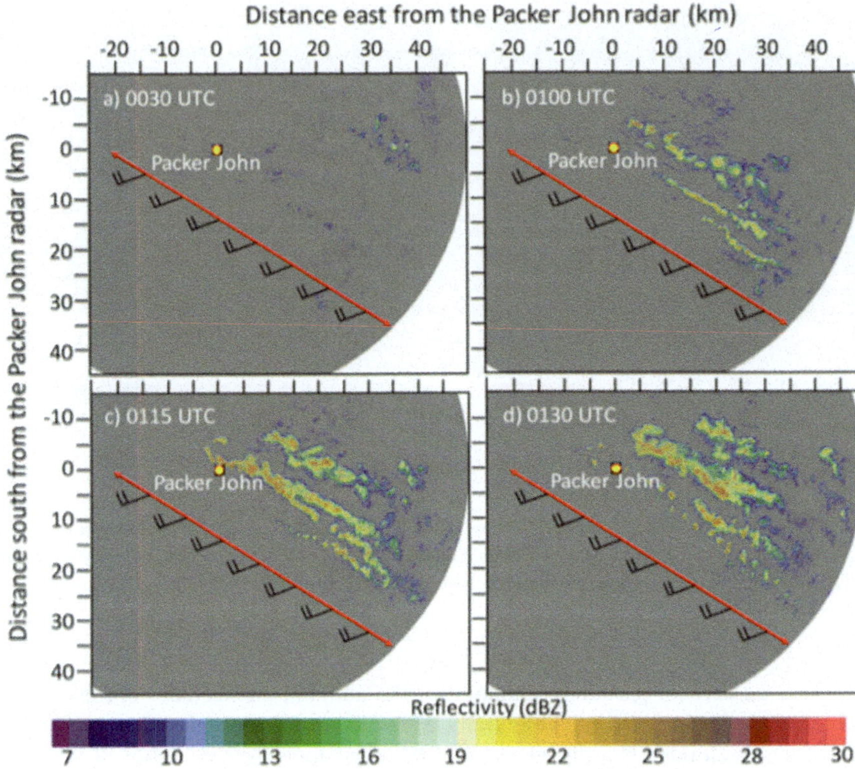

Fig. 2.7 Radar reflectivity from the ground-based Doppler radar located at Packer John mountaintop for four time periods more than 60 min after beginning seeding: **a** 0030, **b** 0100, **c** 0115, and **d** 0130 UTC. The scans were conducted at a 2° elevation angle and, therefore, show the reflectivity just above ground level close to the radar to roughly 1.7 km above ground level at 50-km range. The red line indicates the track of the seeding aircraft, which was repeated eight times. Wind barbs plotted on the aircraft track illustrate mean wind direction and speed (in m per sec) at flight level. Each barb is 10 m s^{-1}. (*Source* French et al. (2018), used with permission from PNAS)

is that the physical processes extend across an extremely large range of spatial and temporal scales. The spatial scales range from submicron to synoptic scale, while the temporal scales can range from microseconds to several hours or longer. One of the often-neglected issues in RE is the scaling up from small to larger scales. This relates to the consideration, explanation, and provision of proof through each link in the chain of events, from the seeding intervention to more precipitation on the ground, in such a way that the result has an acceptable impact with a desirable benefit–cost ratio. This challenge should be viewed in tandem with all the practical and logistical considerations when scaling up from single-cloud experiments to area-wide impacts. The recent observations of French et al. (2018) provide substantial evidence of this chain of events for wintertime orographic seeding. This issue is especially critical and difficult when dealing with convective clouds. Some of these issues were studied by Terblanche et al. (2000) during a semi-operational RE experiment in South Africa. The authors attempted to link the apparent positive storm-scale seeding effect to an observed larger-scale rainfall anomaly observed in the rainfall records in the area of seeding. However, simple calculations proved that there was a "two orders of magnitude challenge" between what could have been realistically expected from the seeding interventions on a storm scale and what was observed in the area-wide rainfall records for the rainfall season. They concluded that the interventions and observations were probably unrelated. In a similar scenario, Terblanche et al. (2005) attempted to calculate the cost–benefit ratio of additional rainfall in a continuation of semi-operational experiments in South Africa. For this purpose, he studied the storm climatology to estimate how many storms will have to be treated in a rainfall season to have the desired area effect if the storm effect they observed could be used as the basis for calculation. From these studies, it became evident that rainfall enhancement could be more favorable than other options to address water stress in South Africa, but there could be several logistical challenges in treating the number of storms required, even though there appear to be sufficient candidate storms for treatment. As most storms develop in a short period of time in the afternoon, the authors concluded that new, more efficient ways to deliver SM (other than standard aircraft) will have to become a priority for the future. Even for winter orographic CS, careful calculation is needed to ensure that the benefit outweighs the cost.

For both the Snowy Rain Enhancement Research Project (Manton et al. 2011) in Australia and the Wyoming Weather Modification Pilot Project (Breed et al. 2014) in the USA, the fractional increase in precipitation for seedable events is on the order of 15%. But seedable events generally make up only a fraction of the overall annual or seasonal precipitation. Moreover, the transformation of precipitation on the ground into hydrological streamflow incurs losses from evaporation and recharging of groundwater, as well as delays, as the water passes through the complex hydrological system.

The relatively small precipitation signal and the complexity of the hydrological system mean that it is very unlikely that the impact of CS could be detected directly in measurements of streamflow or dam volume. Detailed rainfall–runoff modelling is needed to estimate the actual increase in annual streamflows due to CS. On the other hand, the increasing scarcity of potable water around the world means that the potential benefits of CS will continue to increase, while the costs should remain constant or decrease due to technological and scientific advances.

It is not uncommon for communities to seek relief from drought through CS activities. Indeed, Yoshida et al. (2009) found that seeding in Japan may be effective for drought relief. On the other hand, in many countries especially those affected by the El Nino-Southern Oscillation phenomenon (Nicholls and Wong 1990), the variability of precipitation is so high that periods of drought are associated with an essential absence of clouds suitable for seeding. Thus, the cost–benefit analysis for a project needs to account for the prevailing climatic conditions. On the other hand, despite the lack of evidence of area-wide and seasonal-scale impacts, seeding is often carried out on convective clouds based on the potential for a significant benefit at a relatively low cost (Bruintjes 1999). Such strategies are viewed as a component of an overall approach to risk management of water resources, bearing in mind the substantial scientific uncertainties.

2.7 Redistribution and "Negative Enhancement" of Precipitation

The areas affected by CS remain an open question, especially with regard to convective systems. Related uncertainties pertain to the issue of "extra-area" effects, that is, whether seeding can affect the weather beyond the targeted temporal or spatial range. The persistent effects of CS claimed by Bigg (1995) should be carefully assessed, as should the statistical results from experiments in Thailand (Woodley et al. 2003) and Israel (Brier et al. 1973), which claimed effects beyond a few hours.

Some professionals argue that increasing precipitation in one region could reduce precipitation downwind (by "stealing" the atmospheric water vapor); for example, recent modelling studies by Geresdi et al. (2017) suggest that in some circumstances there may be a decrease in precipitation on the leeside of a mountain, even when there is an overall increase over the whole domain. On the other hand, analysis by Long (2001) suggests that enhanced downwind precipitation may be promoted by the transport of ice nuclei and ice crystals or by the dynamic invigoration of clouds through the release of latent heat. Overall, further quantitative studies are needed to resolve these issues, bearing in mind the uncertainties in assessing the impacts of seeding in a designated area.

Givati and Rosenfeld (2004) suggested that urban air pollution in California and Israel may reduce annual rainfall by about 15–25%. According to Khain et al. (2005), small CCN may produce small droplets, which have small collision efficiency, thereby reducing precipitation from deep convective clouds. Introducing superfine hygroscopic SM into the clouds would then initiate the formation of small droplets that compete with existing cloud droplets in the water-vapor absorption process within the cloud. This method may prevent the development of precipitation in some cases. On the other hand, introducing giant hygroscopic SM into clouds can increase the collision efficiency of droplets and lead to the rapid development of rain. This mechanism can be applied to developing upwind clouds with the potential to produce rain over a substantial target area. This "jumping process mechanism" can then reduce the potential of the cloud to develop rain over a target area.

During the last 30 years, considerable work has been done in Russia on precipitation redistribution above megalopolises and neighboring areas to prevent rain occurrences during important local events. More than 80 projects have been completed on national holidays (Koloskov et al. 2011), with various (cold and warm) types of clouds (stratus and convective) being seeded by 6–12 aircrafts dispersing AgI, liquid nitrogen, solid carbonic acid, coarse-dispersion powders, and hygroscopic particles. Four different methods are commonly used, depending on the synoptic situation: (a) dispersion of stratiform clouds; (b) destruction of convective clouds or reduction of the intensity of shower rains and thunderstorms; (c) premature initiation of precipitation from clouds on the upwind side of the target area to create a "precipitation shadow" over the given site; (d) reduction of rainfall intensity over the target area by intensive seeding of the rain-producing clouds moving toward it; this is aimed at weakening the mechanism of precipitation formation through "over-seeding", i.e., by creating excessive concentrations of ice crystals. However, further quantitative substantiation of precipitation redistribution is needed, especially for convective rainfall.

Debates about the effects of seeding beyond the target area point to the fact that WM can be viewed as more than just a means to increase local precipitation. Rather, it can be viewed as a means to alter natural hydrological cycles by increasing the number of times that atmospheric water is recycled at the Earth's surface. As more is learned about the global water balance, and as new tools enable cloud scientists to better understand clouds and their response to seeding, the question of extended area effects will likely become better defined and understood. All these effects will have to be considered against the background of climate change and the associated changes in precipitation in time and space globally.

2.8 Environmental Issues

By design, precipitation enhancement aims to alter the natural environment, and so there is a potential for undesirable changes to the environment. Dennis (1980) gave a critical analysis of the risks associated with CS, including toxicological, ecological, sociological, and legal challenges. He noted an absence of evidence of environmental hazards, and this conclusion has been confirmed over the ensuing decades. Lincoln-Smith et al. (2011) studied the potential risk of the SM (particularly silver and indium) to the overall environment of the Snowy Mountains in Australia. Observations at the generator sites showed that levels of seeding chemicals were well below any trigger levels for health concerns and there was no indication of accumulated impacts over a five-year period. Abshaev et al. (2014) analyzed the results of measurements of AgI and PbI_2 in air, soil, water reservoirs, and precipitation in regions with long-term (more than 40 years) implementation of artillery and rocket HS technology in specific areas (Northern Caucasus, Moldova, and Georgia), and found that the maximum concentration of these hazardous pollutants is several orders below the maximum allowable concentration specified by the World Health Organization. Moreover, Korneev et al. (2017) showed that the utilization of AgI in aircraft seeding in Russian investigations did not lead to measurable increases in the level of these chemicals due to natural and anthropogenic sources. They suggested that seeding has extremely low impacts on the environment, and they did not observe any extra-area effects. Therefore, studies conducted in Russia suggest that direct delivery of SM into clouds should cause no ecological concern even after decades of implementation. On the other hand, Fajardo et al. (2016) used laboratory studies to suggest that some biota could be adversely affected if "large amounts of SM accumulated in the environment".

2.9 Planning Rain Enhancement Projects

Given the state of knowledge about WM, it is important to consider how to plan activities so that they might gain both social and scientific acceptance. The preference of the scientific approach for WM experiments is to carry out well-designed long-term experiments involving proper physical and statistical controls and cloud-physics measurements prior to and during the operations. At each stage in the planning, execution, and evaluation of a WM experiment, it is necessary to consider meteorological and cloud-physics aspects, statistical approaches, and economic, social, and environmental aspects.

Economic analyses have shown that successful RE operations could have real economic benefit, but the impacts of operations have still not been properly quantified. Despite some of the uncertainties of WM, RE remains a potential option and should be viewed as a part of an integrated water-resource management strategy. Each project should be treated as a possible tool among others for water-resource management and should be considered as a scientific project with the inclusion of these four phases: (a) feasibility study using the climatology of clouds and precipitation in the process of site selection; (b) design of the experiment as a function of this climatology and the present knowledge of cloud physics and WM; (c) implementation of an experiment with randomization, using extensive physical measurements and statistical controls; and (d) objective and independent evaluation of the results. Government decision-makers and funding agencies should be aware that such projects need considerable funding and well-trained personnel, as well as time to obtain conclusive results.

It is necessary to consider systematically all aspects of the set-up and proper conduction of the project, consistent with a desirable location, climatology, season, SM, delivery criteria, etc. The results of several "rain-making" projects have been inconclusive because of the lack of sound scientific planning, operation, and evaluation.

The pre-eminent need for a WM project to be planned free of non-scientific influences is very important and must be observed in the procedure for selecting the location and the season for conducting the project. Detailed feasibility studies of meteorological, climatological, hydrological, social, and environmental considerations, including the availability of logistical support are the prerequisites for the selection of an appropriate site and season to maximize the chances of achieving the objectives of RE. The RE site should be in a relatively homogeneous area for two reasons: to minimize errors in estimates of mean precipitation in the target and control areas, and to decrease the natural spatial rainfall variability, which is very important for the assessment of the results. It should be acceptable to have at least one control area with characteristics like those of the target area.

The basic measurements to evaluate and support a seeding hypothesis should be vigorously carried out. Operations should include measurements of physical response variables and should be randomized when possible to allow for an independent evaluation. Finally, attention should be given to the participants' education and training in cloud physics and associated sciences, which should be an essential component of any RE project. The RE may be economically viable and may contribute to alleviating the adverse effects of severe water shortages, but this is still to be demonstrated to the national water managers and policymakers.

2.10 Conclusions

From numerous reports, there is probabilistic evidence that RE can work with reasonable cost–benefit ratios. However, many components of RE processes must be clarified and proven. Experience in applying CS in numerous countries has shown that precipitation can be increased, ranging from essentially zero to more than 15% of the annual norm, depending on the available cloud resources, reagents, and delivery methods. Higher values tend to be associated with direct delivery of SM to the clouds utilizing aircrafts and rockets. However, the reasons for the large variation in impacts are not well understood, and estimates of impacts are sensitive to the estimation of the natural precipitation in the target area.

The current challenge is to provide a credible scientific footing for the planning of WM experiments. To be successful, any WM project should combine the efforts of funding administrators, scientists planning and working as seeding operators, and the scientific committee that assesses the scientific integrity of the project and/or evaluators.

Focusing on better understanding of the aerosol–cloud interactions (natural, anthropogenic, and seeded aerosols) and cloud microphysical and dynamical processes will bring progress also in solving problems related to the role of clouds in precipitation forecasting and climate change. Advances in cloud physics could clarify the human influence on the environment (e.g., brown-cloud, precipitation change issues), and they are not only beneficial to the field of WM but also to weather forecasting and climate issues.

It is important to emphasize that CS technology is not a "cure" for droughts. The technology requires suitable clouds for the success of a program. Unfortunately, when a drought is in progress, these types of clouds are often not available.

Well-conducted RE programs should have a systematic approach covering a wide range of factors, among which are:

- continuously evolving understanding of cloud and precipitation processes emerging from ongoing studies;
- adequate seeding hypotheses, subjecting refined seeding techniques to rigorous field testing with numerical modeling, where appropriate;
- sophisticated measuring instruments (radars, rain gauges, satellite data);
- proper SM and their cloud delivery devices;
- proven methods for assessing physical and economic efficacy, and
- regular personnel training.

Without stable, long-term support, relevant expertise, as well as a sound and systematic scientific approach, such programs will not succeed. The sustained documentation, experimentation of emerging technologies, better reagents and dispersal methodologies, and more accurate forecasting facilities to support decision making for the RE are the key factors.

Scientific understanding of cloud processes continues to increase around the world through the sharing of data and knowledge. The design, implementation, and evaluation of a catchment-scale RE experiment require a major investment of funds and resources. The sharing of the data and results of these experiments through publication in the international scientific literature provides the feedback that will help resolve the remaining uncertainties associated with CS science.

The WM has seen more than 70 years of development from Schaefer's use of dry ice pellets to produce holes in supercooled stratus by snow-out, to the present-day use of digitized radar networks and sophisticated seeding methods showing success in the RE and HS when scientific designs are applied, and proper cloud conditions exist. In the long run, the future of science-based weather modification is real. There is a growing evidence that the basic concepts are correct and that successful implementation is feasible.

The achievements are the product of international collaboration between hundreds of enthusiastic scientists working in the field. An example is to the thorough planning of the WMO Rain Enhancement Project (REP) in Spain which produced more than 30 reports on various aspects of rain enhancement. The facilitating role of the already ten quadrennial WMO Scientific Conferences on WM for the exchange of information and establishment of collaborative studies must be noted. The Indian Government has conducted a detailed science experiment named CAIPEEX to cater to the scientific evaluation of cloud seeding by both statistical and physical means. Two key national cloud-seeding projects are being carrying out in China (Yao 2006; Guo 2015), one consists of the orographic cloud-seeding experiment in six provinces of Northwest China, and the other is the Chinese Randomized Precipitation Enhancement Experiment (CRPEEX) (2014–the present) in four provinces. The experiences and results from these key projects guide and support local operational cloud-seeding activities in China. In recent years, the UAE has made an additional contribution to the scientific and practical development of rainfall-enhancement technology through the UAE Research Program for Rain Enhancement Science (Al Mazroui et al. 2017). Under the UAE's National Center of Meteorology, the program has brought together a cohort of leading scientists and institutions from around the world to untangle the complexities of the natural rainfall process and augment rainfall amounts (http://www.uaerep.ae/). A total of nine three-year distinct research projects have been funded, leveraging multiple disciplines, including material science, bio-geo-engineering, artificial intelligence, and unmanned aircraft systems. In less than five years, the program has stimulated wide research interest in the field of WM, specifically rainfall enhancement, among the international scientific community. The projects have produced state-of-the-art measurements, advanced numerical models, innovative prototypes, and proof-of-concept field demonstrations. Furthermore, the use of field operational programs as a platform for basic research efforts continues to be an excellent opportunity for both scientists and students to conduct a broad range of studies and enhance the knowledge base of RE, as well as the aerosol–cloud interaction which are among the least understood physical processes in weather and climate models.

Confronting already existing or imminent water shortages, meteorologists and their national meteorological services in semi-arid regions can explore the scientific basis for carrying out RE projects. In this connection, they can inform their governments that the United Nations Convention to Combat Desertification (Art.17) also mentions RE as one of the methods for water management. Advances and limitations of RE should be clearly stated. Only well-informed governments and/or communities will provide adequate funding needed for the beneficial application of WM projects.

References

Abshaev MT (1966) Concentration of hailstones and hail embryos. Proc High Mountain Geophys Inst 3(5):191–196

Abshaev MT (1994) A new concept of hailstorm modification. Paper presented at 6th WMO Conference on Weather Modification, 30 May–4 June 1994, Paestum, Italy

Abshaev MT, Malkarova AM, Tebuev AD (2003) Radar estimation of hail damage. Paper presented at the 8th WMO science conference on weather modification, 7–12 April 2003, Casablanca, Morocco

Abshaev AM, Abshaev MT, Sadykhov YA (2004) Diffusion of artificial aerosol in Cu cong clouds. Russ Meteorol Hydrol 9:18–24

Abshaev MT, Sulakvelidze GK, Burtsev II, Fedchenko LM, Jekamukhov MK, Abshaev AM, Kuznetsov BK, Malkarova AM, Tebuev AD, Nesmeyanov PA, Shakirov IN, Shevela GF (2006) Development of rocket and artillery technology for hail suppression, pp 109–127. Al Mandous A, Bojkov R, Al Mazroui A, Al Muhairi M (eds). Achievements in weather modification. Department of Atmospheric Studies, Abu Dhabi, UAE

Abshaev MT, Abshaev AM, Malkarova AM et al (2009) Radar estimation of water content in cumulonimbus clouds. Izv Atmos Ocean Phys 45:731–736. https://doi.org/10.1134/S0001433809060061

Abshaev MT, Abshaev AM, Malkarova AM et al (2010) Automated radar identification, measurement of parameters, and classification of convective cells for hail protection and storm warning. Russ Meteorol Hydrol 35:182–189. https://doi.org/10.3103/S1068373910030040

Abshaev MT, Abshaev AM, Kuznecov BK, Kotelevich AF, Guzoev TH, Balakova N, Chochaev HH, Yakushkin BV, Ponomarenko RN, Trifonov VS (2011a) New advances in automation of anti-hail rocket technology. Paper presented at Tenth WMO conference on weather modification, 4–7 October 2011, Bali, Indonesia

Abshaev AM, Sadykhov YA, Malkarova AM (2011b) On the choice of diffusion schemes in numerical simulation of crystallizing aerosol propagation in the cloud medium. Russ Meteorol Hydrol 36:737–746. https://doi.org/10.3103/S1068373911110057

Abshaev MT, Abshaev AM, Barekova MV, Malkarova AM (2014) Manual on organization and execution of hail suppression projects, Nalchik, 508 pp

Al Mandous A, Bojkov R, Al Mazroui A, Al Muhairi M (2006) Achievements in weather modification. Department of Atmospheric Studies, Abu Dhabi, P. 186

Al Mazroui A (2017a) Advancing the science, technology and implementation of rain enhancement. Project of water security solutions in arid and semi-arid regions and beyond. https://doi.org/10.13140/RG.2.2.34696.21762

Belyaeva MV, Drofa AS, Ivanov VN (2013) Efficiency of stimulating precipitation from convective clouds using salt powders. Izvestiya Atmosph Oceanic Phys 49(2):154–161. https://doi.org/10.1134/S0001433813010039

Bengtsson L (2010) The global atmospheric water cycle. J Environ Res Lett 5:025202. https://doi.org/10.1088/1748-9326/5/2/025202

Bigg EK (1995) Tests for persistent effects of CS in a recent Australian experiment. J Appl Meteorol Climatol 34(11):2406–2411. https://doi.org/10.1175/1520-0450(1995)034%3c2406:TFPEOC%3e2.0.CO;2

Borland SW, Browning KA, Changnon SA, Cooper WA, Danielsen EF, Dennis AS, … Young KC (1977) Hail: a review of hail science and hs. hail: a review of hail science and HS. J Amer Meteorol Soc. https://doi.org/10.1007/978-1-935704-30-0

Braham RR (1986) Precipitation enhancement—a scientific challenge. Precipitation enhancement—a scientific challenge. Amer Meteorol Soc 43:1–5

Breed D, Rasmussen R, Weeks C, Boe B, Deshler T (2014) Evaluating winter orographic CS: design of the wyoming weather modification pilot project (WWMPP). J Appl Meteorol Climatol 53(2):282–299. https://doi.org/10.1175/JAMC-D-13-0128.1

Brier GW, Grant LO, Mielke Jr. PW (1973) An evaluation of extended area effects from attempts to modify local clouds and cloud systems. Proc WMO/IAMAP scientific conference on weather modification, Tashkent, Uzbekistan, WMO, 439–447

Bruintjes RT (1999) A review of CS experiments to enhance precipitation and some new prospects. Bullet Amer Meteorol Soc 80:805–820

Bruintjes RT, Salazar V, Semeniuk TA, Buseck P, Breed DW, Gunkelman J (2012) Evaluation of hygroscopic cloud seeding flares. J Weather Modificat 44(1):69–94. Available at http://journalofweathermodification.org/index.php/JWM/article/view/85

Burtsev II, Ugryumov AI, Dyadyuchenko VN, Martanov VV, Zakharov VM, Shulyakovsky GE, Stasenko VN,…Korneev VP (2018) Essays on the history of weather modification in the USSR and the post-soviet territory. St.Petersburg, RSHMU, 2017. 352 pp. Retrieved from http://mig-journal.ru/toauthor?id=4644

Cooper WA, Bruintjes RT, Mather GK (1997) Calculations pertaining to hygroscopic seeding with flares. J Appl Meteorol Climatol 36(11):1449–1469. https://doi.org/10.1175/1520-0450(1997)036%3c1449:CPTHSW%3e2.0.CO;2

Dennis AS (1980) Weather Modification by CS. Academic Press, 267 pp., 1980. Retrieved from https://digitalcommons.usu.edu/water_rep/670

Dessens J, Sánchez JL, Berthet C,Hermida L, Merino A (2016) Hail prevention by ground-based silver iodide generators: results of historical and modern field projects. Atmosph Res 170:98–111. https://doi.org/10.1016/j.atmosres.2015.11.008

Dinevich LA, Shalaveyus SS (2010) Experience in the use of tracers to study the spread of chemicals in the artificial effect on convective clouds, Modern problems of science and education, No. 12, pp 64–86

Drofa AS, Eran'kov VG, Ivanov VN, Shilin AG, Iskevich GF (2013) Experimental investigations of the effect of cloud-medium modification by salt powders. Izvestiya Atmosph Oceanic Phys 49(3):298–306. https://doi.org/10.1134/S0001433813030043

Eke J, Yusuf A, Giwa A, Sodiq A (2020). The global status of desalination: an assessment of current desalination technologies, plants and capacity. J Desalination 495:114633. https://doi.org/10.1016/j.desal.2020.114633

Fajardo C, Costa G, Ortiz LT, Nande M, Rodríguez-Membibre ML, Martín M, Sánchez-Fortún, S (2016). Potential risk of acute toxicity induced by AgI cloud seeding on soil and freshwater biota. Ecotoxicol Environ Saf 133:433–441. https://doi.org/10.1016/j.ecoenv.2016.06.028

Findeisen W (1938) Die Kolloid-meteorologische Vorgänge bei der Neiderschlags-bildung. Meteorologische Zeitschrift. Meteorol 55:121–133

Flossmann AI, Manton M, Abshaev A, Bruintjes R, Murakami M et al (2018) Peer review report on global precipitation enhancement activities. Research Report. WMO, Geneva

Flossmann AI, Manton M, Abshaev A, Bruintjes R, Murakami M, Prabhakaran T, Yao Z (2019). Review of advances in precipitation enhancement research. Bullet Amer Meteorol Soc 100(8):1465–1480. https://doi.org/10.1175/bams-d-18-0160.1

French JR, Friedrich K, Tessendorf SA, Rauber RM, Geerts B, Rasmussen RM, Xue L, Kunkel ML, Blestrud DR (2018) Precipitation formation from orographic CS. Proceedings of the national academy of sciences of the United States of America, 115(6):1168–1173. https://doi.org/10.1073/pnas.1716995115

Friedrich K, Ikeda K, Tessendorf SA, French JR, Rauber RM, Geerts B,... Parkinson S (2020). Quantifying snowfall from orographic cloud seeding. Proceedings of the national academy of sciences of the United States of America 117(10):5190–5195. DOI https://doi.org/10.1073/pnas.1917204117

Geresdi I, Xue L, Rasmussen R (2017). Evaluation of orographic CS using a bin microphysics scheme: two-dimensional approach. J Appl Meteorol Climatol 56(5):1443–1462. https://doi.org/10.1175/JAMC-D-16-0045.1

Givati A, Rosenfeld D (2004) Quantifying precipitation suppression due to air pollution. J Appl Meteorol Climatol 43(7):1038–1056. https://doi.org/10.1175/1520-0450(2004)043%3c1038:QPSDTA%3e2.0.CO;2

Guo XL, Fu DH, Li XY, Hu ZX, Lei HC, Xiao H, Hong YC (2015) Advances in cloud physics and weather modification in China. Adv Atmosph Sci 32(2), 230–249. https://doi.org/10.1007/s00376-014-0006-9

Huggins AW, Kenyon SL, Warren L, Peace AD, Billish SP, Manton MJ (2008) The Snowy rain enhancement research project. A description and preliminary results. J Weather Modificat 40(1), 28–53. Retrieved from https://journalofweathermodification.org/index.php/JWM/issue/view/55

Khain A, Rosenfeld D, Pokrovsky A (2005). Aerosol impact on the dynamics and microphysics of deep convective clouds. J Royal Meteorol Soc 131(611):2639–2663. https://doi.org/10.1256/qj.04.62

Koloskov BP, Korneev VP, Petrov VV, Beryulev GP, Danelyan BG, Shchukin GG (2011) Some results of activities on the improvement of weather conditions over metropolises. J Russian Meteorol Hydrol 36(2):117–123. https://doi.org/10.3103/S1068373911020063

Korneev VP, Potapov EI, Shchukin GG (2017) Environmental aspects of cloud seeding. Russian Meteorol Hydrol 42(7), 477–483. https://doi.org/10.3103/S106837391707007X

Langmuir I (1950) Control of precipitation from cumulus clouds by various seeding techniques. Science 112(2898):35. https://doi.org/10.1126/science.112.2898.35

Liang H, Abshaev MT, Abshaev AM, Huchunaev BM, Griffiths S, Zou L (2019) Water vapor harvesting nanostructures through bioinspired gradient-driven mechanism. Chem Phys Lett 728, 167–173. https://doi.org/10.1016/j.cplett.2019.05.008

Lincoln-Smith M, Dye A, Kemsley K, Denholm J (2011) Environmental monitoring and assessment. A statistical analysis of concentrations of silver and indium at generator locations. J Weather Modificat 43(1):1–8. From https://www.journalofweathermodification.org/index.php/JWM/article/view/145

Long AB (2001) Review of downwind extra-area effects of precipitation enhancement. J Weather Modif 33(1):24–45. From https://journalofweathermodification.org/index.php/JWM/article/view/237

Manton MJ, Warren L, Kenyon SL, Peace AD, Bilish SP, Kemsley K (2011) A Confirmatory snowfall enhancement project in the snowy mountains of Australia. Part I: project design and response variables. J Appl Meteorol Climatol 50(7):1432–1447. https://doi.org/10.1175/2011JAMC2659.1

Manton MJ, Peace AD, Kemsley K, Kenyon SL, Speirs JC, Warren L, Denholm J (2017) Further analysis of a snowfall enhancement project in the Snowy Mountains of Australia. Atmosph Res 193:192–203. https://doi.org/10.1016/j.atmosres.2017.04.011

National Research Council (2003) Critical issues in weather modification research. Washington, DC: The National Academy Press 131. DOI https://doi.org/10.17226/10829

Nicholls N, Wong KK (1990) Dependence of rainfall variability on mean rainfall, latitude, and the southern oscillation. J Climate 3(1):163–170. https://doi.org/10.1175/1520-0442(1990)003%3c0163:DORVOM%3e2.0.CO;2

Rasmussen RM, Tessendorf SA, Xue L, Weeks CE, Ikeda K, Landolt S, Lawrence B (2018) Evaluation of the wyoming weather modification pilot project (WWMPP) using two approaches: traditional statistics and ensemble modeling. J Appl Meteorol Climatol 57:2639–2660. https://doi.org/10.1175/JAMC-D-17-0335.1

Rauber RM, Geerts B, Xue L, French J, Friedrich K, Rasmussen RM, Tessendorf SA, Blestrud DR, Kunkel ML, Parkinson S (2019) Wintertime orographic cloud seeding-a review. J Appl Meteorol Climatol 58:2117–2140. https://doi.org/10.1175/JAMC-D-18-0341.1

Reinking RF, Martner BE (1995) Feeder-cell ingestion of seeding aerosol from cloud base determined by tracking radar chaff. J Appl Meteorol Climatol 35(9):1402–1415. https://doi.org/10.1175/1520-0450(1996)035%3c1402:FCIOSA%3e2.0.CO;2

Rosenfeld D, Lensky IM (1998) Satellite—based insights into precipitation formation processes in continental and maritime convective clouds. Bullet Amer Meteorol Soc 79(11):2457–2476. https://doi.org/10.1175/1520-0477(1998)079%3c2457:SBIIPF%3e2.0.CO;2

Rosenfeld D, Axisa D, Woodley WL, Lahav R (2010) A quest for effective hygroscopic CS. J Appl Meteorol Climatol 49(7):1548–1562. https://doi.org/10.1175/2010JAMC2307.1

Segal Y, Khain A, Pinsky M, Rosenfeld D (2004) Effects of hygroscopic seeding on raindrop formation as seen from simulations using a 2000-bin spectral cloud parcel model. Atmospher Res 71:3–34. https://doi.org/10.1016/j.atmosres.2004.03.003

Segal Y, Pinsky M, Khain A (2007). The role of competition effect in the raindrop formation. Atmosph Res 83:106–118. https://doi.org/10.1016/j.atmosres.2006.03.007

Semeniuk TA, Bruintjes R, Salazar V, Breed D, Jensen T, Buseck PR (2014) Individual aerosol particles in ambient and updraft conditions below cloud bases in the Oman mountain region. J Geophys Res Atmosph 119:2511–2528. https://doi.org/10.1002/2013JD021165

Shilin AG, Fedorenko AI, Ivanov VN, Savchenko AV (2015) A study of ageing processes in ice-forming aerosols containing silver iodide and organic ice-forming substances, Collected papers "Problems of cloud physics" in the memory of N.O. Journal of Plaude, Russian Hydrometeorological Service, Central Aerological Observatory, Moscow, p 342–347, ISBN: 978-5-901579-62-6

Sulakvelidze GK, Bibilashvili NS, Lapcheva VP (1965) Formation of precipitation and modification of hail processes. Leningrad, Hydrometizdat 203. (Translation available from National Technical Inform. Service, Springfield, VA, USA)

Tai Y, Liang H, Zaki A, El Hadri N, Abshaev AM, Huchinaev BM, Griffiths S, Jouiad M, Zou L (2017). Core/shell microstructure induced synergistic effect for efficient water-droplet formation and cloud-seeding application. J ACS Nano 11(12):12318–12325. Retrieved from http://pubs.acs.org/doi/abs/10.1021/acsnano.7b06114

Terblanche DE, Mittermaier MP, Burger RP, De Waal KJP, Ncipha XG (2005) The South African rainfall enhancement programme: 1997–2001. Water SA 31(3):291–298. Retrieved from http://hdl.handle.net/10520/EJC116273

Terblanche DE, Steffens FE, Fletcher L, Mittermaier MP, Parsons RC (2000) Toward the operational application of hygroscopic flares for rainfall enhancement in South Africa. J Appl Meteorol Climatol 39(11):1811–1821. https://doi.org/10.1175/1520-0450(2001)039%3c1811:TTOAOH%3e2.0.CO;2

Tessendorf SA, French JR, Friedrich K, Geerts B, Rauber RM, Rasmussen RM, Bruintjes R (2019) A transformational approach to winter orographic weather modification research: the SNOWIE project. Bullet Amer Meteorol Soc, 71–92. https://doi.org/10.1175/BAMS-D-17-0152.1

Wegener A (1910) On the condensation processes in the atmosphere. Meteorol. Zeitschrift, p 354–374

Wegener A (1912) Thermodynamik der Atmosphäre. Nature 90:31. https://doi.org/10.1038/090031a0

Woodley WL, Rosenfeld D, Silverman BA (2003) Results of on-top glaciogenic cloud seeding in Thailand. Part I: The demonstration experiment. J Appl Meteorol Climatol 42(7):920–938. https://doi.org/10.1175/1520-0450(2003)042%3c0920:ROOGCS%3e2.0.CO;2

World Meteorological Organization (1985) Synopsis of the WMO Precipitation Enhancement Project. Precipitation Enhancement Project Report No. 34, Geneva, Switzerland

Xue L, Hashimoto A, Murakami M, Rasmussen R, Tessendorf SA, Breed D, Parkinson S, Holbrook P, Blestrud D (2013a) Implementation of a silver iodide cloud-seeding parameterization in WRF. Part I: model description and idealized 2D sensitivity tests. J Appl Meteorol Climatol 52:1433–1457. https://doi.org/10.1175/JAMC-D-12-0148.1

Xue L, Tessendorf SA, Nelson E, Rasmussen R, Breed D, Parkinson S, Holbrook P, Blestrud D (2013b) Implementation of a silver iodide cloud-seeding parameterization in WRF. Part II: 3D simulations of actual seeding events and sensitivity tests. J Appl Meteorol Climatol 52(6):1458–1476. https://doi.org/10.1175/JAMC-D-12-0149.1

Yao Z (2006) Review of weather modification research in chinese academy of meteorological sciences. J Appl Meteorol Sci 17(6):786–795

Yoshida Y, Murakami M, Kurumisawa Y, Kato T, Hashimoto A, Yamazaki T, Haneda N (2009) Evaluation of snow augmentation by cloud seeding for drought mitigation. J Japan Soc Hydrol Water Resour 22(3):209–222. https://doi.org/10.3178/jjshwr.22.209

Zhekamukhov MK, Abshaev AM (2009a) Simulation of rocket seeding of convective clouds with coarse-dispersion hygroscopic aerosol. 1. Condensational growth of the cloud droplets at the salt crystals. Russ Meteorol Hydrol 34:228–235. https://doi.org/10.3103/S1068373909040050

Zhekamukhov MK, Abshaev AM (2009b) Simulation of rocket seeding of convective clouds with coarse-dispersion hygroscopic aerosol. 2. Condensation and coagulation in a cloud seeded with hygroscopic particles. Russ Meteorol Hydrol 34:293–300. https://doi.org/10.3103/S1068373909050045

Zhekamukhov MK, Abshaev AM (2012) Dispersion of crystallizing reagents by detonation method. Part I: Detonation product expansion and kinetics of critical-size embryo droplet formation. Russ Meteorol Hydrol 37:448–454. https://doi.org/10.3103/S1068373912070035

Chapter 3
Fog Harvesting

Rebecca L. Farnum

Abstract Capturing atmospheric water vapor for domestic and agricultural use is an ancient practice. Contemporary fog harvesting science dates to 1900 and owes a great deal to indigenous knowledge and biomimicry (nature-inspired design). The most significant fog harvesting operations have been developed for remote communities (< 1000 residents) in Chile, Morocco, and South Africa, but viable sites have been examined in over 70 places on every continent—including Antarctic. The low-impact technology uses material such as mesh nets to capture water droplets from the air, relying on weather systems and physics to collect water rather than requiring energy or other inputs. Efficient systems can yield > 20 L/m^2/d for more than a decade and cost < \$250/m^2 of mesh. This chapter traces the history of fog harvesting science and practice to review technological innovations in fog harvesting and consider its status as an unconventional water source. Fog harvesting is shown to hold great potential for addressing not only water insecurity but also broader sustainable development needs, providing an entry point for gender equity projects, education initiatives, and enhanced livelihoods. While a growing body of work addresses the technical and mechanical aspects of fog harvesting, impact assessments based on holistic monitoring and evaluation studies are key to demonstrating the effectiveness of fog projects in building community resilience and resource security. Fully tapping into fog as a water source will require greater political and economic buy-in, but the success of organizations like FogQuest and Dar Si Hmad indicate that it is an investment well worth making.

Keywords Fog harvesting · Fog water collection · Indigenous knowledge · Biomimicry · Sustainable development

R. L. Farnum (✉)
Syracuse University, London, UK
e-mail: rlfarnum@syr.edu

M. Qadir et al. (eds.), *Unconventional Water Resources*,
https://doi.org/10.1007/978-3-030-90146-2_3

3.1 Introduction

Perhaps the most well-known image associated with collecting atmospheric moisture is found in the 2009 film *Avatar*, which features a "quiet, provocative shot" of an indigenous woman on a fictional planet drinking water from a leaf—something director James Cameron staged with actress Zoë Saldaña in Hawaii to "get the feel" for, even though what audiences see is computer-generated imagery (Lawrence 2020). The scene emphasizes two key aspects of fog harvesting for water access: one, the practice's relationship with indigenous knowledge, and two, the biomimicry principles behind contemporary advancements.

Hydrologically speaking, fog occurs when air at ground level contains suspended water droplets averaging diameters of 1–50 μm that impair visibility (Ritter et al. 2015)—more colloquially, fog is a cloud lying very close to the ground. Tiny drops of moisture in the air can add up to a significant volume of water that fog harvesting seeks to capture for use. Traditional methods for fog water collection include carefully positioned rock piles encouraging condensation runoff; buckets placed at the base of natural barriers such as flora to collect drip; and honeycomb-shaped walls to facilitate mist and dew buildup around cultivated plants in places as widespread as Gibraltar, Israel–Palestine, and Peru (Dower 2004; Cisneros 2009). Modern techniques are built on these concepts to capture water vapor more effectively. The most common system employs vertical meshes to intercept suspended droplets from the air, which coalesce and fall into a trough below the mesh that leads into a storage tank or distribution system (Abdul-Wahab and Lea 2008). The mesh nets and other materials used in contemporary fog harvesting draw inspiration from the engineering and chemistry of natural water-vapor collectors, such as spider webs, the Namib desert beetle, and cacti (Brown and Bhushan 2016).

The potential of fog to serve as an unconventional water resource is particularly exciting for several reasons. One is simply that fog water collection is underexplored, and thus there are many questions worth answering. Another is that contemporary fog harvesting operations combine a simple concept rooted in traditional practice with advanced materials science. This knowledge blending creates a template for sustainable development and community-based engineering, producing benefits beyond the water itself. Perhaps most significant, though, are the geographic locations of possible application. While only 0.04% of the world's freshwater exists as atmospheric water vapor (Graham et al. 2010), water vapor is present in locations where other forms of freshwater are not. Fog often occurs in arid and semi-arid coastal climates, where proximity to the ocean results in high humidity, but temperatures and wind patterns prevent frequent precipitation. Fog harvesting effectively bypasses the need for airborne water droplets to grow heavy enough to fall to the ground as rain, instead collecting them from the air before they evaporate in the desert heat or their cloud system shifts elsewhere. It can thus serve as an important source of freshwater for communities in certain water-stressed environments, supporting their water security.

This chapter provides an overview of fog water collection, including the history of its development (Sect. 3.2), the status of related technologies and operations (Sect. 3.3 and 3.4), and its future potential amid existing challenges (Sect. 3.5).

3.2 History

Though classified as an "unconventional" water resource, the art of fog harvesting is in truth an ancient practice—as traditional in some parts of the world as drawing buckets of water from a river. Shallow ponds fed by fog drip enabled animal grazing in southern England as early as the Bronze Age (Costley 2016), while archaeological evidence suggests that communities across the world have placed containers at the base of flora to collect condensed water for millennia.

Complicating a clear history of fog water collection is its overlap with dew collection and rainwater harvesting. The three are technically different because they draw on hydrologically distinct phenomena; however, ancient communities were not necessarily concerned with tracking or recording specific water sources. When thousands of shallow reservoirs were constructed in Victorian England, inspired by Bronze Age structures, they were called "dew ponds". A scientist in the 1910s determined that the ponds are fed by drip from fog (water droplets suspended in the air at or just above ground level) rather than formed from dew (water condensed on thin objects on the ground), but the "dew pond" misnomer has stuck (Martin 1915; Frazer 1931; Costley 2016). The source of water in *Avatar*'s leaf-drinking scene is not made explicit and might be rain, fog, or dew—especially since seasonal variations mean that rainwater harvesting devices sometimes collect fog or dew by default, and vice versa. Rainwater harvesting is addressed elsewhere in this volume (see Chap. 4). Less explicit attention is paid to dew collection, though many of the principles transfer from fog water collection.

The modern history of fog harvesting is somewhat easier to track and may be classified in three progressive waves. Evidence of intentional fog harvesting can be found in indigenous engravings and written records from the 15th–19th centuries (Fig. 3.1a). Fog became an object of scientific interest in the twentieth century, with various studies seeking to determine whether fog harvesting might serve as a viable water source (Fig. 3.1b). In 2000, a nonprofit organization dedicated to fog harvesting was formed, signaling a new period in fog harvesting history, this one focused on implementation and technological advancements for large-scale operation (Fig. 3.1c).

Key moments from each of these three waves are plotted on the timelines in Fig. 3.1 and detailed in Table 3.1. Highlights include a quantitative study of fog harvesting potential as early as 1901, conducted on Table Mountain in South Africa by hanging two rain gauges, one with a bunch of reeds suspended just above to serve as a vessel for fog drip (Marloth 1904). The first fog harvesting project using modern technology for harvesting was launched in South Africa nearly seven decades later, providing water for South African Air Force personnel stationed at the Mariepskop

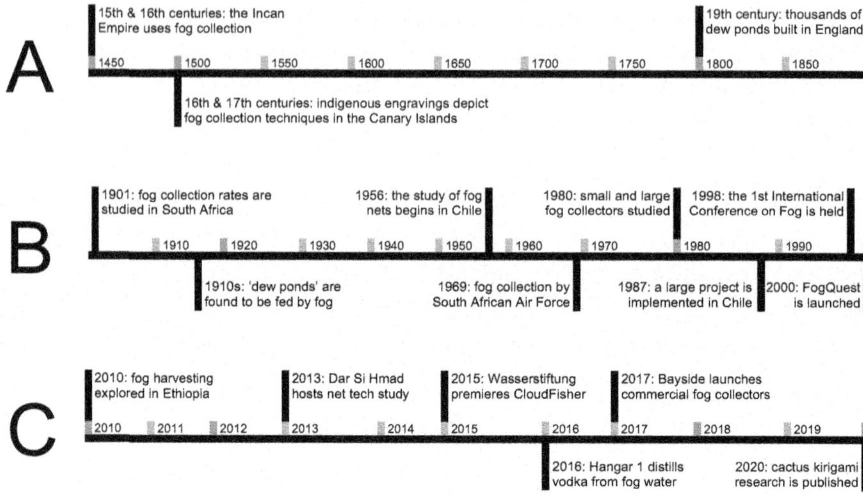

Fig. 3.1 Partial timeline of fog harvesting: **a** early evidence (15th –19th centuries); **b** scientific investigation (20th century); and **c** material advancements (21st century)

radar station (Olivier 2002). In the 1980s, a larger project in Chile developed the standard and large fog collectors that are today's industry mainstreams (Sect. 3.3) to supply water for a coastal community using the mountain fog system (Schemenauer and Cereceda 1994a; FogQuest 2009). More recent developments like the CloudFisher have vastly improved water collection efficiency rates (Trautwein et al. 2016), while fog harvesting has become a media curiosity and whimsical hobby, as well as a life-giving practice—consider, for example, Hangar 1 Distillery's vodka produced from San Francisco's fog (Steinmetz 2016).

Today, fog water collection projects have been undertaken around the world, in places as diverse as Azerbaijan (Maunier and Beysens 2016), Colombia (Garcia-Ubaque et al. 2013), Egypt (Harb et al. 2016), Namibia (Shanyengana et al. 2002), Nepal (Schemenauer et al. 2016), Oman (Abdul-Wahab and Lea 2008), and Saudi Arabia (Gandhidasan and Abualhamayel 2007; Al-Hassan 2009), among many others (Table 3.1 and Sect. 3.4 for a partial account of additional operations). This geographical diversity is matched by the diversity of the materials, engineering strategies, and technologies utilized, as discussed in Sect. 3.3.

3.3 Technological Interventions

Even a cursory review of fog water collection's history showcases that contemporary fog harvesting owes a great deal to indigenous knowledge and practice. Examining modern techniques for fog water collection reveals a similar debt to nature, in the use of biomimetic designs for material technologies. Engineers and material scientists

Table 3.1 An overview of fog harvesting developments

Year	Fog Harvesting Development
Circa 1500	The Incan Empire collected water from the base of trees and plants at high elevations in Peru (Cisneros 2009).
16th–17th centuries	Indigenous communities on the island of El Hierro in the Spanish Canary Islands engraved depictions of fog harvesting techniques (Marzol 2005; Cho 2011).
1800s	Thousands of saucer-shaped 'dew ponds' were built in England for animal grazing, inspired by similar structures dating back to the Bronze Age (Costley 2016).
1901–1904	The first known study of fog harvesting potential was conducted on Table Mountain in South Africa (Marloth 1904).
1910s	Funded by the Royal Society of London, E. A. Martin determined that the misnamed 'dew ponds' in southern England are not fed by dew, but rather by fog drip originating from the English Channel (Martin 1915; Frazer 1931).
1956	The Department of Physics at the Catholic University of the North in Antofagasta, Chile, began working with material nets (Gioda et al. 1993).
1969	The first modern fog harvesting project provided water for the South African Air Force personnel stationed at the Mariepskop radar station (Olivier 2002).
1980	A study in El Tofo, Chile, was initiated to examine fog water collection potential and the efficiency of fog collectors (Schemenauer et al. 1988)
1987	Building on the success of the El Tofo study, a large operational fog harvesting project was implemented on the mountain to supply water to the nearby coastal community of Chungungo (Schemenauer and Cereceda 1994a); the project retains both scientific and media significance but is not currently in use (FogQuest 2009).
1998	The 1st International Conference on Fog and Fog Collection was held in Canada.
2000	FogQuest, a nonprofit supporting fog harvesting for communities in desert climes, was launched; it has supported projects in the Dominican Republic, Ecuador, Eritrea, Guatemala, Haiti, Israel, Peru, Yemen, and more (FogQuest 2020a).
2010	The Mt. Zuquala Monastery received Chilean mesh to evaluate the potential for fog harvesting south of Addis Ababa, an example of FogQuest's facilitation of international cooperation for research and implementation (FogQuest 2010).
2013	German engineering foundation Wasserstiftung piloted a study with Dar Si Hmad, a fog harvesting nonprofit in southwest Morocco, to identify more efficient forms of nets that are resistant to high winds (Jewell 2018).
2015	Wasserstiftung premiered their CloudFisher, a durable fog harvesting net capable of withstanding 120kph winds (Trautwein et al. 2016).

(continued)

regularly borrow ideas from the environment to solve complicated problems, effectively copying evolution's innovations for human use. The most common setup for fog harvesting operations involves mesh netting, by which thin strands of material catch water droplets as a fog system passes through (Fig. 3.2 and Box 3.1)—much like can be seen on a spider's web. This imagery is explicitly acknowledged as the inspiration for a recent advancement in mesh technology (Trautwein et al. 2016).

Table 3.1 (continued)

Year	Fog Harvesting Development
2016	Fog water collection technologies were used to produce a specialty batch of vodka from water originating from San Francisco fog (Steinmetz 2016).
2017	Daniel Fernandez, a professor at California State University, launched a company to make Standard Fog Collectors, and the Marienberg Malla Raschel mesh used by FogQuest became commercially available (Bayside Fog Collectors 2020).
2020	Biomimicry—inspired research into fog harvesting materials—such as the cactus kirigami—continues, with attention to cost, durability, and efficiency (Bai et al. 2020).

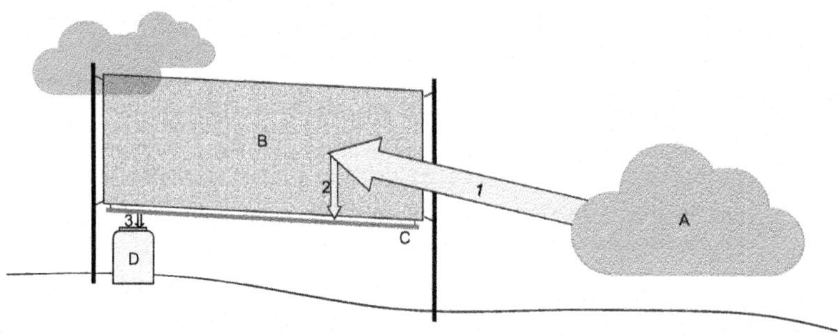

Fig. 3.2 Generalized model of a net-based fog water collection system (Box 3.1)

Box 3.1 Generalized Overview of Net-Based Fog Water Collection Systems

While communities collect fog and dew in any number of ways, many of them using local materials and ancient techniques (Fessehaye et al. 2014; Dower 2004), and most of the contemporary fog harvesting studies and operations utilize a mesh net set up on poles (Fig. 3.2).

Net-based fog water collection relies on three simple steps, which effectively mimic the process of precipitation by using a combination of water's natural cohesive properties and gravity to harvest water vapor from the atmosphere:

1. A fog system (A) passes through a mesh net (B). As the fog goes through the net, water droplets suspended in the air collide with the net's thin strands, catching on the material.
2. Droplets caught on the net coalesce, becoming heavier until gravity naturally pulls them down into a collection trough (C).
3. A pipe from the trough allows water to run into a storage tank (D) or distribution system. A gauge is often set up at this point to measure the quantity of water collected.

This section reviews the technological interventions behind modern fog water collection. The most widespread makes use of a polypropylene Raschel mesh net developed in Chile, found in both Standard Fog Collectors (SFCs) and Large Fog Collectors (LFCs) (Sects. 3.3.1–3.3.2), while the most productive employs a spacer fabric and bungee system in a similar configuration (Sect. 3.3.2). Additional designs are addressed in Sect. 3.3.3 and include everything from mesh materials in alternative structures to wax-infused paper constructions.

3.3.1 Testing Feasibility: Standard Fog Collectors

The first project collecting fog water for human consumption was pioneered in South Africa using two plastic mesh screens measuring 28 × 3.6 m each. The Mariepskop operation successfully collected around 800 L per day when fog was present from October 1969 to December 1970 (Schutte 1971). As science became more interested in the possibility of fog as a water resource, several mesh materials and net set-ups were trialed around the world (Olivier 2002; Klemm et al. 2012; Fessehaye et al. 2014). By the early 1990s, scientists connected with the then largest operational fog harvesting project in Chile were suggesting an industry standard to quantify fog deposition and harvesting potential rates around the world (Schemenauer and Cereceda 1994b).

The SFC is a one m^2 net of polypropylene mesh, double layered, raised 2m off the ground by support posts with a collection trough and outlet tube directly below the mesh (ibid.). Even slight variations in the mesh material could impact collection efficiency, though early work recognized that local materials would help keep costs down, and thus focused on the style and setup more than the precise supplier in detailing the parameters for their proposed SFC.

The nonprofit FogQuest has organized triannual conferences on fog water collection since 1998. The organization is involved with most large-scale fog harvesting projects and suggests the use of an SFC to determine the potential before investing in a larger operation. Fog is a seasonal phenomenon; an SFC should thus be used and monitored for at least a year on site to project the likely yield. Combined with wind speed meters and rain gauges, SFCs also serve a scientific purpose in estimating fog and precipitation rates in an area—useful for building a comprehensive understanding of a watershed's hydrology.

An SFC costs around $ 100 to construct, with the exact cost mostly determined by the site location, given its impact on transportation expenses and possible installation challenges. Since 2017, Bayside Fog Collectors in California have made prebuilt SFCs, required materials, and construction manuals commercially available for hobbyists as well as researchers and community users.

3.3.2 Operating Projects: Large Fog Collectors

True to their name, LFCs are an expansion of SFCs, generally installed after an SFC has demonstrated high potential yield for fog water collection at a particular site. Unlike the SFCs, which is standardized to m^2 of a particular mesh style to allow for comparative data collection, LFCs do not have set dimensions or required materials. However, the majority of LFCs are supported or modelled on the work of FogQuest and thus use the same polypropylene mesh supplied by Marienberg S.A. in Chile, as the SFCs do. Today, the Marienberg Raschel mesh has been used for fog water collection studies and operations in over 35 countries on 5 continents (Klemm et al. 2012).

This mesh is used primarily because of its success in El Tofo, Chile, home to the first major study on fog water collection rates and the world's largest fog harvesting operation during the 1980s and 1990s. After a number of small collector designs were tested at the site, larger collectors were installed with the locally available Raschel mesh (Schemenauer et al. 1988).

Interestingly, the first LFC constructed at El Tofo was 90 m^2, with four horizontal troughs spaced evenly across 3 m of vertical height and running the full length of 30 m of mesh—but a smaller LFC measuring 4×10 m with a single trough at the bottom collected more water per unit of mesh (ibid.). While the researchers posited that favorable exposure likely had more to do with the greater efficiency than design particulars, the 4×10 m collector carried the day. Some 100 LFCs were installed at the site, providing a nearby coastal community with some 15,000 L of potable water each day for ten years (FogQuest 2009).

The El Tofo design continues to be the most common, with LFCs usually constructed with dimensions of 4 m high and 10–12 m wide (a height-to-width ratio similar to the visual in Fig. 3.2). A fog harvesting system costs \$25–50/m^2 of mesh (requiring around \$1,500 on the average to construct an LFC; Qadir et al. 2018) and collects an average of 5 L/m^2/day (roughly 200 L/day per LFC; FogQuest 2020b). Both the cost and the water yielded vary widely across contexts due to factors such as material availability, fog intensity, and storage procedures. A feasibility study in Cape Columbine, South Africa, for instance, estimated collection volumes of 2.5 L/m^2/day (Olivier 2002), while LFC harvesting rates at the El Tofo site varied from 25 to nearly 47 L/m^2/day during peak fog season (Schemenauer et al. 1988).

3.3.3 Considering Efficiency: The CloudFisher

The success of the El Tofo project, combined with growing concern over global water availability, resource security, and sustainable development, has led to the propagation of fog water collection feasibility studies and operational projects around the world. The LFCs used in Chile were readily adopted, thanks in large part to FogQuest and the connected fog harvesting scientific conference series. Fog harvesters were installed at sites in Ecuador, Haiti, Nepal, Yemen, and many more locations, generally targeting small rural communities.

The most common alteration to LFCs made by projects in the 1990s and early 2000s was the net mesh, often due simply to the availability and cost of materials rather than design intent. But mesh style was also an object of targeted interest because scientists quickly became interested in maximizing yield. The "Van Schoor collectors" in South Africa, named after the mine ventilation equipment specialist who designed and donated them, used a carbon impregnated polypropylene mesh from Kimre, Inc. The Kimre mesh had a three-dimensional structure meant to help rotate the air as it passed through the mesh, enhancing water droplet cohesion (Olivier 2002).

The 2010s witnessed a growing number of studies aimed at understanding how mesh style impacts fog water collection rates. The general set-up for these experiments involved installing various materials as the nets on LFCs or similar structures at the same site, and monitoring water collection from each along with wind speed, precipitation rates, and other meteorological data. Alternative materials that have been tested included, metal meshes coated with a hydrophobic compound; high-density polyethylene (HDPE) hail nets; HDPE shade nets; Enkamat, a three-dimensional material used in anti-erosion agricultural turfing; and a number of polyester weaves (Fernandez et al. 2018; Trautwein et al. 2016).

Of course, mesh material and weave style are not the only features contributing to fog water collection rates. Wind speeds and directions, fog cover intensity and humidity, and temperature are among the meteorological factors impacting fog harvesting; while the net angle, size, and proximity to other structures are engineering aspects that have influence. Nor are there simple relationships between these elements: For instance, a study in California found that the mesh material that collected the most water at low wind speeds did not perform as well at high winds (Fernandez et al. 2018). Wind helps push fog through the net, allowing a greater amount of water vapor to contact the material, but heavy winds may also carry the droplets away from the mesh before they are heavy enough to drop into the collection trough. In addition to the physical experiments exploring mesh material, fog harvesting science also makes use of fluid mechanics and aerodynamic principles to mathematically model the most appropriate setups for a collection system (Azeem et al. 2020).

In 2013, German engineers began a study in partnership with Dar Si Hmad, a non-profit in southwest Morocco that had successfully been operating a FogQuest-inspired system for several years. Wasserstiftung sought to address a major fault of the Raschel mesh witnessed at several sites, including those in Eritrea and Morocco, where high wind speeds regularly tore the nets, as well as billowing them away from collection poles such that a great deal of the water that had coalesced on the net never found its way into the collection trough or distribution system. One of the Amazigh villagers working with Dar Si Hmad's system was hiking the mountain and repairing nets on a nearly daily basis, greatly reducing the operation's efficiency and sustainability (similar concerns have been reported from Peña Blanca, Chile; Holmes et al. 2015).

The Wasserstiftung–Dar Si Hmad study led to the creation and launch of the CloudFisher, the most productive fog collector to date, especially at high wind speeds. The CloudFisher uses a three-dimensional spacer fabric made of polyester whose monofilaments were developed with food safety (ensuring potability) and extreme ultraviolet radiation (preventing sun-based deterioration) in mind. Resembling an array of interconnected double helixes, the various strand sizes of the net allow fog to pass through easily, while providing several points at which water droplets might catch. The three-dimensional structure also serves to facilitate the coalescing of droplets and their gravity-led run into the collection trough while making it more difficult for wind to carry them off a flat surface of the net.

Much of the innovation is in the chosen mesh style, but several other improvements were also made to the engineering of a traditional LFC, including decreased net surface and the addition of bungee holders, a flexible gutter, and a support frame (Trautwein et al. 2016). Rather than one large net of 4×10 m with two support poles, a CloudFisher unit is four equally sized panels with full poles between each, creating a total net surface area of 54 m^2, but greatly reducing the size of any one piece of net, thus decreasing wind-billowing effects. This setup also allows for the smaller single panels to be installed for pilots, demonstrations, and hobby use.

The spacer net pieces are fastened to the support poles with rubber expanders, a bungee system allowing the net to move with extreme winds rather than trying to stand against it. This innovation is critical in reducing tears and other wear to the net and is another example of biomimicry—inspired by plant life that flourishes in extreme wind conditions thanks to its flexibility. The system's collection trough is a polyethylene gutter attached to the bottom of the net rather than the support poles, allowing it to move with the mesh to maximize capture rates. Another advancement made by the CloudFisher is its high-density polyethylene support rod, a frame of plastic triangular tiling behind the mesh that allows fog to move through but prevents the wind from billowing the net. Together, these innovations allow the CloudFisher to operate in winds of up to 120 km/h.

While LFCs collect on average 5 L/m^2/day (FogQuest 2020b), the CloudFisher generally captures 10–22 L/m^2/day and has been recorded to harvest more than 65 L/m^2 on particularly fog-intense days at the Mount Boutmezguida study site in southwest Morocco (Jewell 2018). But while the CloudFisher is the most productive fog harvesting device currently available, it is not necessarily the most efficient, depending on site geography and cost considerations. Both the spacer fabric and support grid are far more expensive materials than FogQuest's Raschel mesh. A 40 m^2 LFC will cost about $1,500 and produce about 200 L/day of water, while a 54 m^2 CloudFisher unit runs about $13,000 and can yield approximately 1,200 L/day (FogNet Alliance 2020). Dar Si Hmad has found that the CloudFisher system requires far less maintenance than their LFCs, suggesting that the CloudFisher's heightened up-front costs are a sound investment for long-term operations, but fog harvesting projects facing calmer wind speeds may face different trade-off parameters.

3.3.4 Exploring Designs: Alternative Collectors

Thus far, this chapter has focused on variations to a basic net-and-poles structure for fog water collection, but that setup is far from the only configuration that has been developed by communities and/or tested, both theoretically and physically, by scientists.

Biomimetic design is behind many advancements in fog harvesting nets; nature has also inspired most other collection structures. A particularly poetic example is from the 1970s, when scientists created a tree-like structure made from bamboo modelled after the fountain tree depicted on El Hierro's coat of arms (Gioda et al. 1993). El Hierro is the smallest of the Canary Islands; for centuries, communities have placed buckets at the bottoms of agave, olive, laurel, and juniper trees to harvest fog and dew collected by their branches and leaves. That imagery also encouraged a variety of cylindrical and rectangular prism designs, with a mesh shape setup to drip into a bucket in Chile as early as the 1950s (Schemenauer et al. 1988).

Trees continue to serve as models for other styles of atmospheric water collectors, including Warka Water, a tall vase-shaped tower named after an Ethiopian fig tree. The Warka tower structure is made from juncus (an elastic grass-like flowering plant) or bamboo stalks; inside is a nylon mesh reminiscent of a Chinese lantern (which might also be made from polypropylene), the droplets from which drip into a container inside the tower (Nguyen 2014). As with LFCs, the cost of a Warka unit depends greatly on the location of construction, but it is intended to be simple to build and relatively inexpensive—in ideal conditions, costing around $500 for a tower that can yield water up to 115 L/day. The non-profit launched to support further testing and expansion distinguishes the Warka tower from LFCs by its community-centric design, with the structure intended to blend into the environment and attract public gatherings (Warka Water 2020).

Other examples of biomimicry in fog harvesting can be found in the continuing search for an ideal but cheap mesh. The combination of hydrophilic and hydrophobic components on the back of the *Stenocara gracilipes,* more commonly known as the Namib beetle, helps guide material scientists. A recent breakthrough uses zinc oxide-silver hierarchical nanostructures to mimic the beetle's back, though large-scale field testing and commercial production have yet to be explored (Kim et al. 2019; Dizikes 2011).

Like the CloudFisher, several contemporary developments in fog harvesting engineering address structural flaws in LFCs. Some address the issue of wind damage by altering the shape rather than the material, suggesting a multi-modular, accordion-like structure in which funnel-style net pieces are set at oblique angles to the wind on a tensile structure, or a macro-diamond shape able to collect fog regardless of wind direction (Holmes et al. 2015; Trevino 2020; Fernandez et al. 2019).

Another disadvantage of conventional LFCs is the likelihood of the mesh becoming clogged since the coalesced droplets take time to drip into the bottom trough. Before they drop, their area of the net is at a saturation point, effectively forgoing water from the fog still passing through and around. Once again, material scientists turned to nature for the answer and found it in *Tillandsia landbeckii,* a bromeliad endemic to Peru and Chile. The plant grows on the desert soil and relies on fog for its water source. It's group-positioning to ensure each organism has direct access to a fog stream, its thin filaments organized in a three-dimensional mesh, and its intake pores between leaves prompted scientists to model a structure with corresponding length scales and structures. The result is a multi-layered harp design with vertical filaments rather than a mesh structure, found to be four times more effective than the two-ply Raschel mesh LFC—though, as with the CloudFisher, more expensive and not necessarily optimal for all operations (Azeem et al. 2020). Others have proposed making use of electrostatic charge and water's natural chemistry to overcome aerodynamic drag forces, helping direct water droplets toward a collector to improve efficiency (Damak and Varanasi 2018).

This chapter has repeatedly emphasized the importance of both biomimicry and indigenous knowledge for fog harvesting. Up until now, the role played by indigenous knowledge has primarily been to provide the idea of atmospheric water collection in the first place. But one of the more creative proposals for a new approach to fog harvesting blends material science with a traditional craft, showcasing the value of integrating modern technology with cultural practices. Cacti needles have long been noted for their efficient water collection, but their conical shape is difficult (and expensive) to replicate in three dimensions. However, the water-absorbing spines can be simplified into a two-dimensional triangle through kirigami, a variation of origami that involves cutting the paper. Early tests show water collection rates for the cactus kirigami to be moderately higher than those of harp designs, and far cheaper, with the wax-infused paper-based substrate costing around $0.50/m^2 (Bai et al. 2020). While far more testing is needed, this type of study speaks to the potential of fog water collection both as an affordable water source and as a model for engineering that takes nature and indigenous knowledge seriously as sources for solutions.

This chapter now turns to the status of fog water collection projects, with a focus on cases where fog has shown its potential as a solution to water insecurity that can meet the needs of sustainable development.

3.4 Status

Fog harvesting has been explored as a water source in some 70 locations around the world (Fig. 3.3; Kaseke and Wang 2018; Trautwein et al. 2016). Though it is important to note that the vast bulk of these locations remain in exploratory or experimental stages, Fig. 3.3 highlights the truly global potential for fog water collection, with possible operation sites on every continent—including Antarctica.

As illustrated in Fig. 3.3, most sites considered for fog water collection are coastal. This is due primarily to the climatic and topographic conditions that result in heavy fog systems. These include persistent winds from one direction; mountains or dunes to intercept the fog that are not blocked upwind; and altitudes above sea level to reach the cloud cover (UNEP 1997).

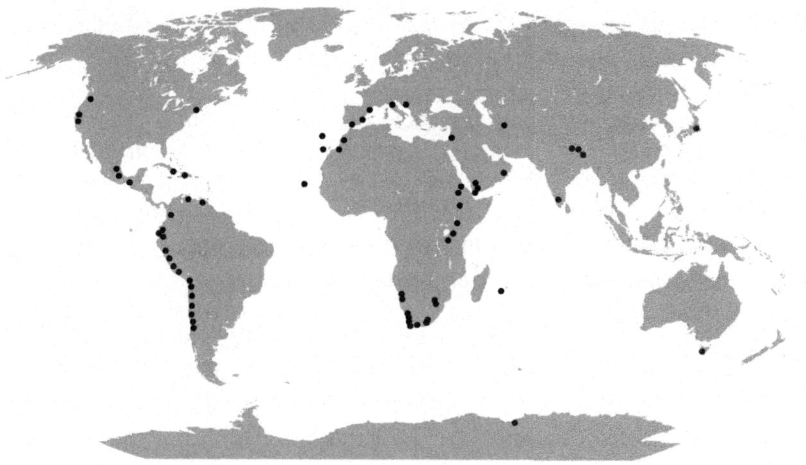

Fig. 3.3 Sites of fog harvesting exploration or operation (based on data drawn from Fernandez et al. 2019; Kaseke and Wang 2018; Trautwein et al. 2016)

The specific local factors that are best for a net-based fog harvesting system are generally not those best for human settlements, with communities more likely to be based at the foot of mountains than atop them, seeking shelter from the elements and greater land area. While recent studies are examining possible adjustments to make fog harvesting tenable in urban environments (Fernandez et al. 2019), existing projects are predominantly rural, requiring significant storage and distribution infrastructure. El Tofo, the Chilean site where LFCs were developed and whose success effectively launched modern fog water collection, is inland from Chungungo, the coastal community that made use of the water. A buried pipeline runs for seven km between El Tofo and Chungungo, transporting stored water collected from the nets to a distribution system in the village (FogQuest 2009).

During the decade of the El Tofo operation, Chungungo doubled its permanent residents as living standards improved. Fog water collection effectively reversed urban migration for the Chilean community—one of the early indicators of the potential for fog harvesting to produce benefits beyond the water itself.

While Latin America and the Caribbean continue to dominate fog water science in terms of the number of experimental sites and published studies, the El Tofo operation has been suspended since 2003. While fog collectors are passive systems, not requiring active energy or other material inputs, they do involve general maintenance. LFC mesh nets will last around 10 years and maintain the general structure for about 20 years (Batisha 2015), but regular attention to the system is needed. At the last turn of the century, local politicians decided to stop maintaining the fog harvesting operation, instead seeking investment in a conventional pipeline or desalination plant. As with most water security problems around the world, barriers to access water are far more political than hydrological or technical (Mehta 2003). FogQuest has found similar issues with projects in Haiti and Yemen, where fog conditions are favorable but national instability and challenges with local buy-in prevent large-scale operations.

Today, several fog harvesting projects exist and successfully provide water for communities, agriculture, and hobbies. The Canary Islands, Ethiopia, Eritrea, Ghana, Guatemala, Nepal, and the US (California) are sites of proven success, with the largest operation currently in southwest Morocco, birth site of the CloudFisher.

Dar Si Hmad, a local nonprofit, began an evaluation period on Mount Boutmezguida in 2006. LFC construction began in 2011, with taps turned on in nearby villagers' homes on March 21, 2015. As of today, CloudFishers have replaced the LFCs to supply 16 villages in Tnine Amellou, with local production for 12 new villages in progress (Dar Si Hmad 2020). Beyond water provided for drinking and domestic use, the quantity of fog produced has enabled its use in the construction of new buildings, sanitation systems, and agricultural production.

Like the El Tofo–Chungungo project, Dar Si Hmad has demonstrated the many positive social and economic impacts that fog water collection projects can have, especially when communities are engaged in all aspects of the project. These include reduced out-migration, improved sanitation, resilient livelihoods, improved numeracy, and literacy (Dodson 2014), enhanced gender parity (Dodson and Bargach 2015), environmental awareness and educational access (Farnum and Moss 2016), international diplomacy (Farnum 2018), among others.

The holistic success of Dar Si Hmad's operation was recognized by the UN through a UNFCC Momentum for Change Award in September 2016 at COP22. Together with the emergence of networks like the FogNet Alliance, the launch of companies such as Bayside Fog Collectors, and continued investment in material and social science research on fog harvesting, Dar Si Hmad's operation signals the status of fog water collection as a serious player in sustainable development. Perhaps most excitingly for unconventional water resources work, the proliferation of fog harvesting operations and research has encouraged several scientific advancements for water production from atmospheric vapor—even in areas where fog is not present and humidity is very low. Metal–organic frameworks (MOFs) have shown promise in absorbing liquids and releasing them when exposed to sunlight. Specially constructed plastic boxes placed in extremely arid deserts can be opened at night, allowing MOFs to absorb atmospheric water, and closed in the morning so that the MOF releases the water vapor to condense and collect (Fathieh et al. 2018).

3.5 Major Barriers and Response Options

Several challenges to fog water collection have already been discussed in this chapter or are well addressed elsewhere in the literature (Table 3.2). Today, the most pressing concerns are: (1) the imbalance between the extent of experimental studies and the number of actualized operations; (2) holistic determinations on what makes a water project "efficient", combined with the need for upfront investment in under-resourced

Table 3.2 Some concerns about, challenges to, and responses for fog water collection

Barrier	Responses
Site selection	Hydrological modelling can predict likely sites for optimal productivity, and SFCs standardize feasibility tests. Previous operations have informed established guidance for site selection, which includes attention to political and community aspects as well as climatic factors (Batisha 2015)
Storage and distribution	As systems are usually set at high altitudes away from settlements, storage and distribution are challenges for water use. Cisterns connected to gravity-fed pipelines are a common, low-impact solution. It is important to note that any water source will require storage and distribution; while these add to the costs and considerations of a fog project, they are not a unique challenge
Cost	There are a few challenges for fog harvesting related to cost. One of the most significant barriers to fog harvesting for communities is that the bulk of the cost is an upfront investment of materials and installation, something under-resourced remote communities are unlikely to be able to afford on their own Another challenge is related to the question of "cost". Because fog harvesting plays out at a far smaller scale than most other water sources, it requires upfront costs but minimal inputs, and often involves a lot of volunteer labor, there is not a clear answer to the "actual" cost involved, with projects estimating \$1.4–16.6/ m^3 (Qadir et al. 2018). This is generally higher than desalination, but lower than options like water trucks—but these figures are not readily comparable as pricing methods vary and do not reflect the same inputs. Unreported costs involved in various water source options include everything from gendered labor inequities to environmental damage
Wind damage	One of the greatest technical challenges to fog harvesting has been the physical setup itself since wind damage has been frequently observed to mesh or other elements of the collector. Engineers have found a variety of solutions, including alternative materials, shapes, and structures (Sect. 3.3)
Material availability	High-efficiency fog collectors make use of specialty materials that may not be readily available, especially since harvesting often occurs in remote locations. Local materials may be used, though there will be trade-offs with capture rates—but a less 'efficient' material may be more effective in the long run if local capacity is able to maintain one but could not fix or replace the other
Energy requirements	Fog harvesting is a passive technology, using physics to capture water. Some projects may use energy for purification or distribution, but it's not required—making fog harvesting one of the most environmentally friendly water sources
Maintenance	Fog systems require few material inputs but do require care, most notably the cleaning of collection troughs/gutters and regular checking of distribution pipes. As with storage and distribution, routine maintenance is an aspect every water system will face, rather than anything exclusive to fog harvesting

(continued)

Table 3.2 (continued)

Barrier	Responses
Water quality	Atmospheric water is normally free of bacteria and contamination, especially in rural sites where fog harvesting is generally conducted (i.e., there is no 'acid rain' effect). Extensive testing has shown fog water to meet WHO drinking water standards at sites around the world. Some projects mix fog water with groundwater or another source to keep supply steady across seasonal variation, mineralize the water distributed, and/or acknowledge community perceptions about water safety (Wei-Haas 2017)

areas to enable projects yielding non-financial benefits; and (3) climate change's possible impact on fog patterns. The first two issues are related and boil down almost entirely to the question of political will, while combatting the third is also a matter of political will, albeit at a far greater scale.

Though technical advancements can (and should) continue to be made in fog harvesting to address factors such as collection rates and material costs, the general premise has been well-established: in certain identifiable climatic conditions, fog water is a viable resource. Whether fog water collection is a viable operation is more uncertain, as the predominant challenges are political and economic rather than hydrological or mechanical.

While reviewing the literature on fog harvesting, it is readily apparent that there is more published work on the technical science of collection than the practical operation of projects, and far more sites have been identified as productive than fog harvesting systems have been installed. For fog water to be taken seriously as an unconventional water resource, the fog harvesting community needs to push beyond feasibility studies and niche material questions to more widespread actualization. Social scientists have acknowledged policy, institutional, and community constraints as well as gender inequality and perceived costs as non-technical challenges to more widespread implementation (Qadir et al. 2018).

As suggested in Sect. 3.3.3, there is not an immediate, objective answer to the most "efficient" water collection and distribution scheme in an area. Which solution is the most cost-effective depends greatly on which costs are taken into account and the timespan considered. The most productive fog harvesting systems generally cost the most upfront but require the least maintenance over time. Fog harvesting has limited scalability beyond small communities and will only collect as much water as the fog cover allows, but it is a passive technology that does not require energy input and can operate in remote areas with little infrastructure.

Ultimately, all water resources have advantages and disadvantages: environmental, economic, and social. Policymakers and communities make decisions about priorities and trade-offs when selecting which water resources to collect and distribute. Fog harvesting is one of the most environmentally friendly forms of water collection with the potential to generate rich dividends for all beneficiaries in the form of enhanced capacities, improved livelihoods, and greater resilience. Designed holistically, a fog water collection operation is not only an engineering system, but a high-impact sustainable development program. The fog harvesting research community should turn its attention more fully to monitoring and evaluation to build a robust evidence base demonstrating the full range of costs and benefits involved in fog water collection. Completing targeted impact assessments not only of fog water's direct use but also of the various community programs related to collection systems could go a long way in demonstrating just how "effective" fog harvesting can be.

One wider concern is the effect that anthropogenic climate change will have on weather patterns that have enabled fog harvesting in known locations for centuries. Climate change is, of course, a far broader challenge to overcome, but this barrier to fog harvesting is yet another example of the greatest disparity in today's world. The populations and ecosystems already most vulnerable are those in line to be most harmed by climate change. Even as communities creatively build resilience, their solutions are being negated by the colossal danger. Yet once again, biomimicry and indigenous knowledge can provide answers. Indigenous peoples have made decisions about agriculture, migration, and shelter for millennia by following the example of other species. While the exact effects of climate change on weather patterns and fog cover can be hard to predict (Qiao et al. 2020; Kawai et al. 2016; Torregrosa et al. 2014), the non-human animals and plants on which communities have based their practices can serve as indicator species. Tracking the *Tillandsia landbeckii* in the Atacama Desert suggests that the fog harvesting plants are moving uphill—a sign that the fog is too (Trevino 2020). Fog harvesting projects would do well to take note of this migration and follow suit.

3.6 Conclusions

Fog harvesting holds great promise for community-level water security, resilience, and sustainable development. Advancements in material science making use of biomimicry and indigenous knowledge have developed highly productive, relatively low cost, and environmentally friendly designs for capturing atmospheric water vapor fit for all forms of human use. While variations in fog patterns, local geography, and community context mean that there is no one-size-fits-all answer to the most productive system for capturing fog water around the world, the cross-pollination of ideas from numerous sites has led to engineering breakthroughs with applications beyond fog harvesting.

Today, sites experiencing extensive fog cover (suggested conditions are visibility < 100 m for at least half the year) can harvest water from fog for well under $\$10/m^3$. The greatest barriers to adopting fog water collection as a solution to resource needs are tied to economic decision-making and community perceptions, rather than technical limitations. Tapping into fog harvesting's potential requires the political will and economic ability to invest resources in projects when the benefits will not be quickly realized and the majority of them are intangible. Unfortunately, how to capture political will is a question far less easy to answer than how to capture water vapor.

This chapter concludes by suggesting five key recommendations for future fog water research and operations:

- present fog water collection projects as a template for blending technological engineering with indigenous knowledge and biomimetic design;
- develop holistic impact assessments capturing the full range of costs and benefits from fog harvesting as a sustainable development practice;
- advocate for financial investment in the form of grants or interest-free loans to support the operationalization of fog water collection at viable sites;
- build recognition for fog harvesting as a scientific discipline with great potential, rather than as a curiosity, to encourage greater political buy-in; and
- investigate further material advancements in fog water collection technologies, but not at the expense of the political and economic work that needs to be done.

References

Abdul-Wahab SA, Lea V (2008) Reviewing fog water collection worldwide and in Oman. Int J Environ Stud 65:487–500. https://doi.org/10.1080/00207230802149983

Al-Hassan GA (2009) Fog water collection evaluation in Asir Region-Saudi Arabia. J Water Resour Manag 23:2805–2813. https://doi.org/10.1007/s11269-009-9410-9

Azeem M, Guérin A, Dumais T, Caminos L, Goldstein R, Pesci AI, de Dios Rivera J, Dumais J (2020) Optimal design of multilayer fog collectors. J ACS Appl Mater Interfaces 12(6):7736–7743. https://doi.org/10.1021/acsami.9b19727

Bai H, Zhao T, Wang X, Wu Y, Li K, Yu C, Jiang L, Cao M (2020) Cactus kirigami for efficient fog harvesting: simplifying a 3D cactus into 2D paper art. J Mater Chem A 27(8):13452–13458. https://doi.org/10.1039/D0TA01204A

Batisha A (2015) Feasibility and sustainability of fog harvesting. J Sustain Water Quality Ecol 6:1–10. https://doi.org/10.1016/j.swaqe.2015.01.002

Bayside Fog Collectors (2020) About us. Bayside Fog Collectors website. Retrieved August 25, 2020, from https://www.baysidefogcollectors.com/about-1

Brown PS, Bhushan B (2016) Bioinspired materials for water supply and management: water collection, water purification and separation of water from oil. J Philos Trans R Soc A, 374(2073). https://doi.org/10.1098/rsta.2016.0135

Cho R (2011) The Fog collectors: harvesting water from thin air. Columbia University, Earth Institute. Retrieved August 25, 2020, from https://blogs.ei.columbia.edu/2011/03/07/the-fog-collectors-harvesting-water-from-thin-air/

Cisneros LJ (2009) Fog-catching in a Peruvian slum. Daily Telegraph, November 12. https://www.telegraph.co.uk/expat/expatnews/6550384/Fog-catching-in-a-Peruvian-slum.html

Costley J (2016) The Earth's eyes: restoring the lost dew ponds. BBC, Springwatch blog. Retrieved August 25, 2020, from https://www.bbc.co.uk/blogs/natureuk/entries/7ad7f541-f496-4689-8440-dc4c5a29a0f8

Damak M, Varanasi KK (2018) Electrostatically driven fog collection using space charge injection. J Am Assoc Sci Adv 4(6). https://doi.org/10.1126/sciadv.aao5323

Hmad DS (2020) A ground-breaking project: harvesting water from fog. Retrieved August 25, 2020, from http://darsihmad.org/fog/

Dizikes P (2011) Out of thick air: MIT graduate student working to make water available for the world's poor by refining the tools and techniques of fog harvesting. MIT News website. Retrieved August 25, 2020, from https://news.mit.edu/2011/fog-harvesting-0421

Dodson LL, Bargach J (2015) Harvesting fresh water from Fog in Rural Morocco: research and impact Dar Si Hmad's Fogwater Project in Aït Baamrane. J Proc Eng 107:186–193. https://doi.org/10.1016/j.proeng.2015.06.073

Dodson LL (2014) A foggy desert: equitable information flow for a fogwater system in Southwest Morocco. Dissertation, University of Colorado Boulder

Dower S (2004) Clouds on tap. The water wheel, Special edition for national water week, 3–5. Retrieved from http://www.wrc.org.za/wp-content/uploads/mdocs/WaterWheel_Archive_water-week%202004.pdf

Farnum RL (2018) Drops of diplomacy: Questioning the scale of hydro-diplomacy through fog-harvesting. J Hydrol 562:446–454. https://doi.org/10.1016/j.jhydrol.2018.05.012

Farnum RL, Moss A (2016) Water school by Dar Si Hmad: STEAM Curriculum for ages 6–14. Morocco: Dar Si Hmad. Retrieved from http://darsihmad.org/wp-content/uploads/2016/10/water-school-sc.pdf

Fathieh F, Kalmutzki MJ, Kapustin EA, Waller PJ, Yang J, Yaghi OM (2018) Practical water production from desert air. Sci Adv 4(6). https://doi.org/10.1126/sciadv.aat3198

Fernandez DM, Torregrosa A, Weiss-Penzias PS, Zhang BJ, Sorensen D, Cohen RE, McKinley GH, Bowman M (2018) Fog water collection effectiveness: mesh intercomparisons. J Aerosol Air Quality Res 18(1):270–283. https://doi.org/10.4209/aaqr.2017.01.0040

Fernandez G, Morichi G, Calixto LB (2019) Exploring fog harvesting nature-based solution tensile membrane structures towards sustainable development in the Italian urban context. In: Proceedings of the TensiNet symposium, Milan, Italy, pp 535–546

Fessehaye M, Abdul-Wahab SA, Savage MJ, Kohler T, Gherezghiher T, Hurni H (2014) Fog-water collection for community use. J Renew Sustain Energy Rev 29:52–62. https://doi.org/10.1016/j.rser.2013.08.063

FogQuest (2020a). About FogQuest. FogQuest: sustainable water solutions. Retrieved August 25, 2020, from http://www.fogquest.org/about-fogquest/what-we-do/

FogQuest (2020b) FogQuest: sustainable water solutions. Retrieved August 25, 2020, from http://www.fogquest.org/f-a-q/

FogQuest (2009) Chile – El Tofo/Chungungo 1987 – 2002 Fog collection proof of technology, Scientific and Operational Projects. FogQuest: Sustainable Water Solutions. Retrieved August 25, 2020, from http://www.fogquest.org/project-information/projects/chile-el-tofo-chungungo/

FogQuest (2010). Ethiopia. FogQuest: sustainable water solutions. Retrieved August 25, 2020, from http://www.fogquest.org/project-information/pending-projects/ethopia/

FogNet Alliance (2020) FAQ – frequently asked questions. FogNet alliance website. Retrieved August 25, 2020, from https://www.fognetalliance.org/faq

Frazer C (1931) "Fog-drip" may hold key to drought relief. Modern Mech Invent 62–65

Gandhidasan P, Abualhamayel H (2006) Fog collection as a source of fresh water supply in the Kingdom of Saudi Arabia. Water Environ J 21:19–25. https://doi.org/10.1111/j.1747-6593.2006.00041.x

García-Ubaque CA, Vaca-Bohórquez ML, García-Ubaque JC (2013) Factibilidad técnica y de salud pública de la recolección de aguas nieblas: Estudio de caso. Revista De Salud Pública 15:366–373

Gioda E, Espejo GR, Acosta BA (1993) Fog collectors in tropical areas. In: Proceedings of the symposium on precipitation and evaporation, vol 3. 20–24 September 1993, Bratislava, Slovakia, pp 20–24

Graham S, Parkinson C, Chahine M (2010) The water cycle. NASA, Earth Observatory. Retrieved August 25, 2020, from https://earthobservatory.nasa.gov/features/Water

Harb OM, Salem MS, Abd EL-Hay GH, Makled KM (2016) Fog water harvesting providing stability for small Bed we communities lives in North cost of Egypt. J Ann Agricult Sci 61(1):105–110. https://doi.org/10.1016/j.aoas.2016.01.001

Holmes R, de Dios Rivera J, de la Jara E (2015) Large fog collectors: New strategies for collection efficiency and structural response to wind pressure. J Atmos Res 151:236–249. https://doi.org/10.1016/j.atmosres.2014.06.005

Jewell C (2018) Pioneering fog-harvesting technology helps relieve water shortages in arid regions. WIPO Magazine, June. Retrieved August 25, 2020, from https://www.wipo.int/wipo_magazine/en/2018/03/article_0003.html

Kaseke KF, Wang L (2018) Fog and dew as potable water resources: maximizing harvesting potential and water quality concerns. J GeoHealth 2(10):327–332. https://doi.org/10.1029/2018GH000171

Kawai H, Koshiro T, Endo H, Arakawa O, Hagihara Y (2016) Changes in marine fog in a warmer climate. J Atmos Sci Lett 17(10):548–555. https://doi.org/10.1002/asl.691

Kim NK, Kang DH, Eom H, Kang., H. W. (2019) Biomimetic fog harvesting surface by photo-induced micro-patterning of zinc-oxide silver hierarchical nanostructures. J Appl Surface Sci 470:161–167. https://doi.org/10.1016/j.apsusc.2018.11.132

Klemm O, Schemenauer RS, Lummerich A, Cereceda P, Marzol V, Corell D, van Heerden J, Fessehaye GM (2012) Fog as a fresh-water resource: overview and perspectives. AMBIO: J Human Environ 41:221–234. https://doi.org/10.1007/s13280-012-0247-8

Lawrence D (2020) How Avatar pulled off the award-winning shot that captured the 'spirit of the film'. Entertainment Weekly. Retrieved August 26, 2020, from https://ew.com/movies/2020/02/03/avatar-the-shot/

Marloth R (1904) Results of experiments on table mountain for ascertaining the amount of moisture deposited from the south-east clouds. Trans R Soc South Africa 14:403–408

Martin EA (1915) Dewponds: history, observation and experiment. Werner Laurie, London

Marzol MV (2005) La Captación del Agua de la Niebla en la Isla de Tenerife. Servicio de Publicaciones de la Caja General de Ahorros de Canarias: Las Palmas de Gran Canaria, Spain. 220 p

Mehta L (2003) Contexts and constructions of water scarcity. Econ Pol Wkly 38(48):5066–5072

Meunier D, Beysens D (2016) Dew, fog, drizzle, and rainwater in Baku (Azerbaijan). J Atmos Res 178–179:65–72. https://doi.org/10.1016/j.atmosres.2016.03.014

Nguyen TC (2014) This tower pulls drinking water out of thin air. Smithsonian Magazine. Retrieved August 25, 2020, from https://www.smithsonianmag.com/innovation/this-tower-pulls-drinking-water-out-of-thin-air-180950399/

Olivier J (2002) Fog-water harvesting along the West Coast of South Africa: a feasibility study. J Water SA 28(4):349–360. https://doi.org/10.4314/wsa.v28i4.4908

Qadir M, Jiménez GC, Farnum RL, Dodson LL, Smakhtin V (2018) Fog water collection: challenges beyond technology. J Water 10(4):372. https://doi.org/10.3390/w10040372

Qiao N, Zhang L, Huang C, Jiao W, Maggs-Kölling G, Marais E, Wang L (2020) Satellite observed positive impacts of fog on vegetation. J Geophys Res Lett 47(12). https://doi.org/10.1029/2020GL088428

Ritter A, Regalado CM, Guerra JC (2015) Quantification of fog water collection in three locations of Tenerife (Canary Islands). J Water 7(7):3306–3319. https://doi.org/10.3390/w7073306

Schemenauer RS, Cereceda P (1994a) A proposed standard fog collector for use in high-elevation regions. J Appl Meteorol Climatol 33(11):1313–1322

Schemenauer RS, Cereceda P (1994b) Fog collection's role in water planning for developing countries. J Nat Resourc Forum 18(2):91–100

Schemenauer RS, Cereceda P, Fuenzalida H (1988) A neglected water resource: the Camanchaca of South America. Bull Am Meteor Soc 69(2):138–147. https://doi.org/10.1175/1520-0477(1988)069%3C0138:ANWRTC%3E2.0.CO;2

Schemenauer RS, Bignell B, Makepeace T (2016) Fog collection projects in Nepal: 1997 to 2016. In: Proceedings of the 7th international conference on fog, fog collection and dew. Wroclaw, Poland

Schutte JM (1971) Die Onttrekking van Water uit die Lae Wolke op Mariepskop. Technical note, 20. Division of Hydrological Research, Department of Water Affairs, Pretoria

Shanyengana ES, Henschel JR, Seely MK, Sanderson RD (2002) Exploring fog as a supplementary water source in Namibia. J Atmos Res 64(1–4):251–259. https://doi.org/10.1016/S0169-8095(02)00096-0

Steinmetz K (2016) There is now vodka literally made from San francisco Fog. TIME Magazine. Retrieved August 25, 2020, from https://time.com/4317269/fog-vodka-san-francisco-hangar-one/

Torregrosa A, O'Brien T, Faloon I (2014) Coastal fog, climate change, and the environment. J Eos, Trans, Am Geophys Union 95(50):473–474. https://doi.org/10.1002/2014EO500001

Trautwein P, Frost E, Jutzi S, Hruschka H (2016) Natural wells in the sky. Wasserstiftung, Schäftlarn

Trevino MT (2020) The ethereal art of fog-catching. BBC: Future Planet. Retrieved August 25, 2020, from https://www.bbc.com/future/article/20200221-how-fog-can-solve-water-shortage-from-climate-change-in-peru

UNEP (United Nations Environment Programme) (1997) Source book of alternative technologies for freshwater augmentation in Latin America and the Caribbean. Unit of Sustainable Development and Environment, Washington DC

Warka Water (2020) FAQ: frequently asked. Questions. Warka Water- every drop counts. Retrieved August 25, 2020, from https://www.warkawater.org/faq/

Wei-Haas M (2017) This device collects water from the clouds. Smithsonian Magazine. Retrieved August 25, 2020, from https://www.smithsonianmag.com/innovation/device-collects-water-clouds-180962705/

Chapter 4
Micro-catchment Rainwater Harvesting

Theib Y. Oweis

Abstract Rainwater harvesting is an ancient practice that helped in meeting basic water needs and reduced water shortages mainly in arid and semi-arid regions. Rainfall, through runoff, can be captured downstream of a suitable "catchment" area. The capture and storage of rainwater can be beneficially used. Harvesting water depends not only on the rainfall amount, but also on its pattern and intensity and on the catchment and storage conditions. Storage is a vital component of rainwater harvesting systems and can be surface or subsurface reservoirs or simply a soil profile. Uses include domestic, agriculture, industrial and environment sectors. Micro-catchment rainwater harvesting (MIWH) systems are based on having a small runoff catchment, normally at the household or farm level. In MIWH, runoff flows as sheet flow downstream to a storage facility to be used later for various purposes. Among the most common MIWH types are the Household systems including rooftops and cisterns and the Farm and Landscape systems including contour ridges, bunds, small runoff basins and strips. This chapter provides an overall description of the types, uses and limitations of MIWH. It also presents cases where MIWH plays an important role in providing necessary water for people and agriculture in addition to combating desertification and coping with climate change in dry environments. The implementation of those systems, however, face several technical, social, financial, and environmental constraints. Recommendations to help overcoming those constraints are provided for the rural dry environments where the need for water and food is critical.

Keywords Runoff · Farm reservoirs · Rooftops · Contour ridges · Bunds · Desertification

4.1 Introduction

In many areas of the world, water scarcity is increasing and the demand for additional amounts of water to satisfy the needs of the rapidly growing population is

---

T. Y. Oweis (✉)
International Center for Agricultural Research in the Dry Areas (ICARDA), Amman, Jordan
e-mail: t.oweis@cgiar.org; theib.y.oweis@gmail.com

M. Qadir et al. (eds.), *Unconventional Water Resources*,
https://doi.org/10.1007/978-3-030-90146-2_4

also increasing. At the same time conventional water resources are limited, often fully utilized, or are already overused (UNESCO 2020). Unconventional water resources are gaining more attention to fill the gap that is widening due to increasing demand. Obvious unconventional resources that have received attention in the last decades include desalinated seawater and brackish water, treated sewage water and lower quality agricultural drainage water (Qadir et al. 2007). A less obvious water source, that is often not listed as unconventional, is that of rainwater harvesting. This resource is now gaining more attention worldwide but especially in dry environments where conventional resources are scarce (Oweis 2017). The two types of rainwater harvesting are the macro and micro catchments. This chapter focuses on the role that micro-catchment rainwater harvesting (MIWH) plays to increase water availability and the most relevant methods for various situations.

Although rainwater harvesting is a known old practice, its nature and role are often misunderstood or confused with other water conservation practices. Many definitions can be found in the literature but a generic one is "the concentration/collection and storage of rainwater, through runoff, and use for beneficial purposes". Beneficial use includes domestic, agriculture, environment and industrial (Oweis et al. 2012).

The rainwater harvesting practice was first developed in the arid and semi-arid regions where conventional surface and ground water resources are scarce. In those regions seasonal rainfall is low and is largely lost in evaporation and/or flow through runoff to salt sinks with little benefits. It is estimated that over 90% of rainwater in dry environments is lost in evaporation (Oweis 2017). Yet those areas are in desperate need for water even for domestic use. Rainwater harvesting responds to this loss of rainwater and provides means to capture, at least part of it, for various uses of the society (Critchley and Siegert 1991). A relatively low rainfall amount falling on large agricultural areas may not be enough to support a crop, as plants water requirements are much higher. In this case both water and the crop are basically lost. The concept of rainwater harvesting in agriculture assumes that if we can borrow the rainwater of say half of the land and add it to the other half then one half of the land gets twice as much rain which may be enough to grow some crops. Furthermore, if one concentrates rainwater from three quarters of the land into one quarter then the later quarter will have 4 times of the annual rainfall, plenty of water to grow any crop. If we can do this, or a proportion of it, then part of rainwater and part of land will be productive, otherwise both the rainfall and the land are lost with no benefits. This concept may be applied to many situations not only in agriculture (Oweis et al. 2012).

With MIWH rainwater becomes more productive, it brings more benefits including biophysical, economic, environmental, and social, but also domestic, food and feed products. Other benefits include those to the environment by enhancing biodiversity, carbon sequestration, controlling erosion and reducing the hazardous impacts of dust storms (Molden et al. 2010). MIWH converts 40–50% of the evaporation losses into transpiration by increasing vegetative cover, hence substantially increasing water productivity (Oweis and Hachum 2006).

Considering the above, any rainwater harvesting system should have a "catchment" which is an area where all or part of the rain falling on it is allowed or forced to run off downstream. This is the first and probably most important component of any rainwater harvesting system. The catchment can be as small as a rooftop or as large as a watershed of many square kms. When the catchment is small, we call the system MIWH. When the catchment area is large the system is called macro-catchment rainwater harvesting (MAWH). The catchment can be natural terrain but to induce more runoff, land may be cleared and smoothened, compacted or covered with impermeable material (Fraiser 1980; Critchley and Siegert 1991).

For the rainwater to be concentrated/collected, some "runoff" must occur. Runoff is essential for any rainwater harvesting system and is the second most important component of the system. If there is no runoff, there is no rainwater harvesting. This is not like soil–water conservation where the objective is to encourage water to infiltrate in the soil and stop any runoff. Of course, runoff here occurs in the catchment which is allocated mainly for this purpose (Oweis et al. 2012). Runoff can occur naturally or may be induced. Most runoff occurs on slopes and on solid surfaces than on flat and sandy soils. The ratio of the amount of runoff to the amount of rain is called the "runoff coefficient". It approaches one for the paved surfaces such as rooftops and other impermeable surfaces and may drop to less than 0.01 for natural terrains. Runoff coefficient can be increased by applying permeable materials on the catchment surface. It can be used with other parameters for designing rainwater harvesting systems (Fraiser 1980; Mzirai and Tumbo 2010).

The third component of any rainwater harvesting system is the "storage" facility. As runoff occurs during and after a rain event, a storage is needed otherwise water will run and be lost downstream. We can store runoff water in surface or underground reservoirs where water can be extracted later for any purpose or stored in the soil profile, as soil moisture, where water can be utilized directly by crops. The latter is a common low-cost storage of the MIWH for agriculture, but small reservoirs are also common for other purposes. Storage is usually the highest in cost among all rainwater harvesting components (Guo and Baetz 2007; Oweis et al. 2012).

The last component of the rainwater harvesting system is the "target/use". In agriculture, it is the crop evapotranspiration but also livestock watering, domestic or industrial and environmental uses (Critchley and Siegert 1991; Oweis et al. 2012). Any system should have the 4 components indicated above although sometimes there is unclear distinction between them. As one can notice this practice provides water for various purposes in an unconventional manner and in places where conventional water resources may not be enough or even available to support the livelihoods of the rural communities. In fact, in many towns and cities where conventional water resources are insufficient, authorities add to the building code provisions for rooftop and yards rainwater harvesting system to provide part of the domestic and landscape water needs and reduce flooding the water and sewage systems (Previati et al. 2010).

4.2 Types and Uses of MIWH

Classifications of rainwater harvesting practices are mainly based on the size of the catchment area or the type of storage facility but can also be related to the uses of the harvested water. MIWH systems include small catchments with area from few square meters to thousands square meters (Thomas et al. 2014). Runoff in the catchment is characterized by a sheet flow that travels only a short distance. In sandy soils and other highly permeable surfaces, materials to help sealing the soil surface are used (Boers and Ben-Asher 1982; Fraiser 1980). The runoff is usually stored and used in an adjacent cultivated area. Storage can be in a small reservoir, a cistern or a tank to be used later for humans, animals and other purposes. Also, storage can be in the soil profile to be used directly by plants (Oweis et al. 2012). In the MIWH, the catchment and the cultivated area are usually on-farm, which gives the farmer more control over the system. As the catchment area may not be cultivated, to provide runoff effectively to the cultivated area, farmers in rainfed areas resist giving this part up as catchment. However, in drier environments where rainfall alone, without water harvesting, may not ensure economic agricultural production then they have no choice but to allocate good portion of the land to be a catchment (Oweis 2017). The two major MIWH systems include the household systems and the farm/landscape systems.

4.2.1 Household Systems

Those usually have a small to medium size catchments in the household vicinity with a surface or underground water storage device. Domestic use for drinking and sanitation in addition to livestock watering are major uses of water. It, however, can be used for agriculture especially for cash crops in greenhouses and household's backyard (Oweis et al. 2012).

Rooftop rainwater harvesting: Those systems mainly use the house roofs as a catchment to collect runoff water and store in tanks or reservoirs for later use. Other surfaces, however, can be used as a catchment including greenhouses and parks (Fig. 4.1). Runoff water is usually stored in tanks, jars, or similar devices. Water is

Fig. 4.1 Rooftop rainwater harvesting system (left) with underground storage in dry northeastern Brazil and greenhouse rainwater harvesting system (right) supporting protected vegetables in Alexandria area of Egypt (Credits: Theib Oweis, ICARDA)

then used in the household or for agriculture such as in hydroponic systems or in protected agriculture (Frasier 1980; Jones and Hunt 2010).

Roofs are usually the top of houses with ceilings made of concrete, steel, wood, or any other material. The runoff water is directed by gutters (narrow channels) and pipe networks to a storage tank which can be above or below ground surface. First runoff of the season is usually flushed out to get rid of contaminations accumulated on the rooftop surfaces during the dry season (Falkenmark et al. 2001; Anderson et al. 2011). For drinking, water should be purified and may be chlorinated before supplied to people. In this case, roofs should be frequently cleaned, and a flirtation and chlorination facility should be provided to ensure water quality is high before using. This may not be necessary if the water is used for other purposes such as gardening (Ben-Asher et al. 1995; Vohland and Barry 2009).

The main cost of rooftop rainwater harvesting comes from installing the storage device. Especially in areas where there is a long dry season, large storage volume is needed to accommodate the long dry period before other rain occurs. However, a combination of rooftop water harvesting, and other sources of water may allow smaller-sized storage at lower cost. Runoff, storage, and use efficiencies can be substantially increased with appropriate modelling of the system and optimizing the catchment and the storage sizes relative to the family size and water needs (Abu-Zreig et al. 2019; Ali et al. 2009).

Cisterns: Cisterns can be traced back to before 3,000 BC when the earliest household cisterns were built in Palestine (Wåhlin 1997). In the Negev desert, cisterns were dug in loess soil and lined with large stones as long ago as the Iron Age. Rock-cut cisterns were practiced in Nabatean times, with some still functioning in the iconic stone city of Petra in Jordan (Lenzen et al. 1985). There are many old and new cisterns in the Syrian steppe and some areas of Libya, Tunisia, and Palestine. Most of these cisterns are water sources for nomads, meeting both human and livestock needs (Ali et al. 2009).

A cistern is a subsurface water collection and storage structure that is generally dug at the lowest level of an earthen micro-catchment (Fig. 4.2). To be successful,

Fig. 4.2 Ancient cistern (left) still functioning and supporting communities in Matrouh, northern Egypt, and renovating an old cistern (right) with settling basin and filtration system at Tel Hadya, northern Syria (Credits: Theib Oweis, ICARDA)

a cistern should have an adequate catchment area to generate enough runoff for various rainfall conditions, have a suitable underlying geologic formation to reduce construction costs and should be managed well for efficient use of the stored water. The first seasonal runoff from the catchment carries contaminants and is usually diverted away from the cistern. The ensuing storm runoffs which are cleaner can enter the cistern. A small settling basin is usually constructed to remove sediments before the water enters the cistern. After the cistern has been filled, the surplus water is usually guided past and downstream by a ditch through an outlet, either to other cisterns or to a disposal point. The water from the cistern is lifted manually by a bucket or by using a pump. This water is mainly used for domestic and livestock needs. Supplemental irrigation of crops is possible if people own large or multiple cisterns. These low-cost structures are affordable for poorer households and relatively safe and reliable to store water for the dry season (Wåhlin 1997; Ali et al. 2009).

In areas where alternative water resources are not available, cisterns still play vital role in providing water for people and livestock needs. This is more critical in the dry areas and for rural communities where livelihoods of the poor depend on them. Cisterns provide water where it is collected and hence provide greater security to the household. They can also support the family garden with necessary irrigation. However, the most important constraint is the size limitation that is associated with the cistern digging and construction costs. Proper management of the runoff–storage–use process can help increase the capacity without additional cost (Oweis and Taimeh 2001; Ali et al. 2009). The famous and still common underground cisterns in northern Egypt (Fig. 4.2) are supplied by a relatively larger micro-catchment.

4.2.2 Farm and Landscape Systems

The on-farm and landscape MIWH techniques support agriculture in dry environments and help combating land degradation. Those systems comprise a small, cultivated area supported by a catchment of greater or more in size. Runoff water is usually stored in the soil profile containing the crops' root system, to be consumed during the growing season or stored in small reservoirs to be used later for crops and other uses, especially as supplemental irrigation or for livestock (Oweis and Taimeh 2001). The catchment area is usually located just upstream of the cultivated area and is usually cleared, smoothed, or compacted to increase the runoff coefficient (Oweis et al. 2012). Sometimes, chemicals are used to treat the soil to increase runoff, but also plastic sheets or mulches are used to cover the whole catchment surface (Xiao et al. 2004 and Dutt 1981). Most common on farm micro-catchments utilize soil-based storage and reservoir-based practices.

Soil-based storage practices: Rainwater runoff from a small micro-catchment is collected at the lowest point downstream where water is slowed down by a ridge or a pit which is also the cultivated area. Here, runoff is given enough opportunity, i.e., time, to infiltrate the soil and be stored in the crop-root zone to support crop growth. Runoff water also carries fertile surface sediments together with indigenous seeds that

Fig. 4.3 Semicircular bunds with runoff collected to support shrubs (left) and newly constructed contour ridges (right) at Tel Hadya, in Northern Syria (Credits: Theib Oweis, ICARDA)

are also deposited in the cultivated area. Improved soil water and fertility in the ridges and pits provide better environments for trees, crops, shrubs, and grasses to grow and for biodiversity to increase. Soil-based techniques require contour ridges/bunds, negarims, and run-off strips (Oweis 2017; Lasage and Verburg 2015).

Contour ridges and bunds (Fig. 4.3) are simple low-cost structures built on undulating lands along the contour lines, so that water stays in place and the ridges are kept intact. When contouring is not precise, earthen ties may be placed at short intervals to keep the water from moving (Oweis et al. 2012). To be successful, they should be implemented on medium to deep soils having high water-holding capacity (Critchley and Siergert 1991). To ensure sufficient runoff, ridges should be designed properly with enough spacings between them to be maintained. Design takes into consideration crop-water requirements, the amounts of rainfall and runoff, and soil storage capacity as the main factors in determining the spacings, which usually are in the range of 5–20 m. The height of the ridge/bund should be carefully chosen to hold the peak storm runoff volume from the catchment. They are usually about 30–60 cm in height and in steep slopes they can be supported by stones (Oweis 2017). Implementation can be manual, but the cost may be high and not economical for large-scale development. Mechanizing this operation is now possible, which drastically reduces the construction cost. A package for restoring the degraded dry rangelands has been developed based on this technique and will be presented later (Ali et al. 2010).

Runoff strips (Fig. 4.4) are designed for field crops in areas where seasonal rainfall is not enough to support economic yields and no irrigation water is available. A catchment strip of adequate width is kept uncultivated and/or smoothened to provide runoff to adjacent cultivated strip so that the rainfall plus runoff is stored in the soil profile and is enough for crop needs (Critchley and Siegert 1991). The cultivated strip should not be too wide so that the runoff can be uniformly distributed across the strip. Catchment strip width depends on the run-off coefficient and rainfall amount, in addition to crop-water requirements. The system implies that farmers may sacrifice part of their land to act as catchments to enable the other part to be productive. This option is only feasible in drier environments where all land may be unproductive without rainwater harvesting (Oweis et al. 2012).

Fig. 4.4 Runoff strips (left) supporting cereals and legumes and small runoff basins (negarims) (right) supporting shrubs at ICARDA, Tel Hadya, northern Syria (Credits: Theib Oweis, ICARDA)

The small basin (negarim) system (Fig. 4.4) is suitable for tree orchards and shrubs. The land is divided into square/rectangular or diamond shape plots of 50–500 m^2 and surrounded by a levee (Oweis et al. 2012). The plot is smoothened or treated to induce maximum runoff. Trees are planted at the lowest part of the plot so runoff water flows from all the catchment to be stored in the plant root zone. One issue here is that the structure is permanent, and no plowing can be done to remove weeds. Farmers usually use other means of weeding, including by hand with herbicides (Critchley and Siegert 1991).

Small-farm reservoir-based practices: Micro-catchments of medium to large size are prepared to provide runoff to small surface reservoirs at the farm level. The catchment occupies the less productive part of the farm such as rocky areas and is usually cleared of obstacles to induce more runoff. Reservoirs may be constructed by a small dam blocking a stream path or by digging a pond in flatter areas (Oweis and Taimeh 1996). Most common small-farm reservoirs are used for supplemental irrigation of rain-fed crops and "hafaer", which are ponds used mainly to provide water for livestock in dry environments (Fig. 4.5). Sediment accumulation and water quality are the most important issues of concern.

One of the most important roles of farm reservoirs is providing supplemental irrigation for crops in areas where rainfall in not adequate or there are drought spells affecting crop yield. Those are common conditions of rain-fed agriculture in the semi-arid areas such as the Mediterranean (Oweis and Hachum 2006). To increase the reservoir capacity, since the reservoir is filled at the beginning of the rainy season, water is moved to be stored in the crop-root zone, so that then a refill can occur with subsequent storms (Oweis and Taimeh 1996).

Hafaer or livestock ponds are very common in arid environments, especially in Africa. Usually, natural runoff into a lower point of the landscape allows water to accumulate where the livestock is watered. In many areas those hafaers are improved by providing defined clean catchments and settling basins to reduce sediments and pollutants in the pond. Also, a spillway may be provided to limit the depth of water in a pond that may threaten the lives of people and livestock. Issues of concern include

Fig. 4.5 Small-farm reservoir (left) for supplemental irrigation of rain-fed crops in Tel Hadya, northern Syria, and a runoff pond (Hafaer) for livestock in western Sudan (right) (Credits: Theib Oweis, ICARDA)

the spread of mosquitoes and other insects due to stagnant water (WOCAT 2012; Oweis et al. 2004; Biazin et al. 2012).

4.3 Applications in Dry Environments

4.3.1 Contour Ridges/bunds for the Restoration of Degraded Rangelands

Rangelands in the dry areas occupy over 30 million km^2 and host over 280 million people. The communities generally are poor and primarily practice livestock herding. These systems receive low and nonuniform rainfall amounts that are inadequate for normal rain-fed farming but can support rangeland grasses and shrubs (Karrou et al. 2011). Land degradation is a common process that defines the features of most of these ecosystems. Efforts by national and international institutions to combat desertification and to restore these ecosystems have had limited success. A typical dry and degraded agro-pastoral system occupies most of West Asia and North Africa and is locally called the '*badia*' in reference to hosting *Bedouin* tribes. The badia is a host for the bedouins in the Middle East who practice sheep and goat herding for a living. The badia covers over 70–80% of the region and receives less than 200–250 mm annual rainfall. Much of the rainfall runs off, as soils are generally crusty, and evaporates or joins salt sinks (Taimeh and Hattar 2006). Because of overgrazing, wood cutting, and drought, vegetation and soils have been eroded causing very low carrying capacities, frequent dust storms, and hardship, altogether resulting in rapid migration to cities (Louhaichi and Tastad 2010; Oweis 2017).

Research by ICARDA and national partners has developed, tested, and scaled up a community-based integrated restoration package using MIWH practices including

Fig. 4.6 Semicircular bunds in the Syrian badia (left) and contour ridge planted with Salsola seedlings in the Jordan badia (right) (Credits: Theib Oweis, ICARDA)

contour ridges and bunds. The restoration package has been developed, tested, and scaled up over 10 years and has shown great potential for restoration of the system (Karrou et al. 2011; Oweis 2017).

Experimental fields were set up in representative watersheds with a combination of MIWH techniques for testing various system parameters and design elements for supporting indigenous shrubs. At the beginning, the structures were constructed manually in farmers' fields (Fig. 4.6). A range of design parameters including water, plant, and soil were tested and evaluated for several seasons. Suitable metrics were calculated for the design of MIWH techniques (Karrou et al. 2011). The selected packages were also tested for several seasons and found satisfactory for restoring the ecosystem. The soil profile was nearly full at the end each rainy season owing to sufficient runoff and soil storage capacity, which drops to near wilting point at the beginning of the next rainy season. It was observed that eroded topsoil, together with indigenous seeds, were largely deposited behind the bunds and ridges providing a fertile environment for plant growth, controlling land degradation, and enhancing biodiversity (Karrou et al. 2011). The site was protected for two years, after which a gradual introduction of animals for grazing occurred, depending on the plants carrying capacity. The shrub species yielded 244–428 kg/ha for Salsola and 668–1,378 kg/ha for *Atriplex* (Akroush et al. 2014; Ali et al. 2010). After four years, the landscape was completely changed, with vegetation covering the bunds and ridges and an indigenous habitat established under the shrubs.

The cost of defining and constructing the bunds and ridges manually amounted to about $2,000–3,500 per ha depending on the site conditions. The high labor cost is a limiting factor for large-scale implementation. The process was mechanized using the Vallerani implement. The 'Vallerani' dolphin (Fig. 4.7), a specially designed plow, can create similar ridges and bunds to those constructed manually. It is a hydraulically controlled machine that can work in various soils and landscapes (Antinori and Vallerani 1994). The plow was tested, adjusted, and calibrated to suit the *badia*

Fig. 4.7 Vallerani delfino (left) in operation constructing contour ridges and bunds at the Badia of Syria and the field after construction of intermittent ridges in the Jordan badia (right) (Credits: Theib Oweis, ICARDA)

conditions (Gammoh and Oweis 2011a). The dolphino plow can create continuous ridges or an intermittent bund separated by the natural terrain (Fig. 4.7). The soil is moved to the downstream side, leaving the upstream side clear to receive runoff water. A 70-cm ripper mounted at the plow breaks the hard surface and subsurface soil and soft rocks to allow water to infiltrate deeper into the soil. To automate the contouring process, a special laser guided device was mounted with a receiver and control panel on the tractor (Gammoh and Oweis 2011b). Laser contour guiding speeded up the implementation and drastically cut down on the cost, in addition to improving tractor efficiency and implementation precision. The total cost of implementing the package was about $32/ha. The internal rate of return on the investment was estimated at 13% and increased to 17% when considering environmental benefits such as erosion control (Akroush et al. 2014).

The package was developed and tested in farmers' fields with full farmer participation at all stages. Most important was the protection of the shrub seedlings in the early stages and managing grazing afterwards (Karrou et al. 2011). Within a few years after implementing the restoration package, the vegetation was already established with many indigenous grass seeds germinating and emerging in the well-moistened ridges and bunds (Fig. 4.8). Furthermore, soil erosion outside the field appeared to have halted, while local-level eroded soil settled in the bunds, improving their fertility (Shawahneh et al. 2011). In 2014, the Jordanian government launched a large initiative to start restoring the badia ecosystem using the package. By 2018 over 10,000 ha were already restored, with plans to continue the restoration (JMOE 2014).

Several lessons were learned from this initiative, the most important being that the involvement of the local communities at all stages of development is essential for their adoption. Although MIWH is instrumental in restoring the degraded rangelands, it must be done in integration with other elements of the development. Good agronomic

Fig. 4.8 Mechanized MIWH bunds (left) after establishment, with water collected and infiltrated within the bunds. Bunds and vegetation growth after a few years (right), with sheep grazing in the restored landscape (Credits: Theib Oweis, ICARDA)

practices and controlling grazing are vital for the success and sustainability of the restoration. Especially, open grazing issues require some resolution by enhancing institutional functions to limit open grazing and enable protection from overgrazing (Oweis 2017).

4.3.2 Management of Small Water Harvesting Reservoirs

Small on-farm water harvesting reservoirs can be at the surface, as in the case of tanks and surface reservoirs, or subsurface, as in the case of cisterns. In all cases the cost of the storage is the main constraint for upscaling these systems. People try to minimize the cost, which necessarily limits the size of the reservoir. At the same time, the volume of the seasonal runoff from the micro-catchment can be much higher than the capacity of the reservoir. If not stored, it may be lost without benefits at a time when the household or farm is in great need for every drop of water (Ali et al. 2009).

Reservoir capacity can be substantially increased by understanding hydrology of the area and the probabilities of having multiple storms with enough runoff. This can support multiple filling if the emptying of the reservoir is done in time. One can empty early-season fills by consuming and storing the extra water in the soil profile for agriculture. Emptying the reservoir will allow space for the next storm's runoff to be stored by doubling the actual capacity. If the probability of another storm is high, then the second fill may be emptied and used/stored in soil. Usually, a garden with some fruit trees and vegetables can benefit from the soil-water storage and support the livelihoods of the household. Only the last fill may be kept and used for the rest of the season for domestic and livestock use. The virtual size of the reservoir can be several times the physical one, and the cost–benefit of the reservoir would be maximized without higher initial investment (Oweis and Taimeh 2001; Ali et al. 2009).

There is a risk, however, that if drought occurs while the cistern is empty it would leave the family/farm with no water for the rest of the season. However, this risk can be reduced by understanding the storm system and making conservative decisions regarding when and how many times the reservoir may be emptied. Hydrological and reservoir modeling can optimize this operation and maximize benefits (Oweis and Taimeh 2001; Abu Zreig et al. 2019).

Farm reservoirs for supplemental irrigation of rain-fed systems: The Muaqqar watershed of the University of Jordan was established in 1985 in a typical dryland environment, 50-km east of Amman. Three surface reservoirs were built in sequence by blocking the wadi in the watershed by an earthen dam with a spillway to protect the structure when it overflows. The size of the lakes behind the dam ranges from 10,000 to 15,000 m^3. The first runoff storm fills the three reservoirs, while extra water flows downstream. Rain from subsequent storms usually has little space in the reservoirs to be stored, so water flows downstream at times when the site actually needs more water. Due to costs and technical reasons, there was no option to build additional reservoirs or to increase the capacity of the existing ones.

Managing the stored water in relation to the agricultural fields provided a low-cost option to extend the storage capacity of the reservoirs without increasing their sizes physically. Pumping the reservoir water once full into the nearby cultivated fields and storing the water in the soil profile allows the reservoir to receive and store new inflows, hence multiplying its storage capacity. When the field-soil profile was full, additional land was supplementally irrigated, while multiple filling and emptying continued with more storms. Water applied in supplemental irrigation supported rain-fed crops, such as barley, wheat, forages, and trees, through the cropping season and beyond. It was possible, on average, to fill up the reservoirs three times during the rainy season, tripling the physical capacity. This requires, however, a thorough analysis of the site hydrology and modeling the runoff–storage–use interactions to optimize the management of the reservoirs (Oweis and Taimeh 2001). Developments upstream may have negative impacts on users downstream (Bouma et al. 2011). Currently, with little development upstream, only limited conflicts occur, but in the future with large-scale development, this needs to be addressed in a comprehensive watershed-management approach (Karrou et al. 2011).

Water applied as supplemental irrigation increased the yield of field crops by up to an additional 2.5 kg/m^3 of water. Great benefits were obtained by irrigating fruit trees and forages. Crops in the rain-fed areas suffer from drought spells that can drastically reduce yields. Applying small amounts of harvested water can alleviate stress and substantially increase yields and water productivity (Mechlia et al. 2009; Oweis and Hachum 2006).

Cisterns for households, livestock, and gardening: The Romans first used rain-water harvesting cisterns in the Matrouh area of northwest Egypt about 300 B.C. to secure water for domestic and livestock needs. The storage capacity of these cisterns were in the range of 500–1500 m^3 and were used collectively by the community.

After the Roman era, people continued to dig cisterns but with smaller capacities of 50–250 m^3 to be used by a single household (MRMP 1992; Ali et al. 2009).

The Matrouh region is dry and receives annual precipitation of about 150 mm. People mainly depend on cisterns to store runoff water, from medium-sized catchments during the rainy winter, to be consumed in the long dry summer season. Cisterns are dug in the deep soil laying below the thick hard-pan near the surface. The hard-pan forms the ceiling of the reservoir, which reduces the construction cost. The bottom and sides of the cistern are plastered using local materials to stop seepage. Currently, cisterns may not be able to meet the total water demand of the household, but they play a significant role in complementing other water sources and reducing water shortages. In addition to providing drinking water for the family and for their livestock, some water is used for limited subsistent farming including fruit trees, grains, medicinal plants, and vegetables in the household backyards (MRMP 1992). Because of the limited capacity of the cistern, people usually are conservative in using the stored water to ensure having drinking water for people and livestock during the long dry summer season. Only a large cistern may be capable of supporting limited gardening surrounding the households (Ali et al. 2009; Xiao et al. 2007).

A World Bank project implemented by ICARDA in the Matrouh area investigated the expansion of cistern capacities through improved water management of the harvested water. Analysis of the occurrence of runoff-producing storms in the area showed that three storms might occur, with a confidence level over 75%. Using soil-water storage to utilize early cistern fillings in agriculture would allow expanding the storage capacity substantially. This means that farmers can use the first fills for supplemental irrigation of crops around the house and other needs during the rainy season. They still need to guarantee that the last fill will be kept for use during the summer dry season for drinking and for the livestock. There remains the small risk of not having enough water for the long dry season but fortunately, there are other supplemental sources of water, such as groundwater, that can be used, in conjunction with harvested water. The water-management approach was tested and applied at several households' cisterns in the project area. It showed that farmers utilize significantly more runoff water to support crops around their households. The household cultivated area is determined based on the physical volume of the cistern and the expected number of storms with comfortable confidence. Benefits are maximized by cultivating high-value drought-tolerant crops and supplementing, when needed, with alternative water resources (Ali et al. 2009; Oweis and Taimeh 2001).

4.4 Challenges and Potential Response Options

4.4.1 Climate Uncertainties and Storage Cost

Climate variability and uncertainty is the greatest challenge for planning MIWH, especially in the long term. Rainfall amounts, intensities, and frequency affect runoff occurrence and rates, which are critical in water harvesting development. Variations in rainfall may disrupt water availability for both domestic and agricultural use where development becomes risky for long-term investments such as in fruit trees cultivation (Sowers et al. 2011). Since water storage ensures water supply during drought spells and through the dry season, it needs to be sufficiently large. However, storage is not unlimited, due to the initial cost in the case of surface and subsurface reservoirs and due to limited soil-water storage capacity in the case of cultivating light soils. Prolonged droughts cause severe moisture stress to these plants with subsequent losses of production to the communities. Prolong periods of drought with limited storage have caused the near collapse of systems that are based only on rainwater harvesting (MRMP 1992). The impacts of climate change can already be felt but are complex and vary from one place to the other. In the dry areas, a decrease in annual precipitation is expected with increases in intensity and variability (IPCC 2014). Furthermore, drought frequency and duration may increase, thus affecting rainwater harvesting strategies in dry environments (Oweis 2017; HLPE 2015).

Optimizing storage facilities, understanding historical rainfall patterns, diversifying storage types, and selecting drought-tolerant crops would reduce the risks associated with climate uncertainties. There are several strategies to reduce the risk of failure, including the selection of drought-tolerant crops, ensuring deepest-level and highest degree of soil-water storage, and having a standby water source to supplement rainfall in the case of extreme events. One can enhance soil storage capacity by applying absorbent polymers and organic matter. As the cost of those amendments in high, they are only used in growing cash crops such as trees and vegetables (Somme et al. 2004). Modeling and optimizing the management of water harvesting reservoirs and cisterns can expand their capacity and support more agriculture (Ali et al. 2009; Oweis 2017; Oweis and Taimeh 1996).

There are also opportunities with climate change. More intensive rainstorms will lead to higher runoff rates. MIWH may better allocate runoff water to target areas with increased infiltration and hence soil water storage and groundwater recharge. This would support higher adaptation rates and a more efficient management of available water resources to cope with drought and enhance communities' resilience (Previati et al. 2010; Kato et al. 2011; HLPE 2015).

4.4.2 Modernization and Technical Capacity

The indigenous knowledge behind rainwater harvesting practices is still relevant today and will be in the future. However, indigenous practices are largely outdated, and many are inefficient currently or in the future. Modernization of tools, new materials, and advanced technologies can be utilized in water harvesting based on the same indigenous knowledge. Using ancient rainwater-harvesting knowledge with modern practices is more relevant, efficient, and cost effective (Oweis et al. 2004).

MIWH systems require specialists to plan, design, and implement these systems. Otherwise, the systems may collapse or function with low efficiency. A spillway, for example, is required for surface reservoirs to pass the extra flow without eroding the structure. Establishing contour lines is another example of the need for skillful people. Planning MIWH requires that the potential need is established and that the water harvesting system has the potential to improve agriculture. When this is established, the selection of the site and the techniques that are suitable and complement each other is critical for the proper functioning of the system (Oweis et al. 2012; Kato et al. 2011). Design of the system requires determining the target catchment ratio that will be enough to meet crop-water requirements. Rainfall characteristics and runoff coefficients, in addition to soil water holding capacity, are important for designing the system. Models are used to help simplify the process, and new tools such as GIS and remote sensing and modern equipment can facilitate the establishment of a proper MIWH (Mekdaschi and Linigar 2013). Furthermore, GIS-based similarity and suitability analysis can help in upscaling successful applications of specific MIWH practices (De Pauw et al. 2008; Ziadat et al. 2014: Gammoh 2011a).

Rainwater harvesting is integrative by nature. Poorly functioning projects contribute to focusing only on water harvesting and ignoring other components of the agricultural system. Having a small reservoir functioning properly requires that agricultural land, a community, and crops and livestock are all linked to it. Furthermore, normal agronomic practices in soil, water, crop, and other inputs should be taken into consideration to maximize the system effectiveness and efficiency (Oweis et al. 2017).

4.4.3 Upstream–Downstream Conflicts

As runoff water generated in the watersheds flows downstream, MIWH implemented in the upstream will halt water and reduce runoff downstream (Falkenmark et al. 2001). For downstream users, any reduction in water to their fields may create a conflict (Vohland and Barry 2009). In the beginning of the watershed development, the reduction in flow is usually small with little impact downstream. But, when more projects are implemented, there can be substantial downstream impacts (Ali et al. 2009). If water rights in the watershed are not established, then available water cannot be optimally allocated to maximize the benefits. Agreed-upon allocation mechanisms and criteria together with an integrated watershed-management approach can optimize the process with minimum conflicts (Bouma et al. 2011). Selection and allocation of MIWH projects should follow upstream–downstream allocation sharing for the rainwater resources. This requires involvement of local communities with thorough discussion of their priorities (Oweis et al. 2017; De Pauw et al. 2008).

4.4.4 *Payment for Environmental Services*

The initial cost of rainwater harvesting systems can be an obstacle to wide adoption, especially for poor communities (UNEP 2009; Bouma et al. 2011). People construct the systems and get direct benefits but, in the case of MIWH, large part of the benefits are social and environmental, which are shared by the whole society. Examples of public benefits include reduction in wind and water erosion that benefit reservoirs and soil fertility. Also, reduction in dust storms because of increased vegetation improves health, in addition to other environmental benefits (Dutilly-Diane et al. 2007). Rainwater harvesting also supports groundwater aquifers, reduces flooding, and enhances biodiversity and ecosystem services. Social benefits of MIWH, such as increased employment and reduction in migration from rural to urban centers, apply to the public at large (Liang and Dijk 2011). It is usually unlikely that poor communities can or should fully finance those services on behalf of the society (Power 2010; Oweis 2017).

Payment for environmental services is a mechanism used in the developed countries to finance such services. Public payment for environmental services does not have priority in developing countries. A public-private partnership for long-term investment would help in providing policy reforms to ensure that incentives and enabling environment for environmental services are established (Nkonya et al. 2016).

4.5 Conclusions

- MIWH is a viable option for providing domestic water for the basic needs of rural communities when other sources of water supply are not available or scarce. In urban areas, rooftop water harvesting can reduce pressure on the domestic water supply system and improve landscape extent and function. MIWH is instrumental in drylands agriculture and in combating desertification through converting lost rainfall and runoff into green water, controlling erosion, and enhancing biodiversity. It can be an effective climate-change adaptation response through alleviating the impacts of any reduction in precipitation or increase in intensive storms. Its role in restoring vegetation in degraded ecosystem contributes to climate-change mitigation.
- Achieving the full benefits of MIWH requires addressing several limiting factors and constraints. These include understanding the uncertainties associated with climate variability and change with needed investment in water storage; the limited capacity of the targeted population to modernize and implement proper systems; upstream–downstream conflicts; and the needs for integration to become most effective.
- As the contribution of MIWH to the environment is substantial, it is recommended that a "payment for environmental services" be applied at the national

level to finance MIWH implementation, especially for restoration of degraded ecosystems. In this regard, incentives should be provided to restoring ecosystems instead of supplemental feeds that encourage increased flocks and further land degradation.

- Involvement of local communities from the beginning is vital for appropriate implementation and adoption of the MIWH practices. In this regard, enhancing the capacity of the personnel associated with water harvesting development is essential to overcome the technical constraints. In many systems, collective use of the resources and the services require that local institutions be formed to help streamline the services and avoid conflicts.

Modernizing rainwater harvesting practices is essential to improve efficiency and effectiveness and reduce cost. Research is still needed to adapt indigenous practices to today's advancements in science and to develop new technologies based on this remarkable ancient knowledge. Work on geospatial modeling for the proper selection and upscaling of rainwater harvesting techniques offers great potential to improve the practice.

References

Abu-Zreig M, Ababneh F, Abdullah F (2019) Assessment of rooftop rainwater harvesting in northern Jordan. Phys Chem Earth 114, Parts A/B/C. https://doi.org/10.1016/j.pce.2019.08.002

Akroush S, Shideed K, Bruggeman A (2014) Economic analysis and environmental impacts of water harvesting techniques in the low rainfall areas of Jordan. Int J Agric Resour Gov Ecol 10(1):34–49. https://doi.org/10.1504/IJARGE.2014.061040

Ali A, Yazar A, Aal AA, Oweis T, Hayek P (2010) Micro-catchment water harvesting potential of an arid environment. J Agricult Water Manag 98(1):96–104. https://doi.org/10.1016/j.agwat.2010.08.002

Ali A, Oweis T, Salkini, AB, El-Naggar S (2009). Rainwater cisterns: traditional technologies for dry areas. J ICARDA (International Center for Agricultural Research in the Dry Areas), Aleppo, Syria.

Anderson J, Zehnder A, Rockström J, Yang H (2011) Potential impacts of water harvesting and ecological sanitation on crop yield, evaporation and river flow regimes in the Thukela River basin, South Africa. J Agricult Water Manag 98(7):1113–1124. https://doi.org/10.1016/j.agwat.2011.02.004

Antinori P, Vallerani V (1994) Experiments in water harvesting technology with the dolphin and train ploughs. In: Proceedings of the FAO Expert Consultation, Cairo (Egypt, 21–25 November 1993). Rome, FAO, pp 113–132

Ben-Asher J, Prinz D, Laron J, Abravia I (1995) Greenhouse roof top water harvesting: a case study under Mediterranean conditions. In: Proceedings of the European conference on water use: water resources management in the mediterranean under drought or water shortage conditions: Nicosia, Cyprus, 14–18 March 1995. Balkema, Rotterdam. pp 145–152

Biazin B, Sterk G, Temesgen M, Abdulkedir A, Stroosnijder L (2012) Rainwater harvesting and management in rainfed agricultural systems in sub-Saharan Africa – a review. J Phys Chem Earth, Parts a/b/c 47–48:139–151. https://doi.org/10.1016/j.pce.2011.08.015

Boers TM, Ben-Asher J (1982) A review of rainwater harvesting. Agricult Water Manag 5(2):145–158. https://doi.org/10.1016/0378-3774(82)90003-8

Bouma J, Biggs WT, Bouwer ML (2011) The downstream externalities of harvesting rainwater in semi-arid watersheds: an Indian case study. J Agricult Water Manag 98(7):1162–1170. https://doi.org/10.1016/j.agwat.2011.02.010

Critchley W, Siegert K (1991) Water harvesting: a manual for the design and construction of water harvesting schemes for plant production. Rome, Italy: FAO (Food and Agriculture Organization of the United States). Retrieved from https://agris.fao.org/agris-search/search.do?recordID=XF2016035521

De Pauw E, Oweis T, Youssef J (2008) Integrating expert knowledge in GIS to locate biophysical potential for water harvesting: a case study for Syria. ICARDA (International Center for Agriculture Research in Dry Areas), Aleppo

Dutilly-Diane C, McCarthy N, Turkelboom F, Bruggeman A, Tiedemann J, Street K, Serra G (2007) Could payments for environmental services improve rangeland management in Central Asia, West Asia and North Africa? Washington DC. USA: IFPRI (International Food Policy Research Institute), CAPRI (System-wide Program on Collective Action and Property Rights), Working Paper No. 62

Dutt GR (1981) Establishment of NaCl-treated catchments. In: Dutt GR, Hutchinson CF, Anaya Garduno M (eds) Proceedings of a workshop: rainfall collection for agriculture in arid and semi-arid regions, University of Arizona, Tucson, Arizona: 10–12. Commonwealth Agricultural Bureau, Slough, pp 17–21

Falkenmark M, Fox P, Persson G, Rockström GJ (2001) Water harvesting for upgrading of rainfed agriculture: problem analysis and research needs, Report number 11. Sweden: SIWI (Stockholm International Water Institute), Stockholm

Fraiser GW (1980) Harvesting water for agriculture, wildlife and domestic uses. J Soil Water Conserv 35(3):125–128

Gammoh IA, Oweis TY (2011a) Performance and adaptation of the Vallerani mechanized water harvesting system in degraded *badia* rangelands. J Environ Sci Eng 5(10):1370–1380. Retrieved from https://hdl.handle.net/20.500.11766/8050

Gammoh IA, Oweis TY (2011b) Contour laser guiding for the mechanized 'Vallerani' micro-catchment water harvesting systems. J Environ Sci Eng 5(9). https://doi.org/10.17265/1934-8932/2011.10.011

Guo Y, Baetz BW (2007) Sizing of rainwater storage units for green building applications. J Hydrol Eng 12(2):197–205. https://doi.org/10.1061/(ASCE)1084-0699(2007)12:2(197)

HLPE (High Level Panel of Experts on Food Security and Nutrition) (2015) Water for food security and nutrition. A report by the High-Level Panel of Experts on Food Security and Nutrition of the Committee on World Food Security. HLPE, Rome

IPCC (Intergovernmental Panel on Climate Change) (2014) Fifth assessment report: climate change 2014. IPCC Technical paper VI. Climate change and water. IPCC, Geneva

JMOE (Jordan Ministry of Environment). 2014. *Badia* restoration program 2014. Jordan Ministry of Environment, Environmental Compensation Unit. http://www.badiarp.gov.jo/en/node/103

Jones MP, Hunt WF (2010) Performance of rainwater harvesting systems in the southeastern United States. J Resourc, Conserv Recycling 54(10):623–629. https://doi.org/10.1016/j.resconrec.2009.11.002

Karrou M, Oweis T, Ziadat F, Awawdeh F (eds) (2011) Rehabilitation and integrated management of dry rangelands environments with water harvesting. Community-based optimization of the management of scarce water resources in agriculture in Central and West Asia and North Africa, Aleppo, Syria: ICARDA ((International Center for Agriculture Research in Dry Areas) report no. 9

Kato E, Ringler C, Yesuf M, Bryan E (2011) Soil and water conservation technologies: a buffer against production risk in the face of climate change? Insights from the Nile basin in Ethiopia. J Int Assoc Agricult Econ 42(5):593–604. https://doi.org/10.1111/j.1574-0862.2011.00539.x

Lasage R, Verburg PH (2015) Evaluation of small-scale water harvesting techniques for semi-arid environments. Arid Environ 118:48–57. https://doi.org/10.1016/j.jaridenv.2015.02.019

Lenzen CJ, Gordon RL, McQuitty AM (1985) Excavation at Tell Irbid and Beit Ras, Jordan. Amman, Jordan: Annals of Department of Antiquities, Vol. XXIX
Liang X, van Dijk MP (2011) Economic and financial analysis on rainwater harvesting for agricultural irrigation in the rural areas of Beijing. J Resourc, Conserv Recycling 55(11):1100–1108. https://doi.org/10.1016/j.resconrec.2011.06.009
Louhaichi M, Tastad A (2010) The Syrian steppe: past trends, current status, and future priorities. J Rangelands 32(2):2–7. https://doi.org/10.2111/1551-501X-32.2.2
Mechlia BN, Oweis T, Masmoudi MM, Khattali H, Ouassar M, Sghaier N, Anane M, Sghaier M (2009) Assessment of supplemental irrigation and water harvesting potential methodologies and case studies from Tunisia. Aleppo, Syria: ICARDA (International Center for Agriculture Research in Dry Areas).
Mekdaschi Studer R, Liniger H (2013) Water harvesting: guidelines to good practice. CDE (Centre for Development), RAIN (Environment Rainwater Harvesting Implementation Network), MetaMeta, Wageningen and IFAD (International Fund for Agricultural Development), Rome
Molden D, Oweis T, Steduto P, Bindraban P, Hanjra M, Kijne J (2010) Improving agricultural water productivity: between optimism and caution. J Agricult Water Manag 97(4):528–535. https://doi.org/10.1016/j.agwat.2009.03.023
MRMP (1992) Project preparation report of Matrouh Resource Management Project, Matrouh Northwest Coast of Egypt. Cairo, Egypt: Ministry of Agriculture
Mzirai OB, Tumbo SD (2010) Macro-catchment rainwater harvesting systems: challenges and opportunities to access runoff. J Inst Rural Devel Plan Depart Agricult Eng 7(2):789–800
Nkonya E, Mirzabaev A, von Braun J (eds) (2016) Economics of land degradation and improvement —a global assessment for sustainable development. Springer International Publishing, Cham. https://doi.org/10.1007/978-3-319-19168-3
Oweis T, Taimeh A (2001) Farm water-harvesting reservoirs: issues of planning and management in dry areas. In: Adeel Z (eds) Proceedings of a joint UNU-CAS international workshop: integrated land management in dry areas, Beijing, China, 8–13. United Nations University, Tokyo, pp 165–182
Oweis T, Hachum A, Bruggeman A (eds) (2004) Indigenous water harvesting systems in West Asia and North Africa. ICARDA (International Center for Agriculture Research in the Dry Areas), Aleppo, p 173
Oweis T, Prinz D, Hachum A (2012) Rainwater harvesting for agriculture in the dry areas. CRC Press/Balkema, Taylor & Francis Group, London, p 262. https://doi.org/10.1201/b12351
Oweis T, Hachum A (2006) Water harvesting and supplemental irrigation for improved water productivity of dry farming systems in West Asia and North Africa. J Agricult Water Manag 80(1–3):57–73. https://doi.org/10.1016/j.agwat.2005.07.004
Oweis T, Taimeh A (1996) Evaluation of small basin water harvesting systems in the arid region of Jordan. J Water Resour Manag 10(1):21–34. https://doi.org/10.1007/BF00698809
Oweis T (2017) Rainwater harvesting for restoring degraded dry agro-pastoral ecosystems; a conceptual review of opportunities and constraints in a changing climate. Environ Rev 25(2):135–149. https://doi.org/10.1139/er-2016-0069
Power A (2010) Ecosystem services and agriculture: tradeoffs and synergies. J R Soc 365:2959–2971. https://doi.org/10.1098/rstb.2010.0143

Previati M, Bevilacqua I, Canone D, Ferraris S, Haverkamp R (2010) Evaluation of Soil Water Storage Efficiency for Rainfall Harvesting on Hillslope Micro-Basins Built Using Time Domain Reflectometry Measurements J Agricult Water Manag 97(3):449–456. https://doi.org/10.1016/j.agwat.2009.11.004

Qadir M, Sharma BR, Bruggeman A, Choukr-Allah R, Karajeh F (2007) Non-conventional water resources and opportunities for water augmentation to achieve food security in water scarce countries. J Agricult Water Manag 87(1):2–22. https://doi.org/10.1016/j.agwat.2006.03.018

Rees D, Qureshi Z, Mehmood S, Raza S (1991) Catchment basin water harvesting as a means of improving the productivity of rain-fed land in upland Baluchistan. J Agric Sci 116(1):95–103. https://doi.org/10.1017/S0021859600076188

Shawahneh N, Saoub H, Oweis T, Haddad N, Karrou M (2011) Impact of micro catchment water harvesting on the diversity of the Badia rangelands of Jordan. In: Karrou M, Oweis T, Ziadat F, Awawdeh F (eds) Water benchmarks of CWANA: rehabilitation and integrated management of dry rangelands environments with water harvesting. Community- based optimization of the management of scarce water resources in agriculture in Central and West Asia North Africa report no. 9. ICARDA (International Center for Agriculture Research in Dry Areas), Aleppo, p 208

Somme G, Oweis T, Abdulal A, Bruggeman A, Ali A (2004) Micro-catchment water harvesting for improved vegetative cover in the Syrian Badia. On-farm water husbandry research reports series No 3. ICARDA (International Center for Agriculture Research in Dry Areas), Aleppo, pp 38.

Sowers J, Vengosh A, Weinthal E (2011) Climate change, water resources, and the politics of adaptation in the Middle East and North Africa. J Clim Change 104(3):599–627. https://doi.org/10.1007/s10584-010-9835-4

Taimeh A, Hattar B (2006) Strategy for steppe development in Jordan. In: Oweis T, Benli B, Bruggeman A, Farahani H (eds) Characteristics of benchmark research agro-ecosystems in WANA: rainfed, irrigated, and marginal drylands. ICARDA (International Center for Agricultural Research in the Dry Areas), Aleppo, pp 39–56

Thomas R, Kirisits M, Lye D, Kinney K (2014) Rainwater harvesting in the United States: a survey of common system practices. J Clean Prod 75:166–173. https://doi.org/10.1016/j.jclepro.2014.03.073

UNEP (United Nations Environment Programme) (2009) Rainwater harvesting: a lifeline for human well-being. A report prepared by Stockholm Environment Institute. UNEP, Nairobi. Retrieved from http://hdl.handle.net/20.500.11822/7762.

UNESCO (United Nations Educational, Scientific and Cultural Organization) (2020) The United Nations World water development report 2020: water and climate change. UNESCO, Paris

Vohland K, Barry B (2009) A review of in situ rainwater harvesting (RWH) practices modifying landscape functions in African drylands. J Agricult, Ecosyst Environ 131(3–4):119–127. https://doi.org/10.1016/j.agee.2009.01.010

Wåhlin L (1997) The family cisterns; 3000 years of household water collection in Jordan. In: Sabour M, Vicør KS (eds) Ethnic encounter and culture change. The third Nordic conference on Middle Eastern studies, Joinsuu

WOCAT (World Overview of Conservation Approaches and Technologies) (2012) Global Database on Sustainable Land Management. http://www.wocat.net/en/global-slm-database/

Xiao G, Zhang Q, Xiong Y, Lin M, Wang J (2007) Integrating rainwater harvesting with supplemental irrigation into rain-fed spring wheat farming. Soil Tillage Res 93(2):429–437. https://doi.org/10.1016/j.still.2006.06.001

Xiao-Yan L, Xie Z, Yan X (2004) Runoff characteristics of artificial catchment materials for rainwater harvesting in the semiarid regions of China. J Agricult Water Manag 65(3):211–224. https://doi.org/10.1016/j.agwat.2003.09.003

Ziadat F, Bruggeman A, Oweis T, Haddad N, Mazahreh S, Sartawi W, Syuof M (2012) A Participatory GIS Approach for Assessing Land Suitability for Rainwater Harvesting in an Arid Rangeland Environment. Arid Land Res Manag 26(4):297–311. https://doi.org/10.1080/15324982.2012.709214

Ziadat F, Mazahreh S, Haddad M, Benabdelouahab T, Attaher S, Karrou M, Oweis T, Kandakji T (2014) Similarity and suitability analysis to assist the out-scaling of sustainable water and land management practices in West Asia and North Africa. Research report No. 11. ICARDA (International Center for Agricultural Research in Dry Areas), Beirut, p 79

Part III
Tapping Offshore and Onshore Deep Groundwater

Chapter 5
Offshore Freshened Groundwater in Continental Shelf Environments

Mark A. Person and A. Micallef

Abstract While offshore groundwater has been utilized by coastal communities as far back as 1000 BC, only in the past 10 years has the global volume of fresh-to-brackish water hosted in offshore aquifers been truly appreciated. There are vast quantities (~300–500 × 10^3 km^3) of offshore freshened groundwater sequestered in continental shelf sediments under water depths of less than 60 m within 110 km of the coastline. New marine geophysical methods now make it possible to map and quantify low salinity offshore groundwater bodies. To date, these offshore resources have not been developed. Offshore freshened groundwater could be produced if wells are located close to the shoreline and coastal desalination plants.

Keywords Offshore freshwater · Continental shelf aquifers · Marine geophysics

5.1 Introduction

The existence of offshore fresh water (salinity of <0.5 parts per thousand) and brackish water (salinity of <10 parts per thousand) hosted within continental shelf sedimentary deposits is a global phenomenon (Micallef et al. 2020a; Fig. 5.1). Cohen et al. (2010) and Post et al. (2013) estimated that the global volume of offshore freshened groundwater ranged 300–500 × 10^3 km^3. These estimates are based on continental shelf borehole salinity data, as well as hydrogeologic model calculations (Cohen et al. 2010). Cohen et al. (2010) reported that the volumes of offshore freshened groundwater varied significantly between 0.2 and 12 km^3 per km of coastline. To provide context to these volume estimates, Konikow (2015) reported that the total volume of water extracted from aquifer systems across the United States in the period

M. A. Person (✉)
New Mexico Tech, Hydrology Program, Socorro, NM 87801, USA
e-mail: mark.person@nmt.edu

A. Micallef
Marine Geology and Seafloor Surveying, Department of Geosciences, University of Malta, Msida, Malta
e-mail: aaron.micallef@um.edu.mt

M. Qadir et al. (eds.), *Unconventional Water Resources*,
https://doi.org/10.1007/978-3-030-90146-2_5

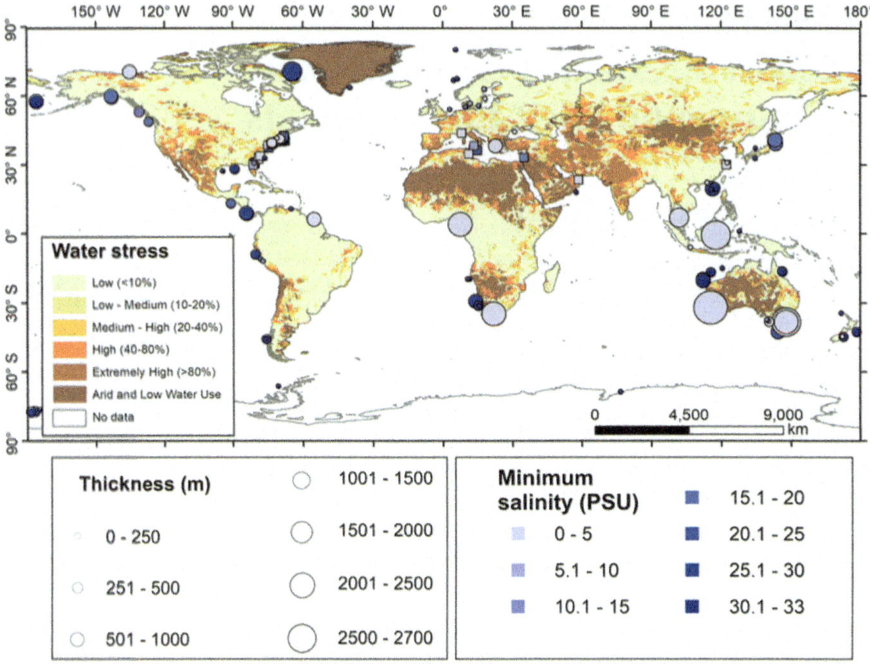

Fig. 5.1 Minimum salinity and thickness of offshore freshened groundwater plotted on a global map of water stress (Micallef et al. (2020a); Hofste et al. (2019)). Practical salinity units (PSU) are more or less equivalent to total concentrations of dissolved solids in parts per thousand (ppt) or 1000 mg/L

1900–2008 was about 10^3 km^3. While these continental-shelf water resource estimates seem vast, they represent only a small fraction of global groundwater storage (Lvovitch 1970). To date, we are unaware of any wells that have been drilled into the continental shelf that have attempted to produce offshore freshened groundwater. However, numerical model calculations presented by Person et al. (2017) suggest that offshore freshened groundwater could be produced from a confined continental-shelf aquifer for at least 30 years, provided that the overlying and underlying confining units are sufficiently tight (permeability $<10^{-16}$ m^2). With growing water shortages projected over the next few decades (Eliasson 2015), offshore freshened groundwater represents a potential unconventional water resource that could be utilized by coastal cities.

The aim of this chapter is to provide an overview of the occurrence, mechanisms of emplacement, volume estimates, recent developments in exploration methods, and production strategies of offshore freshened groundwater resources.

5.2 History of Offshore Freshened Groundwater Discovery and Utilization

The earliest recorded discovery and utilization of continental-shelf, freshened groundwater dates back 3,000 years offshore of Syria. Submarine groundwater discharge was collected using an inverted funnel place over a submarine spring, providing about 1,500 L/s to the city of Tyre (Bakalowicz et al. 2007). During the Bronze Age, the Roman geographer Strabo reported on the collection of offshore freshened groundwater about 3.8 km offshore near the island of Aradus in present-day Syria (Taniguchi et al. 2002). In both cases, the springs issued from karstic aquifers.

In the 1970s, a drilling campaign was undertaken by the U.S. Geological Survey named AMCOR (Atlantic Margin CORing project; Hathaway et al. 1979). The primary goal of this campaign was to obtain information regarding the mineral and hydrocarbon resources of the US Atlantic continental margin. Twenty wells were drilled from Florida to Massachusetts. While the drilling campaign found that the eastern seaboard of the US was hydrocarbon dead, vast quantities of offshore fresh-to-brackish resources were discovered. Offshore fresh and brackish water was found in the Pleistocene (11.7 Ka to 2580 Ka; where Ka equals 1 thousand years before present) to the Miocene (5.3–23 Ma; where Ma equals 1 million years in the past) clastic and limestone formations (Hathaway et al. 1979; Johnston 1983; Lofi et al. 2013).

5.3 Occurrence, Distribution, and Volume Estimates of Offshore Freshened Groundwater

Until recently, most of what we know about the distribution of continental shelf freshened groundwater was derived from a limited number of cross-sectional salinity contour maps constructed using published borehole salinity data and paleo-hydrologic models (Person et al. 2017; Fig. 5.2). Recently, Micallef et al. (2020a) assessed the characteristics of over 300 sites around the world that reported instances of offshore freshened groundwater (Fig. 5.1). These reports were based on borehole and core data, electromagnetic data, onshore indicators, and observations of submarine groundwater discharge. Offshore freshened groundwater has been found on both passive (73%) and active (24%) continental margins (Micallef et al. 2020a). Person et al. (2017) correlated the distribution of offshore freshened groundwater volume with distance from the coastline and seawater depth for five of these cross sections (Fig. 5.3). They found that most of the continental shelf fresh-to-brackish water is typically found in seawater depths of <60 m and distances of <110 km. Offshore freshened groundwater is hosted in both fine (Lofi et al. 2013) and coarse (Hathaway et al. 1979) grained clastic sediments and limestones (Johnston 1983). The offshore

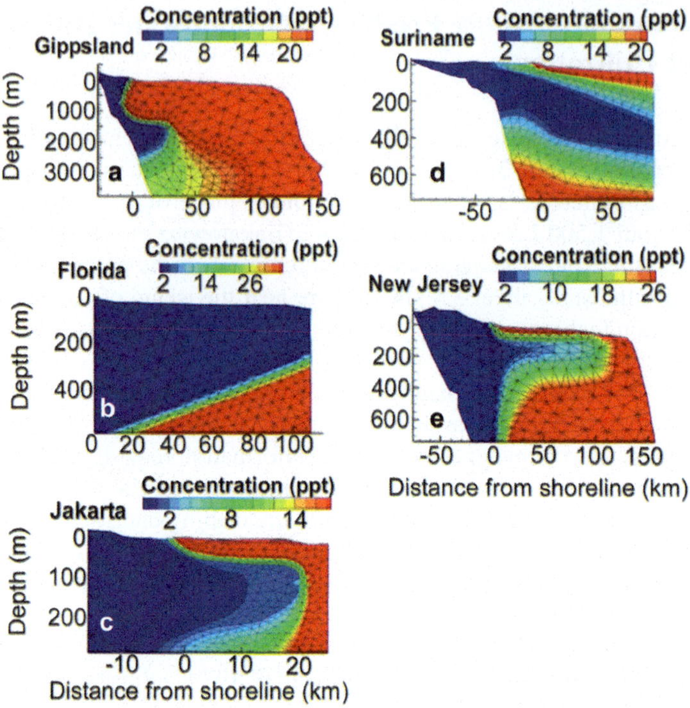

Fig. 5.2 Digital salinity cross sections of offshore freshened groundwater for the following passive margins: (**A**) Gippsland Basin, Australia; (**B**) Florida, US; (**C**) Jakarta, Indonesia; (**D**) Surinam; and (**E**) New Jersey, US. (*Source* Person et al. (2017); Reprinted by permission of the AAPG whose permission is required for further use)

freshened groundwater is typically found below shallow confining units that protect this resource from overlying seawater intrusions.

5.4 Comparison with Onshore Fresh and Brackish Water Resources

How do continental-shelf freshened groundwater resource estimates (~300–500 $\times$ 10^3 km^3) compare to continental groundwater resource estimates? Freeze and Cherry (1979) and Lvovitch (1970) reported that the global volume of continental groundwater is about 60,000 $\times$ 10^3 km^3. This is two orders of magnitude higher than estimates of global offshore freshened groundwater resources, which only represent about 0.7% of all groundwater on Earth. Gleeson et al. (2016), on the other hand, estimated that the global volume of young (<100 years), shallow groundwater resources is about one tenth of offshore groundwater storage (37 $\times$ 10^3 km^3).

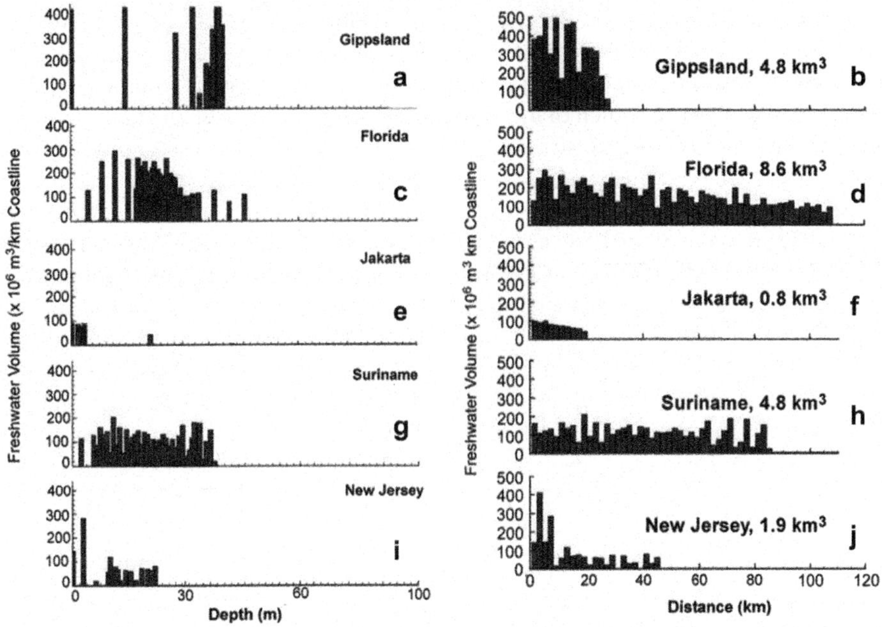

Fig. 5.3 Estimated volume of offshore fresh-to-brackish water ($\times$ 10^6 m^3 per km coastline (**A**, **C**, **E**, **G**, **I**) versus seawater depth and distance (**B**, **D**, **F**, **H**, **J**) from shoreline for the cross sections shown in Fig. 5.2. The cumulative volume of freshwater (km^3) per km of shoreline is listed within each graph. In this figure, freshwater is defined as having a salinity of less than 5,000 mg/L (*Source* Person et al. (2017); Reprinted by permission of the AAPG whose permission is required for further use)

How do continental-shelf freshened groundwater resource estimates compare to continental brackish groundwater resources? The US Geological Survey estimated that the volume of onshore brackish groundwater resources in the US to depths of about 1 km totaled 350 $\times$ 10^3 km^3 (Stanton et al. 2017). Assuming that the length of the US coastline is 7,200 km (neglecting fine-scale fingering features) and each kilometer of coastline contains, on average, 4.4 km^3 of freshwater, there should be about 32 $\times$ 10^3 km^3 of offshore fresh-to-brackish water along the US coastline. Thus, offshore continental shelf groundwater resources would represent about 11% of USA brackish groundwater resources.

5.5 Mechanisms of Emplacement

Several mechanisms have been proposed to account for emplacement of offshore freshened groundwater. Because emplacement of offshore freshened groundwater likely occurred over time scales of 10^5 years, studies of mechanisms for offshore freshened groundwater emplacement have relied heavily on paleo-hydrologic modelling (e.g., Meisler et al. 1984). Michael et al. (2016), on the other hand, argued that present-day meteoric recharge from the onshore portion of continental shelf aquifer can account for much of the observations of offshore freshened groundwater. Indeed, the salinity profiles shown in Fig. 5.2 clearly indicate continuity between onshore and offshore freshwater resources. However, on the US Atlantic continental shelf in New England, offshore freshened groundwater is present (Hathaway et al. 1979; Gustafson et al. 2019), yet the continental-shelf sedimentary deposits do not extend onshore. Micallef et al. (2020b) argued that offshore freshened groundwater emplacement is greatly enhanced during periods of low sea-level stands, such as during the last glacial maximum 21 Ka (Fig. 5.4). Sea-level fluctuations are driven by the waxing and waning of continental ice sheets during the Pleistocene Epoch (Hansen et al. 2013). During glacial periods, the shore-normal hydraulic gradient increased, and local groundwater flow systems developed on the exposed continental shelf (Fig. 5.4). During sea-level rise, shallow sand-dominated aquifers were exposed to seawater and haline convection, causing mixing of seawater with freshwater (Post and Kooi 2003; Thomas et al. 2019).

At high latitudes, continental ice sheets likely overran the continental shelf (Denton and Hughes 1981). Fluid pressures at the base of the ice sheet could have been as high as 90% of the ice sheet height (floating conditions; Person et al. 2007), enhancing infiltration of freshwater into continental shelf aquifers (Person et al. 2003;

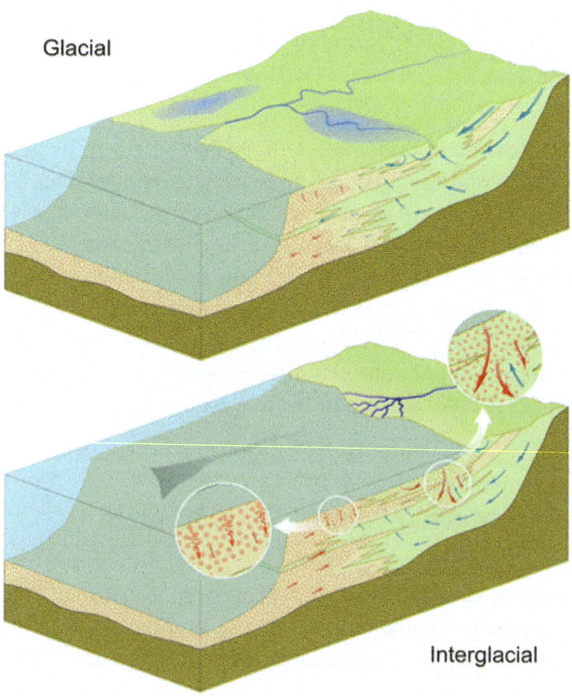

Fig. 5.4 Schematic diagram depicting hydrologic conditions on the continental shelf during glacial and interglacial periods. During glacial periods sea-level was as much as 120 m lower than today. The red arrows denote seawater flow into continental shelf sediments. The blue arrows are associated with freshwater infiltration and flow (Post et al. (2013))

Cohen et al. 2010; Siegel et al. 2012, 2014). In some instances, permafrost formation during glacial periods may have blocked infiltration in the near-shore environment (Edmunds 2001). Infiltration of glacial meltwater from proglacial lakes was proposed by Person et al. (2012) as an additional source of freshwater infiltration on glaciated continental shelves.

It should be kept in mind that while many of the numerical models described in this section have been "calibrated" to borehole salinity observations, paleo-hydrologic models suffer from uncertainty regarding input data (e.g., hydraulic conductivity) and how boundary conditions vary across geologic time (e.g., changes in bathymetry). This renders conclusions from hydrologic modeling uncertain. Geochemical tracers and age dating of groundwater would help resolve this uncertainty, but this would require new dedicated drilling campaigns.

5.6 Recent Developments in Offshore Freshened Groundwater Exploration

Until recently, the discovery of offshore freshened groundwater relied on marine borehole data. Currently, the price for drilling a single offshore well on the continental shelf using a jack-up rig is on the order of several million $. Recent geophysical developments in marine magnetotelluric (MT) and controlled source electromagnetic (CSEM) surveys have provided geophysicists with the ability to image offshore freshened groundwater (Gustafson et al. 2019; Fig. 5.5). MT surveys rely on the natural electromagnetic waves generated by solar winds and lightning storms that penetrate the earth's surface (Simpson and Bahr 2005). These incoming waves generate secondary electromagnetic waves that diffuse back to the seafloor where they are detected by changes in electrical voltage, measured by seafloor MT units. Marine CSEM surveys generate electromagnetic waves using a transmitter deployed in a towed array, behind a marine vessel. Marine MT and CSEM surveys cost a fraction of the price of one offshore well. Inversion programs produce formation resistivity images of continental-shelf sediments (Gustafson et al. 2019; Micallef et al. 2020b). Because freshwater is far more electrically resistive than seawater, continental reservoirs whose pore spaces host freshwater show up as regions of high electrical resistivity (Fig. 5.6). These high resistivity targets must be drilled to confirm that there is sufficient permeability to produce offshore freshened groundwater.

5.7 Development of Offshore Freshened Groundwater

There are issues associated with the development of offshore freshened groundwater including well placement, production rates, pipelines, and desalinization. The offshore freshened groundwater resource should be sufficiently close to the shoreline

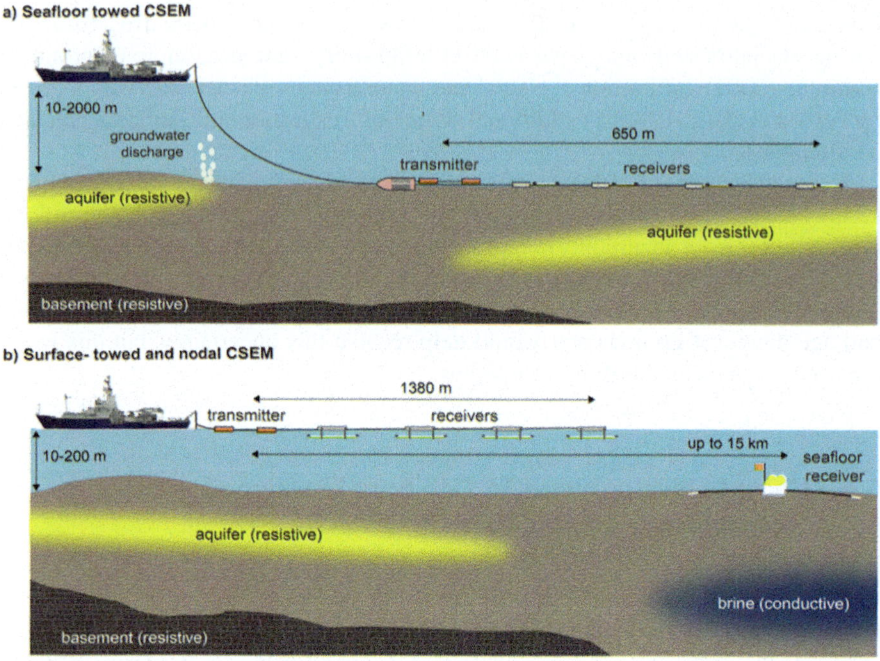

Fig. 5.5 Marine electromagnetic systems for mapping offshore freshened groundwater. **a** Seafloor towed CSEM system. **b** Surface-towed and nodal CSEM system (*Source* Micallef et al. (2020a))

that the construction of a pipeline is economically feasible. On the other hand, the wells should be placed sufficiently far offshore to eliminate significant interaction with the onshore hydrologic system (Yu and Michael 2019). Ideally, this pipeline would come onshore close to an existing desalination facility. The resource must be overlain and underlain by confining units sufficiently tight to prevent seawater from infiltrating the aquifer horizon. Exploration wells must be drilled to verify that the targeted horizon has sufficient porosity and permeability to produce large quantities of water (~60 L/s). Horizontal wells greatly reduce fluid pressure gradients, extending the production lifetime. Several wells should be drilled in different directions from an offshore platform. Orienting some of these wells parallel to the coastline would reduce potential interactions with onshore freshwater pumping. Person et al. (2017) used hydrological modeling of offshore freshened groundwater production using horizontal wells in order to estimate well yields and production lifetime. They found that freshwater could be produced for up to 30 years without seawater encroachment, provided that sufficiently tight (permeability $<10^{-16}$ m^2) confining units are present and that the offshore freshened groundwater aquifer is laterally extensive. Producing offshore freshened groundwater is akin to oil production; it is non renewable on human time scales.

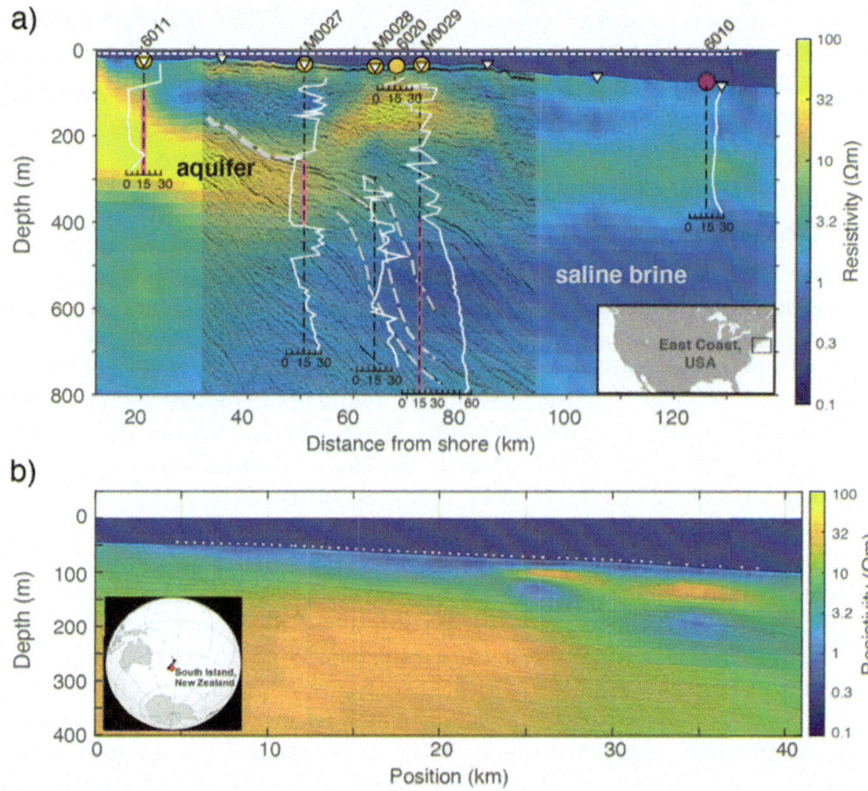

Fig. 5.6 **(a)** Shore-normal resistivity model profile from offshore New Jersey derived from 2D joint inversion of surface towed CSEM and seafloor MT data. The resistivity contours are overlain with a corresponding seismic reflection image. The resistive zones (> 10 Ωm) are interpreted as low-salinity OFG. Observed pore water-salinity profiles (white lines) are presented for boreholes AMCOR sites 6010, 6011, and 6020, and IODP Expedition 313 sites M0027-29 on a linear scale ranging from 0 to 60 PSU. Black dashed lines indicate a salinity value of 15 PSU. Seismically imaged confining units and clinoform structures that influence groundwater salinity distribution patterns are denoted by light-grey dashed lines. Modified from Gustafson et al. (2019). **(b)** Offshore New Zealand resistivity model derived from 2D inversion overlain with the related seismic reflection section (modified from Micallef et al. 2020a). The model contains an extended, seaward dipping, resistive body (> 20 Ωm) at depths of 25-215 m below seafloor, which is interpreted as the main offshore freshened groundwater body. The shallow resistive feature 25-40 km offshore (> 20 Ωm) follows seismic reflectors and is interpreted as freshened groundwater within a fine sand unit. (*Source*: Micallef et al. 2020a)

Offshore produced water may be predominantly brackish. We believe there are synergistic benefits between coastal desalination and production of offshore freshened groundwater. Desalination of brackish water rather than seawater can reduce the costs of freshwater production by as much as a factor of three (Karagiannis and Soldatos 2008), while at the same time decreasing the environmental impacts of desalination by as much as 50% (Muñoz and Fernández-Alba 2008). Because ocean water is vulnerable to surface contamination (e.g., oil spills), using groundwater wells rather than seawater intake pipes would reduce the threat of pollution (Stein et al. 2020). This may not be true for karst systems where contamination can move rapidly through large dissolution features.

5.8 Discussion

This chapter has provided evidence that continental shelf fresh-to-brackish water is vast (~10^5 km^3) and globally distributed. That said, we still know little about the timing of emplacement/replenishment of offshore freshened groundwater. Paleo-hydrologic models that included calculations of groundwater residence times from Micallef et al. (2020a) indicated that freshened groundwater offshore off the South Island of New Zealand (Canterbury Bight) was likely emplaced over the past three glacial cycles (~300,000 yr). If these estimates are correct, offshore freshened groundwater production is not renewable on human time scales.

While recent advances in offshore electromagnetic surveying techniques will undoubtedly lead to additional discoveries of offshore freshened groundwater over the next decade (e.g., Attias et al. 2020), no dedicated drilling campaigns have been carried out to investigate these freshwater reservoirs. What is the permeability and porosity of these offshore aquifers? Can offshore freshened groundwater be produced over a period of decades without the invasion of seawater? A recent IODP drilling campaign provides a cautionary note. IODP expedition 313 off the New Jersey coastline drilled three wells that revealed the presence of freshened groundwater at seawater depths of 33.5–36.0 m. This zone of offshore freshened groundwater was hosted in relatively low-permeability deposits (Lofi et al. 2013). Thomas et al. (2019) argued that this freshened groundwater is in a region where chemical diffusion and haline convection have not yet run their course. A production well installed at these sites would not produce viable amounts of offshore freshened groundwater.

We still know little about the economics of utilizing this unconventional water resource. The capitalization costs of installing an offshore freshened groundwater system will be high. Exploitation of offshore freshened groundwater will likely only be economically viable when developed in combination with a new or existing coastal desalination plant. The cost savings of offshore fresh-to-brackish water production would result from reduced energy costs associated with desalination of brackish water rather than seawater (Karagiannis and Soldatos 2008).

But what is the cost of doing nothing for coastal megacities that have exhausted their shallow water resources? Megacities such as Sao Paulo and Cape Town have both experienced extreme droughts recently and spent considerable funds transporting water. We conclude this section with a quotation from Benjamin Franklin: "When the well is dry, we know the worth of water" (1993).

5.9 Conclusions

This chapter has focused on documenting the global occurrence, volumes, and mechanisms of emplacement of offshore freshwater resources in continental-shelf environments. There are vast quantities (~300–500 × 10^3 km^3) of offshore freshwater hosted in continental-shelf environments globally. Much of this offshore freshwater was likely emplaced during low sea-level stands during the Pleistocene. The chapter also addressed the benefits and possible onshore environmental impacts of developing this resource. Offshore freshwater is likely not a renewable resource. Provided there are relatively low-permeability ($<10^{-16}$ m^2) shallow confining units, modeling studies have shown that production of significant volumes of offshore freshwater from underlying confined aquifers can be sustained for a period of at least 30 years using horizontal drilling technologies. New marine electromagnetic exploration methods have been developed and applied over the past five years and have successfully imaged offshore freshwater in continental-shelf environments.

The following recommendations for future work on this topic are offered:

- Drilling campaigns should be undertaken to assess the feasibility of producing offshore freshwater in continental-shelf environments. As part of this effort, pumping tests should be conducted and water samples collected for ^4He and ^{81}Kr dating of groundwater (Müller et al. 2016; Gerber et al. 2017). Ideally, horizontal drilling technology should be used.
- Marine geophysical studies should continue to be funded to better characterize the lateral variability of offshore freshwater resources on a global scale.
- Large-scale continental shelf paleo-hydrologic models such as Cohen et al. (2010) should be constructed along the margins of other continents.
- Additional studies assessing the impacts of offshore pumping on onshore water quality and land subsidence should be performed (e.g., Yu and Michael 2019). Some insights might be gained by considering the impact of offshore oil production on the onshore hydrogeologic system (Gambolati et al. 2006).

Acknowledgements Partial support from the US National Science Foundation (NSF FRES 1925974) and by the European Research Council (ERC) under the European Union's Horizon 2020 research and innovation programme (grant agreement No. 677898; MARCAN) is acknowledged.

References

Attias E, Thomas D, Sherman D, Ismail K, Constable S (2020) Marine electrical imaging reveals novel freshwater transport mechanism in Hawai 'i. J Sci Adv 6(48):eabd4866. https://doi.org/10.1126/sciadv.abd4866

Bakalowicz M, El-Hajj A, El Hakim M, Al Charideh AR, Al-Fares W, Kattaa B, Fleury P, Brunet P, Dörfliger N, Seidel JL, Najem W (2007). Hydrogeological settings of karst submarine springs and aquifers of the Levantine coast (Syria, Lebanon). Towards their sustainable exploitation. *TIAC ,7*:721–732

Cohen D, Person M, Wang P, Gable CW, Hutchinson D, Marksamer A, Dugan B, Kooi H, Groen K, Lizarralde D, Evans RL, Lane Jr JW (2010). Origin and extent of fresh paleowaters on the Atlantic continental shelf, USA. J Groundw 48(1):143–158. https://doi.org/10.1111/j.1745-6584.2009.00627.x

Denton GH, Hughes TJ (eds) (1981) The last great ice sheets. Wiley-Interscience, New York, pp 437–467

Edmunds WM (2001) Palaeowaters in European coastal aquifers—the goals and main conclusions of the PALAEAUX project. J Geol Soc Lond Spec Publ 189(1):1–16. https://doi.org/10.1144/GSL.SP.2001.189.01.02

Eliasson J (2015) The rising pressure of global water shortages. Nat News 517(7532):6–6. https://doi.org/10.1038/517006a

Franklin B (1993) Poor Richard's Almanac, 1733–1758. *Oxford Text Archive Core Collection*

Freeze RA, Cherry JA (1979) Groundwater. Prentice Hall, 604 p

Gambolati G, Teatini P, Ferronato M (2006) Anthropogenic land subsidence. Encycl Hydrol Sci. https://doi.org/10.1002/0470848944.hsa164b

Gerber C, Vaikmäe R, Aeschbach W, Babre A, Jiang W, Leuenberger M, Lu ZT, Mokrik R, Müller P, Raidla V, Saks T, Purtschert R (2017) Using 81Kr and noble gases to characterize and date groundwater and brines in the Baltic Artesian Basin on the one-million-year timescale. Geochimica et Cosmochimica Acta 205:187–210

Gleeson T, Befus KM, Jasechko S, Luijendijk E, Cardenas MB (2016) The global volume and distribution of modern groundwater. Nat Geosci 9(2):161–167

Gustafson C, Key K, Evans RL (2019) Aquifer systems extending far offshore on the US Atlantic margin. Sci Rep 9(1):1–10

Hansen J, Sato M, Russell G, Kharecha P (2013) Climate sensitivity, sea level and atmospheric carbon dioxide. Philos Trans R Soc A Math Phys Eng Sci 371(2001):20120294. https://doi.org/10.1098/rsta.2012.0294

Hathaway JC, Poag CW, Valentine PC, Manheim FT, Kohout FA, Bothner MH, Miller RE, Schultz DM, Sangrey DA(1979) U.S. Geological Survey core drilling on the Atlantic shelf. J Sci 206:515–527. https://doi.org/10.1126/science.206.4418.515

Hofste RW, Kuzma S, Walker S, Sutanudjaja EH, Bierkens MF, Kuijper MJ, Sanchez MF, Van Beek R, Wada Y, Rodríguez SG, Reig P (2019) Aqueduct 3.0: updated decision-relevant global water risk indicators rep. World Resources Institute, Washington DC. https://doi.org/10.46830/writn.18.00146

Johnston RH (1983) The salt-water-fresh-water interface in the tertiary limestone aquifer, southeast Atlantic outer continental-shelf of the USA. J Hydrol 61(1–3):239–249. https://doi.org/10.1016/0022-1694(83)90251-2

Karagiannis IC, Soldatos PG (2008) Water desalination cost literature: review and assessment. J Desalin 223(1–3):448–456. https://doi.org/10.1016/j.desal.2007.02.071

Konikow LF (2015) Long-term groundwater depletion in the United States. Groundwater 53(1):2–9

Lofi J, Pezard P, Bouchette F, Raynal O, Sabatier P, Denchik N, Levannier A, Dezileau L, Certain R(2013) Integrated onshore-offshore investigation of a Mediterranean layered coastal aquifer. Groundwater 51(4):550–561. https://doi.org/10.1111/j.1745-6584.2012.01011.x

Lvovitch MI (1970, July) World water balance (General report). In: World water balance: proceedings of the reading symposium, pp 401–415. http://hydrologie.org/redbooks/a092/093019.pdf

Meisler H, Leahy PP, Knobel LL (1984) Effect of eustatic sea-level changes on saltwater-freshwater relations in the northern Atlantic Coastal Plain, vol 2255. US Government Printing Office. https://doi.org/10.3133/wsp2255

Micallef A, Person M, Berndt C, Bertoni C, Cohen D, Dugan B, Evans R, Haroon A, Hensen C, Jegen M, Key K, Thomas AT (2020a) Offshore freshened groundwater in continental margins. Rev Geophys 59(1):e2020RG000706. https://doi.org/10.1029/2020RG000706

Micallef A, Person M, Haroon A, Weymer BA, Jegen M, Schwalenberg K, Faghih Z, Duan S, Cohen D, Mountjoy JJ, Woelz S, Tiwari AK (2020b) 3D characterisation and quantification of an offshore freshened groundwater system in the Canterbury Bight. J Nat Commun 11:1372. https://doi.org/10.1038/s41467-020-14770-7

Michael HA, Scott KC, Koneshloo M, Yu X, Khan MR, Li K (2016) Geologic influence on groundwater salinity drives large seawater circulation through the continental shelf. J Geophys Res Lett 43(20):10782–10791. https://doi.org/10.1002/2016GL070863

Muñoz I, Fernández-Alba AR (2008) Reducing the environmental impacts of reverse osmosis desalination by using brackish groundwater resources. J Water Res 42(3):801–811. https://doi.org/10.1016/j.watres.2007.08.021

Müller T, Osenbrück K, Strauch G, Pavetich S, Al-Mashaikhi KS, Herb C, Merchel S, Rugel G, Aeschbach W, Sanford W (2016) Use of multiple age tracers to estimate groundwater residence times and long-term recharge rates in arid southern Oman. J Appl Geochem 74:67–83. https://doi.org/10.1016/j.apgeochem.2016.08.012

Person M, Dugan B, Swenson JB, Urbano L, Stott C, Taylor J, Willett M (2003) Pleistocene hydrogeology of the Atlantic continental shelf, New England. Geol Soc Am Bull 115(11):1324–1343. https://doi.org/10.1130/B25285.1

Person M, McIntosh J, Bense V, Remenda VH (2007) Pleistocene hydrology of North America: the role of ice sheets in reorganizing groundwater flow systems. Rev Geophy 45(3). https://doi.org/10.1029/2006RG000206

Person M, Marksamer A, Dugan B, Sauer PE, Brown K, Bish D, Licht KJ, Willett M (2012) Use of a vertical δ 1^8O profile to constrain hydraulic properties and recharge rates across a glacio-lacustrine unit, Nantucket Island, Massachusetts, USA. Hydrogeol J 20(2): 325–336https://doi.org/10.1007/s10040-011-0795-1

Person M, Wilson JL, Morrow N, Post VE (2017) Continental-shelf freshwater water resources and improved oil recovery by low-salinity waterflooding. AAPG Bull 101(1):1–18. https://doi.org/10.1306/05241615143

Post VEA, Kooi H (2003) Rates of salinization by free convection in high-permeability sediments: insights from numerical modeling and application to the Dutch coastal area. Hydrogeol J 11:549–559. https://doi.org/10.1007/s10040-003-0271-7

Post VEA, Groen J, Kooi H, Person M, Ge S, Edmunds WM (2013) Offshore fresh groundwater reserves as a global phenomenon. J Nat 504:71–78. https://doi.org/10.1038/nature12858

Siegel J, Dugan B, Lizarralde D, Person M, DeFoor W, Miller N (2012) Geophysical evidence of a late Pleistocene glaciation and paleo-ice stream on the Atlantic Continental Shelf offshore Massachusetts, USA. J Mar Geol 303–306:63–74. https://doi.org/10.1016/j.margeo.2012.01.007

Siegel J, Person M, Dugan B, Cohen D, Lizarralde D, Gable C (2014) Influence of late Pleistocene glaciations on the hydrogeology of the continental shelf offshore Massachusetts, USA. J Geochem Geophys Geosyst 15(12):4651–4670. https://doi.org/10.1002/2014GC005569

Simpson F, Bahr K (2005) Practical magnetotellurics. Cambridge University Press. https://doi.org/10.1017/CBO9780511614095

Stanton et al. (2017) Stanton JS, Anning DW, Brown CJ, Moore RB, McGuire VL, Qi SL, Harris AC, Dennehy KF, McMahon PB, Degnan JR, Böhlke JK (2017) Brackish groundwater in the United States (No. 1833). US Geological Survey. https://doi.org/10.3133/pp1833

Stein S, Sola F, Yechieli Y, Shalev E, Sivan O, Kasher R, Vallejos A (2020) The effects of long-term saline groundwater pumping for desalination on the fresh–saline water interface: field observations and numerical modeling. Sci Total Environ 732:139249

Taniguchi M, Burnett WC, Cable JE, Turner JV (2002) Investigation of submarine groundwater discharge. J Hydrol Process 16(11):2115–2129. https://doi.org/10.1002/hyp.1145

Thomas AT, Reiche S, Riedel M, Clauser C (2019) The fate of submarine fresh groundwater reservoirs at the New Jersey shelf, USA. Hydrogeol J 27:2673–2694. https://doi.org/10.1007/s10040-019-01997-y

Yu X, Michael HA (2019) Offshore pumping impacts onshore groundwater resources and land subsidence. J Geophys Res Lett 46(5):2553–2562. https://doi.org/10.1029/2019GL081910

Chapter 6
Continental Brackish Groundwater Resources

Mark A. Person and Nafis Sazeed

Abstract The global volume of continental brackish water is on the order of $5{,}000 \times 10^3$ km^3. This is about ten times the volume of freshwater to brackish water hosted in marine continental-shelf environments. On average, brackish water resources occupy about 11% of aquifer volume. Brackish water resources tend to occur in the upland areas of sedimentary basins close to recharge areas. While brackish water utilization for municipal water supplies is growing at a near exponential rate, economic barriers exist in many countries to the use of desalinated brackish water in the agricultural production of high-value crops. Linking brackish water desalination to geothermal greenhouse and aquaculture facilities in regions of high heat flow may be one strategy for agricultural cost reduction.

Keywords Brackish water · Desalination · Geophysics · Geothermal

6.1 Introduction

Since the 1970s, production of brackish water resources has grown rapidly, providing municipal drinking water for semi-arid regions of Europe, the US, China, India, Africa, the Middle East, and Australia. Here, we define brackish water as having a salinity ranging between 3,000–10,000 mg/L. As seen in Fig. 6.1, the growth in the number of brackish water reverse-osmosis desalination plants across the US is nearly exponential. The use of brackish groundwater in desalination plants is widespread across the US (blue dots in Fig. 6.2).

Increases in the development of brackish water resources is being driven by climate change (Sola et al. 2019) and population growth, as well as the diminished availability of shallow fresh groundwater resources (Aeschbach-Hertig and Gleeson 2012). In the US, water demand is expected to increase to about 12% by 2050 (Stanton et al. 2017).

M. A. Person (✉) · N. Sazeed
New Mexico Tech, Hydrology Program, Socorro, NM 87801, USA
e-mail: mark.person@nmt.edu

N. Sazeed
e-mail: nafis.sazeed@student.nmt.edu

M. Qadir et al. (eds.), *Unconventional Water Resources*,
https://doi.org/10.1007/978-3-030-90146-2_6

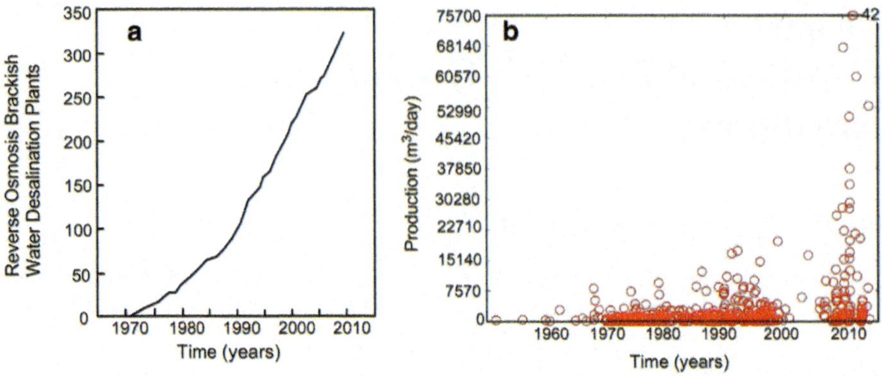

Fig. 6.1 Cumulative number of reverse-osmosis desalination facilities using brackish water in the US (*Source* Stanton et al. 2017)

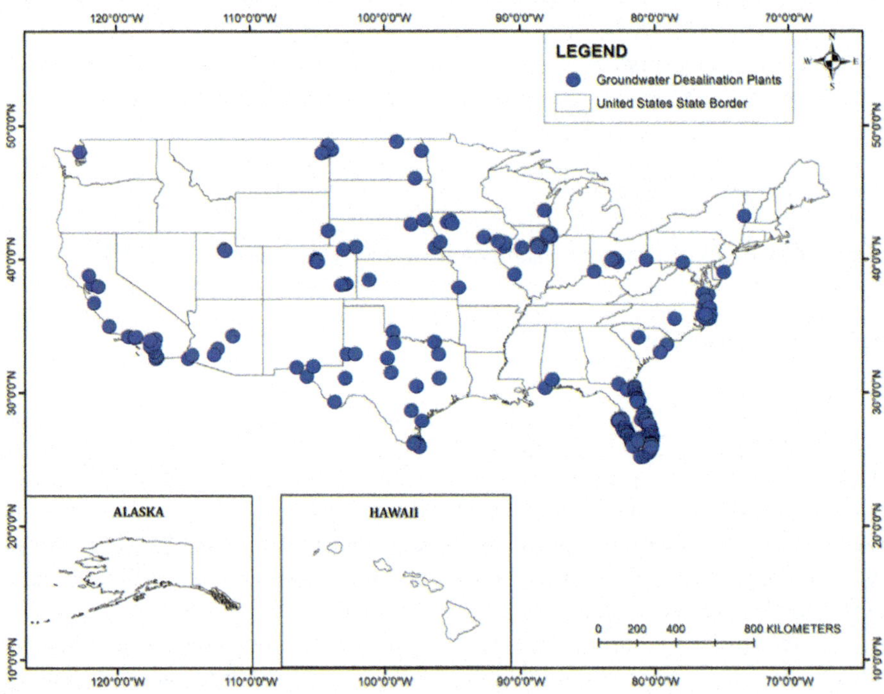

Fig. 6.2 Locations of reverse-osmosis desalination facilities across the US that utilize brackish water resources (*Source* Stanton et al. 2017)

Similar increases are projected for other parts of the globe (De Fraiture and Wichelns 2010). Most of the available shallow freshwater resources are already overexploited (Gleeson et al. 2016). In Israel and Spain, produced water from desalination plants is also being used for agricultural production of high-value crops (Ghermandi and Minich 2017; Aparicio et al. 2017).

Most research on the development of brackish water resources has focused on improvements in membrane-technologies, brine disposal practices, and economic feasibility studies (Morillo et al. 2013; Burn et al. 2015; Barron et al. 2015; Ziolkowska et al. 2016; Ghermandi and Minich 2017). Not much emphasis has been placed on assessments of the volume of continental brackish water resources nor on determination of the optimal locations for citing desalination plants using hydrogeological or geophysical criteria.

The goal of this chapter is to report on the occurrence, volumes, geochemistry, and mechanisms of emplacement of brackish water in continental settings. Geophysical exploration methods that can be used to detect fresh, brackish, and saline groundwater will also be discussed.

6.2 Occurrence and Volume Estimates

There is still uncertainty concerning, or any synthesis study documenting, the global occurrence or volumes of brackish water resources. However, the US Geological Survey (USGS) recently published a comprehensive assessment of the occurrence, volumes, and chemical characteristics of brackish groundwater resources across the US (Stanton et al. 2017). The information from Stanton et al. (2017) was then used to estimate global brackish-water endowments.

Stanton et al. (2017) found that nearly all regional aquifer systems across the US host significant volumes of brackish water (Table 6.1). The authors analyzed the presence of fresh (< 1,000 mg/L), brackish (3000–10,000 mg/L) and saline groundwater (>10,000 mg/L) over depth intervals of 0–152 m, 152–457 m, and 457–914 m. Brackish water hosted in sedimentary basins ranged between 1% to 86% of total aquifer volumes with an average of about 11%. Stanton et al. (2017) reported that about one quarter of the brackish water wells in their study had production rates of greater than 500 m^3/d, and 1% of these wells produced at a rate above 5,000 m^3/day. The total volume of brackish water resources in the US was estimated by Stanton et al. (2017) to be about 300×10^3 km^3. To put this number in perspective, the total groundwater production in the US for the year 2000 was 0.12×10^3 km^3 (Wada et al. 2010). The US represents only about 6% of the global land mass. Thus, an order of magnitude estimate for the global volume of brackish water resources is about 5,000 $\times 10^3$ km^3.

Analysis of salinity data across the US by Ferguson et al. (2018) noted that the salinity-depth relationship varies widely between sedimentary basins. This is likely due to variations in subsurface geologic conditions, available recharge, the presence/absence of evaporite minerals (Hanor 1994), and/or evapo-concentrated paleo

Table 6.1 Volumetric estimates of brackish water of selected aquifers across the US (Stanton et al. 2017)

Principal aquifers system name	Number wells	Percent sampled wells with brackish ground-water (m)	Median depth of sampled wells with brackish ground-water	Volume (km^3)	Percent of total aquifer volume
Eastern midcontinent					
Mississippian aquifers	3,105	13	91	4,393	22
New York and New England carbonate-rock aquifers	906	13	22	596	20
New York sandstone aquifers	280	2	34	83	8
Ozark Plateaus aquifer system	229	11	38	354	20
Ozark Plateaus aquifer	5,662	5	103	3,464	11
Pennsylvanian aquifers	4,043	10	61	3,172	26
Sand and gravel aquifers of alluvial or glacial origin	9,089	5	20	2,576	14
Silurian-Devonian aquifers	2,826	15	102	4,723	25
Valley and Ridge aquifers	6,233	1	76	659	3
Principal aquifer not present or not determined	27,367	7	27	15,351	16
Southwest basins					
Basin and Range basin-fill aquifers	13,874	24	64	14,622	32
Basin and Range carbonate-rock aquifers	335	25	111	817	33
Central Valley aquifer system	6,276	20	131	6,469	37
Rio Grande aquifer system	2,813	33	60	2,989	38
Principal aquifer not present or not determined	7,485	20	53	11,137	26

(continued)

Table 6.1 (continued)

Principal aquifers system name	Number wells	Percent sampled wells with brackish ground-water (m)	Median depth of sampled wells with brackish ground-water	Volume (km^3)	Percent of total aquifer volume
Western midcontinent					
AdaVamoosa aquifer	513	11	43	400	43
Arbuckle-Simpson aquifer	216	6	103	104	15
Blaine aquifer	513	80	32	521	79
Central Oklahoma aquifer	1,524	13	27	980	45
Colorado Plateaus aquifers	4,289	35	134	16,514	41
Denver Basin aquifer system	1,715	27	17	1,392	25
Edwards-Trinity aquifer system	14,162	20	107	21,179	43
High Plains aquifer	17,740	9	30	6,006	13
Lower Cretaceous aquifers	6,431	57	274	35,913	71
Lower Tertiary aquifers	6,127	68	46	19,753	77
Paleozoic aquifers	1,141	39	576	6,786	58
Pecos River Basin alluvial aquifer	1,400	58	55	2,197	73
Roswell Basin aquifer system	18	44	114	54	23
Rush Springs aquifer	440	30	41	704	72
Sand and gravel aquifers of alluvial or glacial origin	15,793	44	22	13,497	69
Seymour aquifer	669	54	14	242	87
Upper Cretaceous aquifers	5,120	72	61	22,675	77
Western Interior Plains aquifer system	707	10	304	1,259	12

(continued)

seawater during basin evolution over geologic time (Hanor and McIntosh 2006). Figure 6.3a presents salinity–depth information for all wells across the US. It is interesting to note that groundwater in numerous wells at depths of 500–1,000 m contain both fresh and brackish water. Figure 6.3b presents salinity–depth data for the Williston and Albuquerque basins. The Williston Basin has, on average, much

Table 6.1 (continued)

Principal aquifers system name	Number wells	Percent sampled wells with brackish ground-water (m)	Median depth of sampled wells with brackish ground-water	Volume (km^3)	Percent of total aquifer volume
Wyoming (Upper) Tertiary aquifers	52	17	30	46	12
Principal aquifer not present or not determined	37,397	42	30	66,245	44
Early Mesozoic basin aquifers	2,139	2	84	654	13
New York and New England crystalline-rock aquifers	3,934	<1	11	38	<1
Piedmont and Blue Ridge carbonate-rock aquifers	8,578	<1	58	304	1
Sand and gravel aquifers of alluvial or glacial origin	3,370	2	22	250	7
Principal aquifer not present or not determined	5,732	3	43	1,246	6
Columbia Plateau basaltic-rock aquifers	2,288	1	66	200	1
Columbia Plateau basin-fill aquifers	1,197	1	13	79	5
Pacific Northwest Volcanic					
Pacific Northwestern volcanic rock or basin-fill aquifers	517	1	63	58	3
Sand and gravel aquifers of alluvial or glacial origin	14	7	2	2	8
Snake River Plain volcanic rock or basin-fill aquifers	1,322	1	34	158	6
Principal aquifer not present or not determined	6,789	2	50	1,263	6

(continued)

Table 6.1 (continued)

Principal aquifers system name	Number wells	Percent sampled wells with brackish ground-water (m)	Median depth of sampled wells with brackish ground-water	Volume (km^3)	Percent of total aquifer volume
Western Mountain Ranges					
Northern Rocky Mountains Intermontane Basins aquifer systems	1,785	2	35	267	6
Sand and gravel aquifers of alluvial or glacial origin	518	3	0	79	6
Willamette Lowland basin-fill aquifers	370	1	10	42	4
Principal aquifer not present or not determined	7,652	3	39	2,222	12
Total Onshore Brackish Water USA (km^3) 294,734					
Total Offshore Fresh-Brackish Water USA (km^3) 53,092					

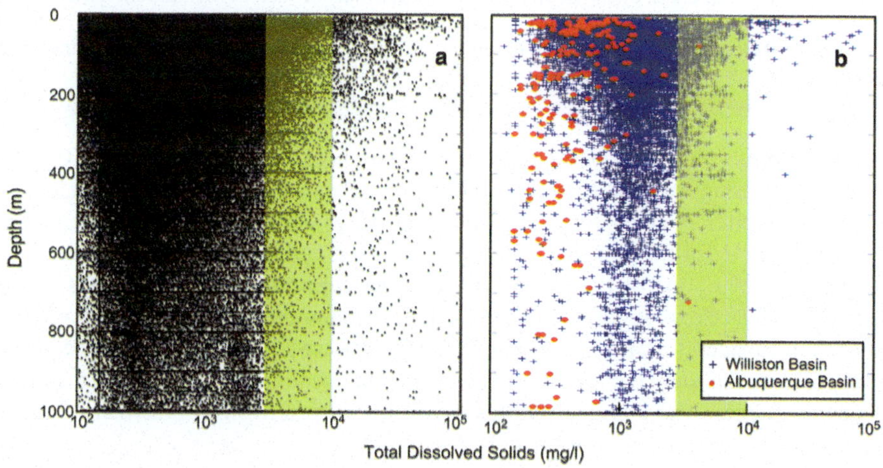

Fig. 6.3 **a** Semi-log plot of salinity versus depth for wells across the US. **b** Semi-log plot of salinity versus depth for the Williston and Albuquerque Basins. The yellow shading denotes brackish water (3,000–10,000 mg/L). (*Source* Stanton et al. 2017)

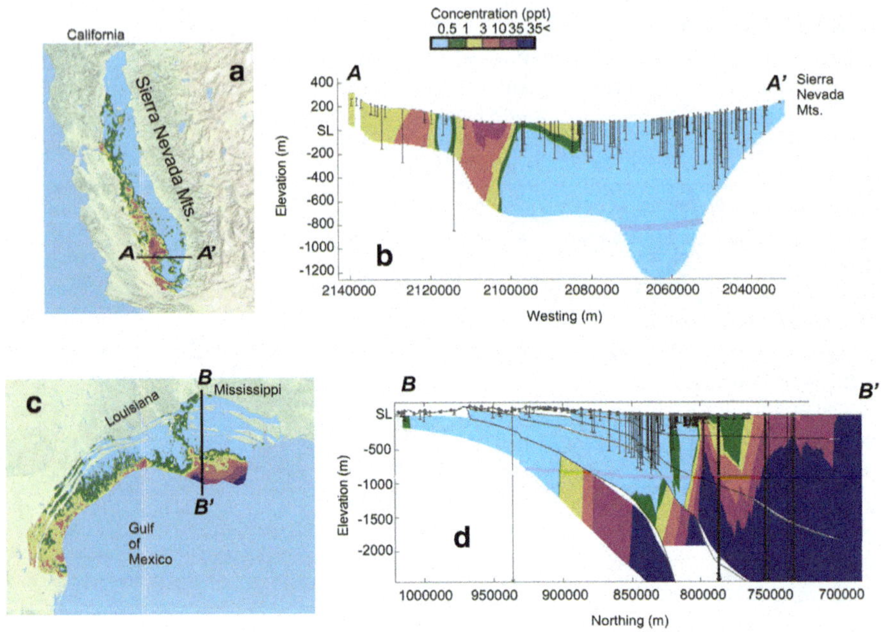

Fig. 6.4 Surface **a**, **c** and cross-sectional **b**, **d** contour plots of total concentrations of dissolved solids for the Gulf of Mexico, Louisiana, and the Central Valley, California. Brackish water is denoted by the blood-orange pattern. The black vertical lines denote wells (*Source* Stanton et al. 2017)

higher salinity conditions, perhaps due to the presence of evaporite minerals (Grasby et al. 2000). Stanton et al. (2017) also constructed salinity contour plots for select, data-rich aquifer systems across the US. Plan-view salinity maps are presented in Figs. 6.4a, 6.4c for the Central Valley of California and Louisiana Gulf Coast. The Central Valley is one of the most important agricultural producing regions of the United States. Cross-sectional salinity contour maps presented in Figs. 6.4b and 6.4d indicate lateral continuity of salinity trends in these aquifer systems. The contour maps reveal consistent patterns of lower salinity waters in the uplands where meteoric recharge occurs. The width of the transition zone from fresh-to-brackish-to-saline water is variable but is on the order of 5–10 km.

6.3 Geochemistry

The geochemistry of brackish water can have an important impact on the cost of desalination. Stanton et al. (2017) classified brackish water resources across the US into four geochemical groups. Group 1 was characterized by $NaHCO_3$ dominated waters with a mean pH of 8.1. Group 2 consisted of $CaSO_4$ dominated fluids.

Group 3 was characterized by NaCl-rich fluids having the highest mean concentration relative to the other groups (~8,400 mg/L; Stanton et al. 2017). The fourth group has no dominant geochemical anions or cations and is characterized by relatively low concentrations (~1,300 mg/L). Solubility analysis carried out by Stanton et al. (2017) found that brackish groundwater was commonly oversaturated with respect to $CaCO_3$, $BaSO_4$, and SiO_2, indicating that these fluids will likely result in scaling issues in conveyance pipes and reverse-osmosis membranes. Stanton et al. (2017) reports that relatively high concentrations of trace metals such as arsenic would need to be removed before they can be used as drinking water sources (Table 6.2) or for livestock (Table 6.3).

Arehart et al. (2003) reported that relatively high-temperature (>100 °C) geothermal systems within the Basin and Range Province in the southwest US are brackish. In New Mexico, low temperature (<100 °C) geothermal systems are also brackish, with relatively high concentrations of silica and arsenic (Pepin et al. 2014). In these crystalline basement-hosted geothermal systems, the dissolution of highly saline fluid inclusions is thought to be the likely source of elevated salinity (Ellis and Mahon 1964, 1967).

6.4 Timing and Mechanisms of Brackish Water Emplacement

Below about a depth of 600 m, groundwater is, on average, older than the Holocene (11,700 years in the past; Jasechko et al. 2017). Brackish water typically forms along the flow path within aquifer systems as groundwater flows down hydraulic gradients, mixing with more saline fluids (Hanor and McIntosh 2006) and/or undergoing water-rock interactions with soluble minerals such as calcite, gypsum, and halite. Meteoric recharge enters the aquifer where it crops out in upland regions of sedimentary basins. Initial recharge concentrations are generally less than 30 mg/L (Blackburn and Maleod 1983). Much of this paleo-recharge occurred during the late Pleistocene when climatic conditions were about 6 °C cooler than the present (Putnam and Broecker 2017). Under these conditions, groundwater recharge rates were likely higher. Zhu et al. (2000) estimated that recharge to the Navajo aquifer in the Black Mesa Basin of Arizona varied by a factor of three over the past 20,000 years. In the Northern Hemisphere, many sedimentary basins at high latitudes (>40° N) were overrun by continental ice sheets and experienced large influxes of glacial meltwater exceeding modern recharge rates (Person et al. 2007; Lemieux et al. 2008). Paces et al. (2020) found geochemical evidence of up to a 60 m increase in water table heights within the present-day recharge area of the Williston Basin during periods of glaciations. This could either be due to elevated recharge or enhanced aquifer pressures as the Laurentide sheet overran the groundwater discharge area of the Williston Basin in Manitoba, Canada, reversing groundwater flow directions (Grasby et al. 2000).

The distribution of brackish water and brines is likely related to regional topographic gradients (Ferguson et al. 2018) and regional geology. To illustrate how

Table 6.2 Trace metal concentrations of brackish water compared to water-quality standards for drinking water. The characteristics of the four geochemical groups are described in the text in Sect. 6.3 (*Source* Stanton et al. 2017)

	Arsenic (10ug/L)		Fluoride (4 mg/L)		Nitrate (10 mg/L)		Selenium (50 ug/L)		Uranium (30 ug/L)	
Group	# wells	Percent exceedance	# wells	Percent exceedance	# Wells	Percent exceedance	# Wells	Percent exceedance	# wells	Percent exceedance
1	1,054	11	3,206	14	2,223	2	1,041	3	508	8
2	2,758	4	3,655	1	2,321	13	2,742	5	1,932	18
3	1,711	19	2,935	6	1,724	7	1,359	3	801	14
4	1,402	14	2,000	2	1,216	17	1,175	1	857	8
Total	6,925	11	11,796	6	7,484	9	6,317	3	4,098	14

Table 6.3 Trace metal concentrations of brackish water compared to water-quality standards for livestock The characteristics of the four geochemical groups are described in the text in Sect. 6.3 (*Source* Stanton et al. 2017)

	Arsenic (10ug/L)		Fluoride (4 mg/L)		Nitrate (10 mg/L)		Iron (2 mg/L)		Selenium (50 ug/L)	
Group	Number wells	Percent exceedance	# wells	Percent exceedance	# Wells	Percent exceedance	# Wells	Percent exceedance	# Wells	Percent exceedance
1	1,054	5	963	4	3,206	35	3,303	7	1,041	2
2	2,758	1	2,511	2	3,655	10	4,977	18	2,742	5
3	1,711	6	1,555	14	2,935	22	2,969	14	1,359	3
4	1,402	2	1,256	1	2,000	7	2,343	14	1,175	1
Total	6,925	3	6,285	5	11,796	19	13,592	14	6,317	3

variations in subsurface permeability and regional water-table gradients control the distribution of brackish water in basins, we constructed a series of numerical simulations for an idealized sedimentary basin using the sedimentary basin model *RIFT2D*. The transport equations solved are presented in Mailloux et al. (1999). An initial linear salinity gradient was imposed such that, at the deepest point in the basin (~3.5 km depth), the solute concentration was about two times greater than seawater salinity (75 ppt or 75,000 mg/L), and hydrostatic initial conditions for groundwater flow were assumed. Along the top boundary, the concentration was set to 0 mg/L, and the hydraulic head was set to equal the elevation of the land surface. No-flux boundary conditions were assigned along the sides and base of the solution domain, both for groundwater flow and solute transport. Rock-water interactions such as the dissolution of evaporite minerals were not considered. Hence, these simulations represent the flushing of an initially saline basin. The models were run between one to five million years and do not reflect steady-state conditions. For all units, a constant porosity, longitudinal, and transverse dispersivities of 0.2, 100 m and 10 m, respectively, were assigned. Two of the models (Figs. 6.5a and 6.5b) were assigned a uniform permeability of 10^{-15} m^2 in both the x- and z-directions. Figures 6.5c and 6.5d considered a layered system consisting of a regional aquifer (permeability of 10^{-15} m^2), sandwiched between two confining units with a permeability of 10^{-18} m^2. Figures 6.5a and 6.5c used a regional hydrologic gradient across the basin of 500 m/200,000 m, or 0.25%. The other two models have a lower hydraulic gradient of 100 m/200,000 m, or 0.05%. The lower hydraulic gradient ($\Delta h = 100$ m) simulations were run for five million years, while the high gradient ($\Delta h = 500$ m) simulations were run for one million years. As can be seen, the distribution of brackish water (green shading) in the upgradient portion of these idealized basins varied across for the four different scenarios. In one scenario (Fig. 6.5c), an overturn in salinity occurs within the confined aquifer. The width of the brackish zone varied within 10–20 km, which is consistent with field observations shown in Fig. 6.4. The formation

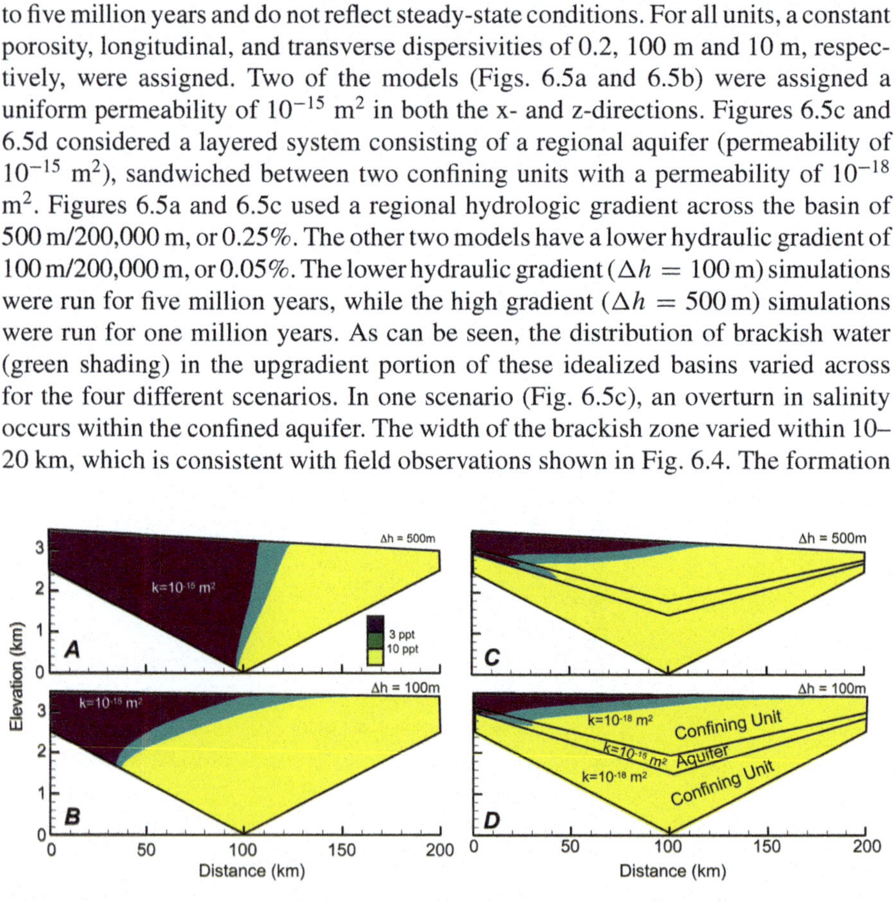

Fig. 6.5 Computed brackish water distribution across an idealized sedimentary basin. The variable "Δh" refers to the total linear head drop across the top surface of the basin. The variable "k" is permeability of a given unit in m^2

of brackish in all simulation runs was due to diffusive–dispersive mixing of younger, relatively fresh recharge with more saline connate fluids.

6.5 Geophysical Exploration Methods for Brackish Water Resources

Because rock formations in sedimentary basins host both fresh and saline water, their formation resistivity can vary between about 1000 and 0.2 Ωm (Fig. 6.6; Peacock et al. 2015). Surface electromagnetic (EM) methods such as time-domain electromagnetics (TEM) can detect changes in subsurface water quality (Simpson and Bahr 2005). A TEM system induces an electromagnetic wave that penetrates the subsurface. The surface EM waves are generated by passing a current through a copper wire loop at various frequencies. The copper wire loop is deployed at the land surface and is typically 40 m × 40 m or 100 m × 100 m. This EM wave interacts with subsurface layers of various formation resistivities spawning secondary electromagnetic waves that propagate back to the land surface. Then, these secondary electromagnetic waves are detected at the land surface using a magnetic antenna. TEM systems can also be deployed in aircraft suspending a rigid transmitter loop (Auken et al. 2009). These noninvasive electromagnetic soundings are used to determine layers having different formation resistivity. TEM soundings typically have penetration depths of 100–200 m.

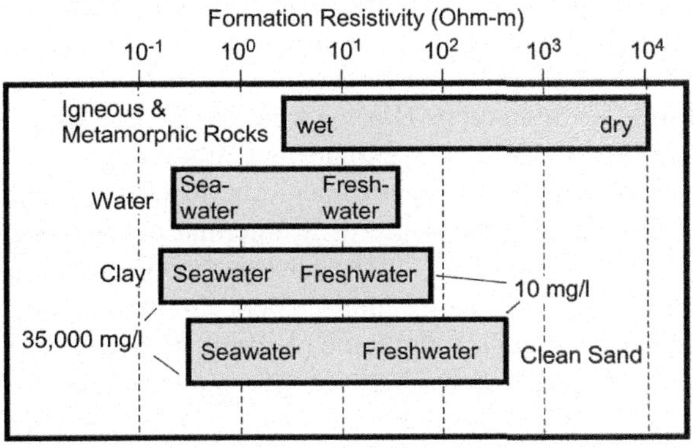

Fig. 6.6 Plot of formation resistivity for various geologic materials, pore water salinity, and air/water saturation conditions. For clean sand deposits, the endmember formation resistivity conditions for freshwater (10 mg/L) and seawater (35,000 mg/L) are plotted for clean sands. We used Archies law to calculate the formation resistivity of a clean sand saturated with seawater and freshwater using a cementation factor of two and porosity of 0.3. Clay resistivity was calculated using Glovers law, assuming a clay resistivity of 10 Ωm. For gravel deposits, wet and dry refer to whether the pores are filled with water or air (*Source* Peacock et al. 2015)

Because brackish water and brines generally conduct electric currents better than freshwater, TEM soundings can be used to image subsurface salinity variations and rock types (Fig. 6.6). Because clay deposits have elevated formation resistivities, interpretation of formation salinity is not unique, and some knowledge of subsurface geology is needed.

Recently, the USGS presented results from an airborne TEM survey (SkyTEM) across Paradox Valley, Utah (Ball et al. 2015). This was done to assess the efficacy of detecting variations in formation water salinity and lithology (Ball et al. 2015). As groundwater flows across the Paradox Valley towards the Dolores River, its salinity increases dramatically due to dissolution of shallow evaporite deposits (Fig. 6.7). By the time groundwater reaches the Dolores River, its salinity far exceeds that of seawater (35,000 mg/L). This can easily be seen in plan-view maps of formation resistivity produced by the SkyTEM survey (Fig. 6.7). The low formation resistivity ($<1\ \Omega$m) near the Dolores River is interpreted to be upwelling saline groundwater. Similar aerial electromagnetic surveys have been performed along the coast line in California to assess saltwater intrusion (Goebel et al. 2019).

6.6 Discussion

The use of brackish water for municipal drinking-water supplies is both economic and growing in a near exponential rate (Fig. 6.1). Siting reverse-osmosis desalination facilities overtop regions where brackish water salinity levels are relatively low could make desalination more cost effective (Burn et al. 2015). Economic barriers currently exist for the use of brackish water resources in agricultural production. Farmers are generally averse to using desalination technologies due to high capitalization costs and transitions to high-value crops (Barron et al. 2015). Improved efficiencies are needed to make the use of desalinated groundwater more attractive (Reca et al. 2018). Improved economic efficiencies might be achieved in regions of high heat flow by combining desalination with direct use of geothermal heating in greenhouses and aquaculture facilities (Goosen et al. 2010; Mahmoudi et al. 2010; Christ et al. 2017).

Owing to the long time scale of emplacement, brackish water is a nonrenewable resource. That said, the global volume of young ($\leq$ 100 years) groundwater is about 350×10^3 km^3 (Gleeson et al. 2016). The total volume of old groundwater (>100 years of age), which includes nearly all brackish fluids, is about $2{,}100 \times 10^3$ km^3 (Gleeson et al. 2016).

Production of brackish water resources will likely have an impact on shallow groundwater conditions. There is a dearth of studies that have considered these potential impacts.

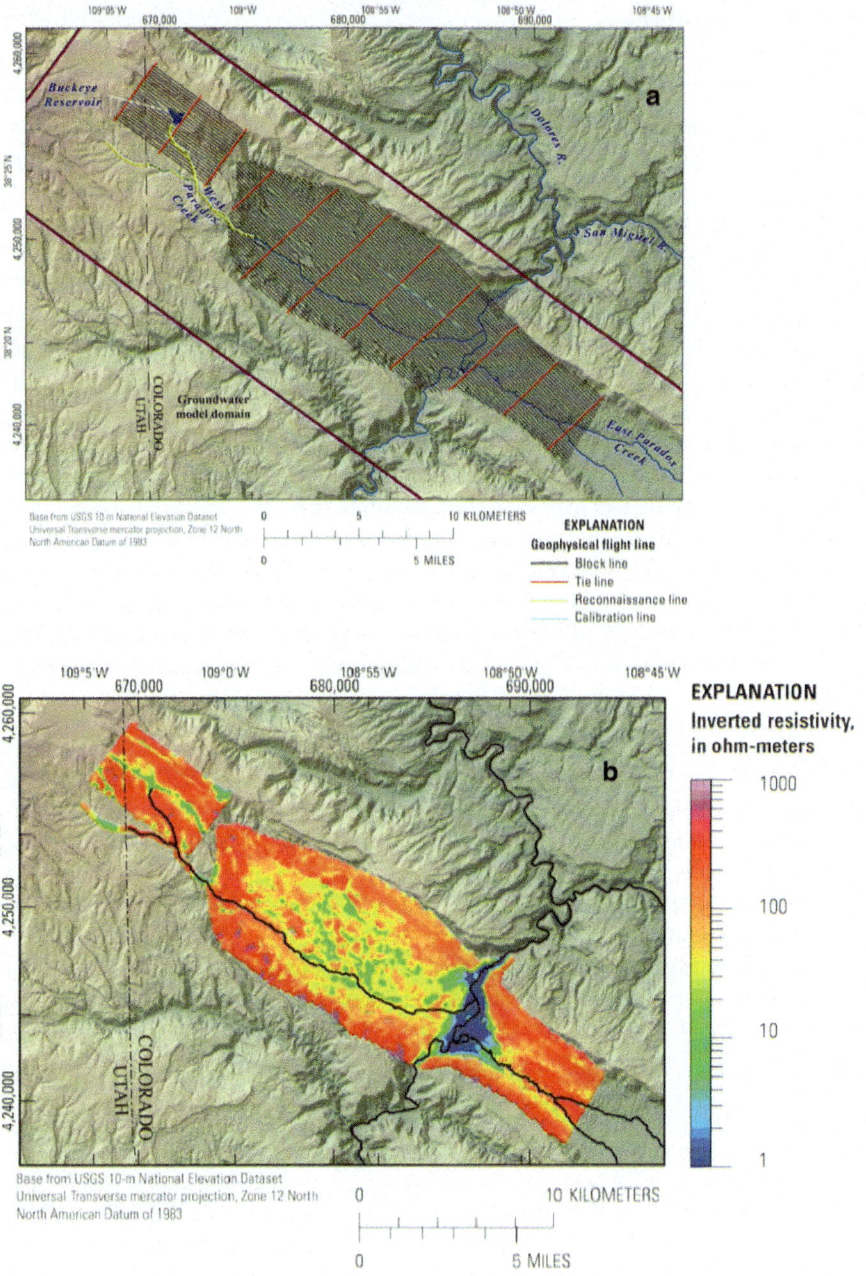

Fig. 6.7 SkyTEM survey lines **a** across Paradox Valley and **b** formation resistivity for depth slices between 21 m and 28 m (*Source* Ball et al. 2015)

6.7 Conclusions

This study has focused on the mechanisms of emplacement, occurrence, geochemistry, electromagnetic exploration methods, and volumes of brackish water resources. Many of our findings are based on the study of Stanton et al. (2017). Vast quantities (estimate ~5,000 × 10^3 km^3 globally) of brackish water hosted in continental sedimentary basin aquifer systems were found within the underlying crystalline basement. This is about one order of magnitude greater than continental-shelf fresh-brackish water resources. Most brackish water resources are relatively old (>10,000 years) and were likely emplaced during the periods of glaciation when temperatures were cooler and groundwater recharge rates were higher. It is likely that the current distribution of brackish water resources may not be in equilibrium with modern recharge conditions. The vertical distribution of brackish water resources varies considerably from basin to basin. Generally, fresh-to-brackish waters are found in the upland regions of sedimentary basins. The use of electromagnetic surveys can optimally locate desalination facilities beneath a relatively fresh-brackish water resource. Siting desalination facilities within upland locations proximal to recharge areas can also help to reduce energy costs. Crystalline-basement hosted geothermal systems with circulation depths of up to 6–8 km (Mailloux et al. 1999; Pepin et al. 2014) contain significant volumes of brackish water resources. Desalination of geothermal fluids may improve the economics of greenhouse and aquaculture facilities.

The following steps can be taken to better characterize continental brackish water resources:

- Quantify global estimates of brackish water resources using additional data sources, such as presented in van Weert et al. (2009), Thorslund and Vliet (2020) and Gleeson et al. (2016).
- Perform systematical SkyTEM surveys across water-stressed regions of the world including the western US.
- Develop hydrologic models and conduct field studies to assess the impacts of brackish water development on shallow water resources.

Acknowledgements The authors benefited from their interactions with Dr. Jennifer Stanton of the US Geological Survey in Marlborough, MA, who published a comprehensive assessment of brackish water resources in the US. Partial support from the US National Science Foundation is acknowledged (NSF FRES 1925974 and EAR 1830172).

References

Aeschbach-Hertig W, Gleeson T (2012) Regional strategies for the accelerating global problem of groundwater depletion. J Nat Geosci 5(12). https://doi.org/10.1038/ngeo1617

Aparicio J, Candela L, Alfranca O, García-Aróstegui JL (2017) Economic evaluation of small desalination plants from brackish aquifers. Application to Campo de Cartagena (SE Spain). J Desalinat 411:38–44. https://doi.org/10.1016/j.desal.2017.02.004

Arehart GB, Coolbaugh MF, Poulson SR (2003) Evidence for a magmatic source of heat for the steamboat springs geothermal system using trace elements and gas geochemistry. Trans Geothermal Resour Council 27:269–274

Auken E, Christiansen AV, Westergaard JH, Kirkegaard C, Foged N, Viezzoli A (2009) An integrated processing scheme for high-resolution airborne electromagnetic surveys, the SkyTEM system. J Explorat Geophys 40(2):184–192. https://doi.org/10.1071/EG08128

Ball LB, Bloss BR, Bedrosian PA, Grauch VJS, Smith BD (2015) Airborne electromagnetic and magnetic survey data of the paradox and San Luis Valleys. Colorado, US: US Department of the Interior; US Geological Survey 19. https://doi.org/10.3133/ofr20151024

Barron O, Ali R, Hodgson G, Smith D, Qureshi E, McFarlane D, Campos E, Zarzo D (2015) Feasibility assessment of desalination application in Australian traditional agriculture. J Desalinat 364:33–45. https://doi.org/10.1016/j.desal.2014.07.024

Blackburn G, McLeod S (1983) Salinity of atmospheric precipitation in the murray-darling drainage division, australia. Austr J Soil Res 21(4):411–434. https://doi.org/10.1071/SR9830411

Burn S, Hoang M, Zarzo D, Olewniak F, Campos E, Bolto B, Barron O (2015) Desalination techniques—a review of the opportunities for desalination in agriculture. J Desalin 364:2–16. https://doi.org/10.1016/j.desal.2015.01.041

Christ A, Rahimi B, Regenauer-Lieb K, Chua HT (2017) Techno-economic analysis of geothermal desalination using Hot Sedimentary Aquifers: a pre-feasibility study for Western Australia. J Desalin 404:167–181. https://doi.org/10.1016/j.desal.2016.11.009

De Fraiture C, Wichelns D (2010) Satisfying future water demands for agriculture. J Agric Water Manag 97(4):502–511. https://doi.org/10.1016/j.agwat.2009.08.008

Ellis AJ, Mahon WAJ (1964) Natural hydrothermal systems and experimental hot-water/rock interactions. J Geochimica et Cosmochimica Acta 28(8):1323–1357. https://doi.org/10.1016/0016-7037(64)90132-2

Ellis AJ, Mahon WAJ (1967) Natural hydrothermal systems and experimental hot water/rock interactions (Part II). J Geochimica et Cosmochimica Acta 31(4):519–538. https://doi.org/10.1016/0016-7037(67)90032-4

Ferguson G, McIntosh JC, Perrone D, Jasechko S (2018). Competition for shrinking window of low salinity groundwater. J Environ Res Lett 13(11). https://doi.org/10.1088/1748-9326/aae6d8

Ghermandi A, Minich T (2017) Analysis of farmers' attitude toward irrigation with desalinated brackish water in Israel's Arava Valley. J Desalin Water Treat 76. https://doi.org/10.5004/dwt.2017.20198

Gleeson T, Befus KM, Jasechko S, Luijendijk E, Cardenas MB (2016) The global volume and distribution of modern groundwater. J Nat Geosci 9(2):161–164. https://doi.org/10.1038/ngeo2590

Goebel M, Knight R, Halkjær M (2019) Mapping saltwater intrusion with an airborne electromagnetic method in the offshore coastal environment, Monterey Bay, California. J Hydrol: Regional Stud 23. https://doi.org/10.1016/j.ejrh.2019.100602

Goosen M, Mahmoudi H, Ghaffour N (2010). Water Desalination using geothermal energy. J Energies 3(8):1423–1442. https://doi.org/10.3390/en3081423

Grasby S, Osadetz K, Betcher R, Render F (2000) Reversal of the regional-scale flow system of the Williston basin in response to Pleistocene glaciation. J Geol 28(7):635–638. https://doi.org/10.1130/0091-7613(2000)28%3c635:ROTRFS%3e2.0.CO;2

Hanor JS (1994) Origin of saline fluids in sedimentary basins. Geolog Soc Spec Publication 78(1):151–174. https://doi.org/10.1144/GSL.SP.1994.078.01.13

Hanor JS, McIntosh JC (2006) Are secular variations in seawater chemistry reflected in the compositions of basinal brines? J Geoch Explorat 89(1–3):153–156. https://doi.org/10.1016/j.gexplo.2005.11.054

Jasechko S, Perrone D, Befus KM, Bayani Cardenas M, Ferguson G, Gleeson T,... Kirchner JW (2017) Global aquifers dominated by fossil groundwaters but wells vulnerable to modern contamination. J Nat Geosci 10(6):425–429. https://doi.org/10.1038/ngeo2943

Lemieux JM, Sudicky EA, Peltier WR, Tarasov L (2008) Dynamics of groundwater recharge and seepage over the Canadian landscape during the Wisconsinian glaciation. J Geophys Res Earth Surface 113(1). https://doi.org/10.1029/2007JF000838

Mailloux B, Person M, Kelley S, Dunbar N, Cather S, Strayer P, Hudleston PJ, Cather S (1999)Tectonic controls on the hydrogeology of the Rio Grande Rift, New Mexico. Water Resour Res 35(9):2641–2659

Morillo J, Usero J, Rosado D, El Bakouri H, Riaza A, Bernaola FJ (2014) Comparative study of brine management technologies for desalination plants. J Desalin 336(1):32–49. https://doi.org/10.1016/j.desal.2013.12.038

Paces JB, Palmer MV, Palmer AN, Long AJ, Emmons MP (2020) 300,000 yr history of water-table fluctuations at Wind Cave, South Dakota, USA-Scale, timing, and groundwater mixing in the Madison Aquifer. Bullet Geol Soc Amer 132(7–8). https://doi.org/10.1130/B35312.1

Peacock JR, Mangan MT, McPhee D, Ponce DA (2015) Imaging the magmatic system of Mono Basin, California, with magnetotellurics in three dimensions. J Geophys Res Solid Earth 120(11):7273–7289. https://doi.org/10.1002/2015JB012071

Pepin J, Person M, Phillips F, Kelley S, Timmons S, Owens L, Witcher J, Gable C (2014) Deep fluid circulation within crystalline basement rocks and the role of hydrologic windows in the formation of the Truth or Consequences, New Mexico low-temperature geothermal system. J Geofluids 15(1–2). https://doi.org/10.1111/gfl.12111

Person M, McIntosh J, Bense V, Remenda VH (2007) Pleistocene hydrology of North America: the role of ice sheets in reorganizing groundwater flow systems. J Rev Geophys 45(3). https://doi.org/10.1029/2006RG000206

Putnam AE, Broecker WS (2017) Human-induced changes in the distribution of rainfall. J Sci Adv 3(5), 1–15. https://doi.org/10.1126/sciadv.1600871

Reca J, Trillo C, Sánchez JA, Martínez J, Valera D (2018) Optimization model for on-farm irrigation management of Mediterranean greenhouse crops using desalinated and saline water from different sources. J Agric Syst 166:173–183. https://doi.org/10.1016/j.agsy.2018.02.004

Simpson F, Bahr K (2005) Practical Magnetotellurics. Cambridge University Press. https://doi.org/10.1017/CBO9780511614095

Sola I, Sánchez-Lizaso JL, Muñoz PT, García-Bartolomei E, Sáez CA, Zarzo D (2019) Assessment of the requirements within the environmental monitoring plans used to evaluate the environmental impacts of desalination plants in Chile. J Water 11(10). https://doi.org/10.3390/w11102085

Stanton JS, Anning DW, Brown CJ, Moore RB, McGuire VL, Qi SL,... Böhlke JK (2017) Brackish groundwater in the United States. Water Availability and Use Science Program In Professional Paper 1833. https://doi.org/10.3133/pp1833

Thorslund J, van Vliet MTH.(2020) A global dataset of surface water and groundwater salinity measurements from 1980–2019. J Scientific Data 7(1). https://doi.org/10.1038/s41597-020-0562-z

van Weert F, van der Gun J, Reckman J (2009) Global overview of saline groundwater occurrence and genesis. Report nr. GP 2009-1. International Groundwater Resources Assessment Centre, Utrecht, 103 pp

Wada Y, Van Beek LPH, Van Kempen CM, Reckman JWTM, Vasak S, Bierkens MFP (2010) Global depletion of groundwater resources. J Hydrol Land Surface Stud 37(20). https://doi.org/10.1029/2010GL044571

Zhu C (2000) Estimate of recharge from radiocarbon dating of groundwater and numerical flow and transport modeling. J Water Resour Res 36(9):2607–2620. https://doi.org/10.1029/2000WR900172

Ziolkowska JR, Reyes R (2016) Impact of socio-economic growth on desalination in the US. J Environ Manag 167, 15–22. https://doi.org/10.1016/j.jenvman.2015.11.013

Part IV
Reusing Used Water

Chapter 7
Municipal Wastewater

Birguy Lamizana, Olfa Mahjoub, Serena Caucci, Clever Mafuta, Edeltraud Guenther, Gueladio Cisse, Kim Andersson, and Francesc Hernández-Sancho

Abstract Municipal wastewater is a major source of water for multiple uses, particularly in water-scarce regions. It is now recognized as a valuable resource rather than a waste stream with a focus toward resource recovery. Tailored technologies, adaptive policies, and regulations, as well as innovative finance mechanisms that create

B. Lamizana (✉)
Marine and Freshwater Branch, Ecosystem Division, UN Environment Programme, Nairobi, Kenya
e-mail: birguy.lamizana@un.org; birguy.lamizana@gmail.com

O. Mahjoub
National Research Institute for Rural Engineering, Water, and Forestry (INRGREF), Laboratory of Agricultural Sciences and Techniques (LR16INRAT05), National Institute of Agronomic Research of Tunisia (INRAT), University of Carthage, Ariana, Tunisia
e-mail: olfama@gmail.com

S. Caucci · E. Guenther
United Nations University Institute for Integrated Management of Material Fluxes and of Resources (UNU-FLORES), Ammonstrasse 74, Dresden 01067, Germany
e-mail: caucci@unu.edu

E. Guenther
e-mail: guenther@unu.edu

C. Mafuta
GRID-Arendal, P.O. Box 183, 4802 Arendal, Norway
e-mail: clever.mafuta@grida.no

G. Cisse
Department of Epidemiology and Public Health (EPH), Swiss Tropical and Public Health Institute (Swiss TPH), University of Basel, Socinstrasse 57, 4051 Basel, Switzerland
e-mail: gueladio.cisse@swisstph.ch; gueladio.cisse@unibas.ch

K. Andersson
Stockholm Environment Institute, Linnégatan 87D, 104 51 Stockholm, Sweden
e-mail: kim.andersson@sei.org

F. Hernández-Sancho
Institute of Local Development (ILD-WATER). Water Economics Group, University of Valencia, Avda. Tarongers s/n, Valencia 46022, Spain
e-mail: francesc.hernandez@uv.es

M. Qadir et al. (eds.), *Unconventional Water Resources*,
https://doi.org/10.1007/978-3-030-90146-2_7

an enabling environment, need to be in place. In doing so, wastewater use should be an essential component in new policies for a circular economy, which aims to decouple economic activity from finite resource consumption. If safely managed, wastewater use can also be an important strategy to alleviate pollution in ecosystems, while producing green business opportunities. There is an increasing diversity of available technologies for resource recovery from wastewater in the presence of major challenges due to a lack of systematic planning and design to identify and implement sustainable solutions in the context of a circular economy and a Nexus thinking approach. Acceptance of reused wastewater by people and policymakers still remains a challenge. This acceptance is linked to many aspects including the general absence of adequate national legislation and the insufficiency of information sharing about the advantages, the progress in technological performance, and the safety regarding the environment and human health. There are still barriers spanning several complex and multiple dimensions that impede, delay, or completely block the expanded use of municipal wastewater. The good news is that options are available and if responsible decision-makers are aware, these barriers can be actively overcome.

Keywords Municipal wastewater · Fit-for-purpose use · Resource recovery · Barriers · Circular economy · Nexus approach

7.1 Introduction

Municipal wastewater is wastewater generated within the limits of a city (urban and peri-urban), mainly, municipal sewerage, non-sewerage wastewater from treatment systems, and combined stormwater and sewer systems. Hence, its quantity and quality depend on the size of the city, the levels of density, population, and development, and industrialization. At the same time, municipal wastewater is the sink and the source of thousands of physical (suspended and organic matter), chemical (nutrients, salts, heavy metals, and organic compounds, including persistent and emerging ones), and biological (pathogens of various sizes and pathogenicity) components. This composition, which strongly defines the quality of treated wastewater for reuse purpose, varies in time and space and under the prevailing conditions of collection from the point of release to the point of treatment and reuse.

Since its very beginning, civilization has strongly influenced the relationship of the human-made environment with wastewater over time. Collection and treatment technologies have also substantially improved. However, the reuse of wastewater has remained closely related to acceptance of the by-products and their quality, the buy-in of societies, and driving forces like health, environment, and economics. The good news is that this is changing rapidly.

Wastewater quality should be guaranteed both for discharge in receiving water bodies and for intended reuse. The reduction of contaminant concentrations is often in compliance with national sanitation and environmental policy and standards, when existing. If collected and reused, wastewater can play a crucial role as an alternative to a constant water-supply source. Indeed, physical water scarcity is a threat for more than half of the worldwide population and is a reality for 20% of human's economic water shortage (WWAP 2017). In places with major competing demands from human settlements and industrial or agricultural activities, water becomes scarce. Wastewater use can be a valid supplement for sustainable and resource-efficient water supplies if facilitated by integrated watershed-planning processes in which both upstream (e.g., source control) and downstream (e.g., collection, treatment, and reuse) wastewater management is considered.

Accordingly, wastewater use should be an essential component in new policies for a circular economy, in which communities should look for synergies with actual productivity but balance, at the same time, the consumption of finite natural resources, thus preserving them in the long term. If safely managed, wastewater use can also be an important part of a strategy to alleviate pollution in ecosystems, while producing green business opportunities. However, on a global scale, we are missing out on these potential additional reuse benefits since a major part of the municipal wastewater generation, estimated to be 354 km^3 per year (FAO 2020), is discharged with limited or no treatment and indirectly used in irrigation of approximately 29.3 million ha (Mha) of land (Thebo et al. 2017), at very high public-health costs.

Constraints to water reuse are well known nowadays and are progressively abridged thanks to advances in treatment technologies and storage to fulfill the reuse purpose and water needs, including agricultural irrigation. The availability of alternative low-cost solutions for safe reuse is yet to be developed to mitigate health and environmental risks. However, advanced technologies to remove persistent pollutants may not be affordable. The multi-barrier perspective is poised to go beyond water quality to address protection of consumers from potential harm caused by the spread of, and exposure to, pathogens and chemicals.

The focus of this chapter is municipal wastewater and its challenges, as well as on the contribution of wastewater to water security when properly managed.

7.2 Technological Interventions

For wastewater to be safely managed, there must be infrastructure for its collection and storage, conveyance and transport, treatment, and resource recovery and reuse (Andersson et al. 2020). Wastewater management systems can be either centralized or decentralized:

- Centralized systems, which can either be combined with sewerage (collecting stormwater also) or separate from sewerage (separate wastewater and stormwater sewers),
- Combined on-site and centralized systems, including on-site septic tanks and off-site treatment plants,

- Semi-centralized systems, including many systems that feed into one, and
- Decentralized on-site systems (no sewerage) that are household based.

A city or town can have a combination of centralized, decentralized, and on-site systems with fecal sludge management, to meet the overall city sanitation requirements. Centralized systems collect and treat large volumes of wastewater for larger communities or whole towns and cities. In contrast, decentralized wastewater management systems treat wastewater ranging from individual houses up to small communities close to the source of wastewater generation. Decentralized systems are characterized by a dispersed siting of treatment facilities and resource-recovery applications within a given geographical boundary.

Once wastewater is collected, there are various technologies and approaches that can be applied for its treatment, including preliminary treatment, primary treatment, secondary treatment, and tertiary treatment, as illustrated in Fig. 7.1.

Preliminary Treatment: This includes manual or mechanical screening to remove large materials such as rags, plastic, and other foreign object that may interfere with or damage treatment equipment. Grit is also removed (WRC 2015).

Primary Treatment: This entails flow balancing/flow equalization, and wastewater treatment at this stage uses the pond system (Goel et al. 2005). There are various

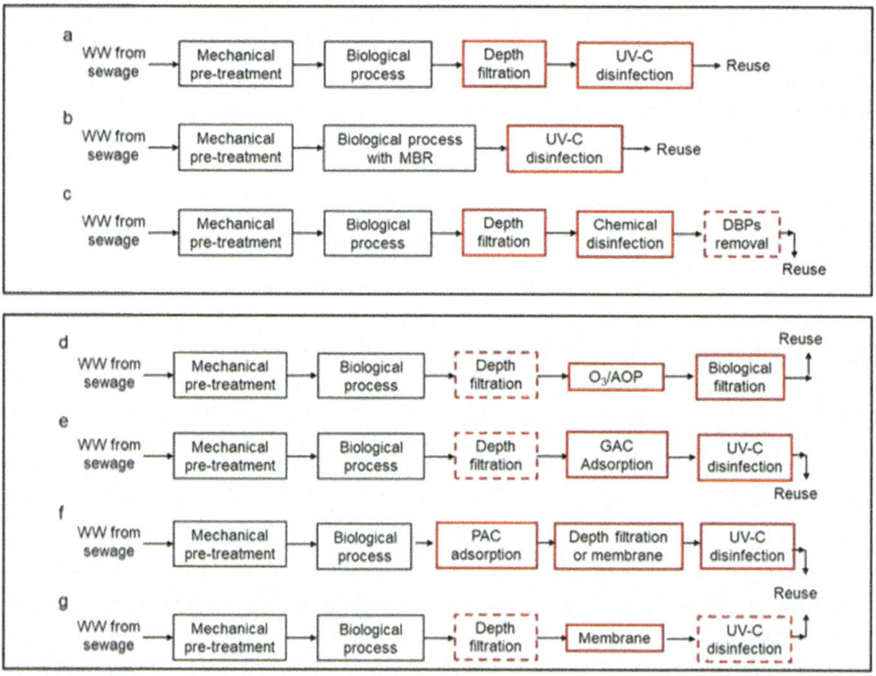

Fig. 7.1 Wastewater treatment technologies and approaches (*Source* Rizzo et al. 2020; used by permission)

pond systems, including anaerobic ponds, conventional ponds, high-performance pond systems, and ponds integrated with advanced technologies. Wetlands (both natural and artificial) can also be used at the primary treatment stage as filters for sediments and nutrients (Mahmood et al. 2013). Primary sedimentation at the primary treatment stage helps separate solids from liquids, while flotation produces buoyant bubble-solids agglomerates that get removed by scrapping (Zaharia 2017).

Secondary Treatment: At this stage, growth technologies, such as rotating biological contactors and trickling filters, are used (WRC 2015). Secondary treatment technologies also include nutrient-removal activated-sludge processes. Other treatment processes considered here are membrane separation biological treatment, decentralized wastewater treatment systems, pond enhanced treatment and operation, packaged plants, including activated sludge plants/extended aeration, and membrane bioreactors (WRC 2015).

Tertiary Treatment: Constructed wetlands can be used in tertiary treatment of wastewater and so are ecosystem technologies (Almuktar et al. 2018). Disinfection is done at the tertiary treatment stage and includes chemical disinfection through chlorination and ozonation and UV radiation. Maturation ponds give a final polish to effluents before they are discharged into the environment (Muralikrishna and Manickam 2017).

Sludge Handling: The previously mentioned processes and technologies treat the liquid phase of wastewater. The solid phase needs different treatments, and this entails thickening, stabilization, and dewatering (WRC 2015).

The selection of the optimal municipal wastewater treatment is based on the intended end use of the reclaimed water. Fit-for-purpose wastewater treatment indeed points to the intended end-use, economic viability, and environmental sustainability of the water reuse activities (Table 7.1). In the context of fit-for-purpose wastewater treatment, the technologies differ mainly in terms of treatment efficiency, cost, energy use, and associated carbon emissions (Chhipi-Shrestha et al. 2017). Establishing the most efficient and suitable type of wastewater treatment system is site-specific, and, as such, it requires capacity-building efforts to assess and select the most appropriate treatment technologies (WHO and UN-Habitat 2018). Further, technological developments are increasingly fostering resource recovery from municipal wastewater and turning it into a new source of water (IWA 2020).

7.3 History

"The history of men is reflected in the history of its sewers," as written by Victor Hugo in 1892. Hence, municipal wastewater management mirrors the societal status, the progress of the technologies, and the economic sectors. Managing the whole water cycle is not the achievement of today. Unveiling the history of the society's relationship to water is therefore fundamental to understand in present-day discussions

Table 7.1 Wastewater treatment technologies (based on the synthesis of information from Andersson et al. 2020; WRC 2015; WWAP 2017)

Treatment by-product	Reclaimed water				Sewage sludge
Treatment stage	Preliminary	Primary	Secondary	Tertiary/advanced treatment	
Processes	Screening	Biological oxidation and nitrogen removal	Chemical coagulation, biological or chemical nutrient removal, filtration, and disinfection	Activated carbon, reverse osmosis, ozone/advanced oxidation processes, disinfection, soil aquifer treatment, UV-C disinfection, chemical disinfection, disinfectant by-products removal, biological filtration, depth filtration, granular activated carbon adsorption	Thickening, stabilization, and dewatering
Treatment technology	Sedimentation	Pond's systems, including anaerobic ponds, conventional ponds, high-performance pond systems, and ponds integrated with advanced technologies	Activated- sludge processes, namely modified Ludzack-Ettinger and oxidation ditch activated sludge with alternating aerobic and anaerobic zones	Construction ponds, maturation ponds	

(continued)

Table 7.1 (continued)

Treatment by-product	Reclaimed water				Sewage sludge
Treatment stage	Preliminary	Primary	Secondary	Tertiary/advanced treatment	
End uses	No use is recommended	Surface irrigation of above-ground fruit orchards and vineyards, and non-food crops. Recharge of non-potable aquifers. Maintenance of wetlands and other wildlife habitats, as well as stream augmentation. Industrial cooling.	Irrigation of landscapes, golf courses, food crops. Flushing of toilets. Car washing. Industrial cooling.	Indirect potable reuses, through groundwater recharge of potable aquifers. Surface-water reservoir augmentation and potable reuse. Recycled water used for irrigation or industry.	Organic fertilizers Energy
Advantages	Protects equipment at later stages of treatment from damage.	Ponds have no clogging risk. Ponds have high nutrient-removal rates	Activated sludge treatment is good at removal of BOD. The treatment plants can also remove N and P. The treatment is rapid and relatively odor-free.	Maturation ponds achieve good bacterial removal. Biogas can be recovered as a source of energy from maturation ponds. Constructed wetlands have low energy requirements; are cheap to maintain; and provide aesthetic, commercial and habitat value; effective disinfection for bacteria, protozoa, and some viruses	

(continued)

Table 7.1 (continued)

Treatment by-product	Reclaimed water				Sewage sludge
Treatment stage	Preliminary	Primary	Secondary	Tertiary/advanced treatment	
Disadvantages	Biological and chemical polluting loads are not removed	Ponds are land-intensive and are unsuitable in very windy regions	The treatment is ineffective in deep water, and under freezing weather conditions. There is little removal of bacterial loads and high sludge production.	Constructed wetlands need large areas; can result in clogging of the system, especially when it is used as primary treatment; if national standards are stringent, UV may not be enough for removal of all bacteria, while some disinfectant such as chlorine cannot be used.	

(Tempelhoff et al. 2009) and the way it is impacting policies and technical developments (Lofrano and Brown 2010). Controlling the collection, discharge, and eventually the reuse of municipal wastewater have considerably evolved through time. However, wastewater has always been considered dirty (Lofrano and Brown 2010); changing the mindset about reuse has taken some time, even centuries. Despite that the Antonine Plague in 165 AD was described to be far deadlier than the current SARS-CoV-2 virus causing the pandemic of COVID-19 (Watts 2020), the transport of this virus via the waterways is raising questions about its occurrence in municipal wastewater and whether it could cause risks to people (Joonaki et al. 2020). Indeed, wastewater is used to monitor the prevalence of COVID-19 infection in communities. Wastewater is recognized as the major source of pathogens that transfer/exposure cause the spread of very common water-borne diseases centuries ago, like typhoid and cholera, among others. Plague bacteria, for instance, that emerged in 541–542 AD (i.e., the Plague of Justinian), is suspected of being protected by some water-borne microorganisms and likely to remerge in our era (Markman et al. 2018). Therefore, the acceptance of reused water at the present time is expected to face new challenges under the approach of the circular economy that governments are tending to promote.

The first successful efforts to control the flow of water were made in Mesopotamia (the north of today's Iraq and Syria) after the agricultural revolution, in the last 9,000–10,000 years, and then in Egypt around 6,000–7,000 years ago (Table 7.2). In Mesopotamia, water technology was not limited to irrigation and pioneered sanitary engineering, with many cities possessing networks of wastewater and stormwater drainage systems (Tamburrino 2010). Around 3500–2500 BC, the Mesopotamian Empire was the first to address sanitation problems and used drains to eliminate wastes (Lofrano and Brown 2010). Unlike civilizations in Mesopotamia and Egypt, based on the exploitation of the water of the large rivers (i.e., the Tigris, Euphrates, and Nile), the Greek civilization has been characterized by limited, and often inadequate natural water resources. During the Classical period, Greeks have had an elaborate channel drainage system to convey stormwater and waste out of the city (Mays 2010). They used combined sewer and drainage systems in alleys or on side streets between houses to remove wastewater released by the public toilets, houses, and other premises, along with stormwater (Antoniou 2010). Around 430–424 BC, typhoid was probably the most devastating epidemic water-borne disease since it wiped out one-third of the population in Athens. By the fourth century BC, malaria was also a common epidemic in ancient Greece, killing a large portion of the population (Adhikari 2019). However, this did not prevent the population from reusing wastewater. Between the 3rd and the 2nd millennium B.C, in the early Bronze Age, communities in several civilizations started using human wastes as fertilizers and domestic wastewater for irrigation and aquaculture (Crouch 1993 and Angelakis et al. 2018). Such developments were driven by the necessity to make efficient use of natural resources, to make civilizations more resistant to destructive natural elements, and to improve the standards of life, both at the public and private levels.

Around 1100 BC, the use of wastes (including human) for aquaculture was practiced in China during the Yin dynasty (Khouri et al. 1994). The very first sewers did see the light in 600 BC as canals dug in the streets. Around 490 BC, the sewerage

Table 7.2 History timeline of wastewater management and reuse since the Bronze Age

Timeline	BC								AD				
	ca 10,000–4,500	ca 4,000–2,500	ca 3,200	ca 2,000	ca 1,000	ca. 600	500–67	100 BC–330 AD	330–1500	1500	1800	1900	2000
Event	Soil treatment	Stormwater drainage system	Sewage and rainwater discharge to agricultural land	Wastewater disposed into surface water or used for agricultural irrigation	Wastes used for aquaculture during the Yin dynasty	Use of water of different qualities for different purposes including irrigation	Stormwater and wastewater collected through a sewerage network to be used for irrigation	Use of pipes to move wastewater and stormwater to a basin outside the city	Use of wastewater in agriculture	Use of untreated wastewater for nutrients and water reuse	Treatment consisted of dilution into receiving waters. First buried sewers for control of disease.	1960s: Development of secondary biological treatment in US. Wastewater irrigation in Mezquital Valley, Mexico	Revision of standards and guidelines: WHO guidelines on wastewater use in agriculture; US-PA guidelines on water reuse to include Indirect Potable Reuse (IPR), French water reuse regulation; Israel regulations
Location/Civilization	n.d.	Mesopotamia	Crete, Greece	Indus civilization/Minoan period in Greece	China, Asia		Ancient Athens, Greece	Greece	Greece	Mexico	US	US and Mexico	Worldwide (WHO), US, France, Israel
Historical period	Prehistory and Bronze Age						Classical time		Middle Ages	Dark Age and early to mid-modern times		Contemporary time	

network was meant to collect human wastes and stormwater, which were both reused for irrigation and fertilization of irrigated crops using brick-lined pipes (Tzanakakis 2014).

In the Middle Ages, hydraulic techniques were better mastered by the Romans (Box 7.1), leading to an increase in water demand and volume of discharge of wastewater coming from public baths later became a source of nuisance and diseases. Since that time, the use of different qualities of water was common, indicating the various purposes intended for use, i.e., the potable, the sub-potable, and the non-potable uses, including the reuse of wastewater combined with runoff (Crouch 1993). During Roman times, collection and treatment for reuse of wastewater could be assumed to today's "decentralized or on-site sanitation", though integration and prevention, as prerequisite, were missing (Huibers et al. 2010). Only about 400 years later sewer systems appeared (Brunet 2006). Despite their complexity, they lacked the basic principle of sanitation, combining domestic wastewater, drainage water, and stormwater. Cloaca Maxima, an open drain designed to carry stormwater in Rome, was one of the earliest sewage systems constructed to collect drainage water and remove wastes (Malissard 2002). Later in Medieval times, waste disposal was so unregulated that European countries witnessed several disease outbreaks (like the Black Plague), due to the lag in water technology and knowledge and poor sanitation in general, resulting in the death of 25% of the population. During the classical age, ancient Greeks used wastewater in agriculture. Wastewater originating from the public toilets and residences and stormwater were collected in combined sewer and drainage systems. At the Acropolis, wastewater from households and workshops was drained into a central sewer made of clay pipes (Angelakis et al. 2018).

Despite the industrial revolution, sewerage systems witnessed great progress only in the second half of 1800s, and sewage farms were adopted for wastewater treatment and disposal in many regions like Europe, North America, and Australia (Khouri et al. 1994). By 1900, in the US, pit privies and open ditches were replaced by buried sewers, and, around 1960, almost 50% of US population had access to wastewater treatment. In Mexico, the Aztecs started using various wastewater components before 1500 AD, and then, around 1890, wastewater was collected in drainage canals for agricultural irrigation in the Mezquital Valley. The use of untreated wastewater contributed to the prosperity of the region up until today (Becerri and Jimenez 2007).

Wastewater reuse since 2000 has been witnessing the promotion and implementation of new approaches and concepts (One Health, One Water). Moreover, the establishment of guidelines and regulations, plus changes in the mindset, have occurred, promoting the acceptance of potable reuse under prevailing climatic constraints; this requires bearing in mind the increasing health and environmental risks and balancing such risks with the advancement of technologies. These concepts arose to address concerns like emerging antibiotic resistant bacteria (ARB) and genes (ARG) that have gained momentum since the beginning of the century due to the overuse of antibiotics. Alternative therapies were used to treat infections in ancient times, but none were as reliably safe and effective as modern antimicrobial therapy (Earla 2014). The overuse and misuse of antibiotics have led to the development of mitigation strategies that are recognized to be fundamental by the UN. Together with

ARB and ARG, COVID-19 is considered as one of 21st century's challenges. The health concerns are based on the transmission routes through aerosols and droplets. Hence, treatment technologies and monitoring tools are expected to provide scientific evidence to mitigate transmission linked to wastewater reuse (Bogler et al. 2020).

Box 7.1 Municipal wastewater management in Africa Proconsularis (Tunisia) under the Roman Empire
Tunisia (known as Africa Proconsularis during the Roman era) was among the most urbanized areas after Rome; its cities experienced exceptional development during the Roman civilization (Mahjoub and Chaibi 2014). Tunisia was one of the most prosperous region of the Roman Empire thanks to the advancement of water engineering and sanitation. The first sewers the Romans built for collecting used/polluted water were canals along the roadways. Liquid and solid wastes produced by households and the runoff and stormwater, today called "municipal wastewater", used to flow in a unique system. Wastewater discharged by craft workshops and 'industrial' activity used to flow in the same canals. In fullers' workshops, the wastewater used to be collected in drains underneath and then discharged into the main drain in the street (De Feo and De Gisi 2013). The sanitation system built by the Romans was meant to facilitate several operations, including collection, transport, treatment, disposal, and/or reuse of wastewater (Mahjoub and Chaibi 2014).

7.4 Status

Municipal wastewater is a sustainable and reliable water supply source to supplement limited freshwater resources in arid and semi-arid regions of the world (WWAP 2017). As the overall water demand for human and economic activities is constantly increasing and thus placing freshwater resources under stress, wastewater use could make water resources more resilient and reduce over-abstraction, pollution, and climate change impacts (WWAP 2017). The use of treated municipal wastewater in productive activities has a vast potential to reverse the worrisome projections that nearly six billion people will suffer from clean water scarcity by 2050 (Boretti and Rosa 2019).

In multiple regions of the world, the use of municipal wastewater is common practice to protect valuable water resources and take advantage of the nutrients contained in sewage for crop production (WWAP 2017). The practice is becoming particularly common in the wastewater-rich urban and peri-urban areas where the resource is readily available, and where there is a market for agricultural products.

Global data show that there is a major gap between the wastewater treatment capacities of high-income and low-income countries. Today, around 70% of the municipal wastewater generated in high-income countries is treated, but the ratio of water treatment falls to 38% in upper-middle-income countries, 28% in lower-middle-income countries, and 8% in low-income countries (WWAP 2017). The main objective for high-income countries to engage in water reuse is to maintain environmental quality and reduce the impacts of water scarcity (WWAP 2017). On the other hand, in low-income countries, the main challenge to the reuse of municipal wastewater is insufficient infrastructure, technical and institutional capacity, and financial resources (Cossio et al. 2020).

Although municipal wastewater treatment helps to reduce the health risks, improved sanitation capacity cannot be equated to improved wastewater management (WWAP 2017). Even in high-income countries, the presence of a sewerage system does not guarantee pollution-free disposal (UN-Water 2015). This is a challenge particularly in urban areas because the data reveal that globally only 26% of urban and 34% of rural sanitation and wastewater services achieved the status of 'safely managed sanitation' (Guy and Mili 2016). The implementation of low-cost solutions and safe water reuse options is a matter of increasing urgency in low-income and lower-middle-income countries since the level of wastewater treatment there is extremely low.

The figures on the global status of 'safely managed sanitation' (Fig. 7.2) show that in 2017, in most countries in sub-Saharan Africa and Latin America and the Caribbean, less than 50% of wastewater was treated (WHO 2019). Wastewater has been increasingly treated to satisfy demand from different sectors, including industry and agriculture, but still large volumes of wastewater are discharged daily into the waterways, where it creates significant pressure on the aquatic environment and causes health, environmental and climate-related risks (IWA 2020).

The improvements in the sanitation and wastewater treatment systems around the world have been followed by a paradigm change triggered by the need for a more sustainable economy and circular use of resources. Water and wastewater are key components of sustainable development and have a critical role in transitioning to a circular economy regarding the energy consumed and produced, and the materials exploited. Another important advantage of integrating wastewater management in circular economy models relates to resource recovery and reuse that could transform wastewater treatment and sanitation from a costly service to a self-sustaining activity that adds value to the economy (Rodriguez et al. 2020). In reaction to the drawbacks of the conventional 'take–make–consume and dispose' model of growth, reuse of municipal wastewater and resource recovery have been increasingly seen as a key opportunity to be seized by society and businesses.

Fig. 7.2 Regional sanitation coverage for household drinking water, sanitation, and hygiene in 2017 (Modified from UNICEF and WHO 2019)

Reuse of municipal wastewater and by-product recovery also have the potential to generate new business opportunities by recovering energy, nutrients, metals and other by-products (WWAP 2017; WHO and UN-Habitat 2018). As such, wastewater represents a widely available and valuable resource that, if sustainably managed, is poised to become one of the central elements in the transformation towards a circular economy. However, integration of wastewater in the circular economy framework requires a strong emphasis not only on the technological aspects but also on the socioeconomic aspects of perceiving municipal wastewater as a resource. The importance of considering wastewater as a resource is well-recognized, and major technological improvements have been made, especially with regard to the recent changes in municipal wastewater composition where the increasing presence of contaminants of emerging concern (CECs) and AMR requires the development of advanced treatment options (Rizzo et al. 2020).

Despite the important technological development for resource recovery, a major hurdle is the lack of systematic planning and design to identify and implement the most affordable and sustainable solutions in the context of a circular economy. A coordinated and pragmatic governmental policy environment combined with the efforts of diverse sectors, such as industry, utilities, health, agriculture, and the environment, is needed to promote innovative safe recycling and reuse of wastewater (WWAP 2017).

With the push for reuse and resource recovery from wastewater and more countries embracing the circular-economy approach, the importance of monitoring wastewater is becoming crucial, particularly during stressful times such as a pandemic. Wastewater monitoring has been used for decades to assess the success of vaccination campaigns against the poliovirus (Metcalf et al. 1995) and more recently as an early warning system for COVID-19, which has thrown up fresh challenges.

Usually, monitoring wastewater takes place throughout the works—from the inlet to the final effluent—and at every stage of the treatment process. In many cities, due to lack of the capacity (financial, technical, institutional, regulatory, enforcement, and stakeholders' engagement), the monitoring system is not as optimal as during the normal times; and monitoring is subject to severe strain during pandemics and disasters when the government focus shifts to providing health services without looking into integrating those with the important segments, such as wastewater that may create a vicious cycle, impacting both public health and the environment.

Wastewater quality monitoring can be time consuming and requires reliable techniques and skilled capacities. The evolution from traditional to modern technologies for water quality monitoring can save time and accelerate interventions, which is not the case in the less developed countries. Real-time detection of physical, chemical, and biological pollutants can be very crucial and may include specific sensors, instrumentation, and signal processing to enhance the detection of trace elements. Wastewater is a complex mixture of compounds, and quality monitoring can strongly support prioritization of contaminants and establishing threshold concentrations for regulations (Korostynska et al. 2013). Water reuse can be significantly facilitated by on-line monitoring technologies. However, the latter need to be recognized by national regulations to improve end-users' trust in the efficiency of the treatment process.

Good examples of wastewater use show that municipal wastewater can be a safe source for multiple purposes. Today, the city of Windhoek (Namibia) treats wastewater to meet drinking water-quality standards, while meeting 25% of the city's potable water supply from wastewater (WWAP 2017). Another successful example is in Chennai, India, where the reuse of 40% of the generated municipal wastewater satisfies 15% of the city's water demand. In Monterey (CA, in the US), a water-scarce region, a large agricultural area is supplied with almost 80,000 m^3 per day of treated municipal wastewater for irrigation and crop production (Kehrein et al. 2020). Water stress is an important trigger to engage in water-reuse activities, and, at the national level, Israel and Singapore lead the trend. In Israel, treated wastewater accounts for almost a quarter of the agricultural water demand, while in Singapore the ratio reaches up to 40% with its NEWater reclamation plant for indirect potable reuse (Kehrein et al. 2020; WWAP 2017).

Important progress on a broad front is happening for non-sewered wastewater management, for example, through many fecal sludge management (FSM) initiatives, on both global and local scales, where services for treatment and reuse are promoted for decentralized or onsite wastewater systems (e.g., the City-Wide Inclusive Sanitation Initiative and the new FSM Alliance). Expanding services for non-sewered areas is critical considering that a third of the human population is currently dependent on onsite sanitation solutions (Hidenori et al. 2016). Resource recovery of fecal sludge for fertilizer or energy is an important driver to facilitate adequate services for non-sewered sanitation. An example is the city of Naivasha, Kenya, where fecal sludge is collected and treated into a very competitive solid fuel in the form of briquettes, with a potential saving of 22 trees per ton of sold briquettes (Andersson et al. 2020).

Wastewater use can provide an important alternative source of freshwater, yet the potential of municipal wastewater as a resource has remained underexploited. As untreated sewage and inadequately treated municipal wastewater continue to deplete the quality of water around the world, the 'resource' should be addressed not as a problem but as a smart solution in the context of a restorative and regenerative circular economy.

7.5 Major Barriers and Response Options

7.5.1 Major Barriers

Barriers to the expanded reuse of municipal wastewater are factors that impede, delay, or completely block this innovative process (Hueske et al. 2015). Only if the responsible decision-makers are aware of potential difficulties when implementing new processes, the barriers can be actively overcome (D'Este et al. 2012). These barriers span several complex and multiple dimensions that are highly interconnected, e.g., technological, regulatory, economic, social and cultural, environmental and health, educational, knowledge, and capacities (Perraton et al. 2014; UN-Water 2020; Ventura et al. 2019). Innovation related barriers involve challenges in policy, leadership, technology, organization, human and financial resources, and individual, state, societal, and private interests (Hueske et al. 2015). The durability of, and interest in, reuse schemes depend on the role of the multiple parties, the required levels of treatment, the ability to fund capital and operational investment, and if operated on a "not-for-profit" or a commercial basis (Perraton et al. 2014).

Technological barriers: There are four main broad categories of reuse of municipal wastewater: direct potable use; indirect potable use; non-potable use; and industrial use (Capodaglio 2020). The technological challenges depend heavily on the purpose of the reuse plans and the local capacities to afford what this purpose entails. Most urban water systems continue using decades- or century-old technologies, while the growing requirements from health and environmental perspectives call for more and more innovation (Kiparsky et al. 2016). There are still limitations in the efficiency of the current technologies and treatment processes for the removal of chemicals, particularly contaminants of emerging concern (UN-Water 2020). The pertinent regulations can increase the challenges for finding and affording the appropriate technologies. The energy demand of some treatment technologies constitutes a major challenge, and the processes can be a significant source of greenhouse-gas emissions. Wastewater treatment facilities are among the major energy consumers at a municipal level worldwide, amounting to about 1% to 3% of the total energy output of a country (Capodaglio and Olsson 2020). There are also several technological limitations in some contexts for measuring pollutants in waters, increasing fears among potential end-users related to the reliability of the water quality monitoring systems. Another area of constraints is the distance between where the wastewater is treated and where the various reuses take place (UN-Water 2020). This could entail additional technological constraints related to the transport of treated wastewater to, and storage at, the place of reuse. In general, there is a continuous need for the development and the diffusion of new technologies responding to new requirements of water-related security and sustainability.

Regulatory barriers: Municipal wastewater uses by low-income urban farmers used to face constraints linked to the availability of and secure accessibility to land (Perraton et al. 2014; UN-Water 2020), which are linked to insufficient land-ownership regulation, institutional and governance arrangements, environmental protection laws, and water quality standards set by authorities. Institutional factors are particularly critical determinants of pathways for sustainable reuse schemes in water systems (Kiparsky et al. 2016).

Economic barriers: While undertaking a water reuse project is fully justified in terms of objectives, it is not always possible to defray costs by charging tariffs. Moreover, who should pay for water reuse projects? Should only water users pay, or should all beneficiaries contribute to the costs? The economics of the water reuse projects differ significantly among the various cases because the water reuse systems are generally designed ad hoc in line with the characteristics of the private users. Considering water regeneration costs and the tariffs paid by water users, in most cases, some degree of subsidy is needed to recover the full costs. The first step to improve the application of this economic principle is to identify the barriers that prevent policy makers from establishing higher water reuse tariffs.

The recycling of water not only increases the availability of water resources, but it creates significant environmental benefits. However, the value of these benefits is often not calculated because they are not determined by the market. Valuation of these benefits nevertheless constitutes a barrier to overcome to justify suitable investment policies and financing mechanisms for promoting water reuse. The benefits of water reuse should be estimated, and the option of water reuse should be compared with other alternatives. In most cases, the alternative to the development of a water reuse project is the status quo, that is, not to implement any alternative. So, the costs of non-action should be established. From the perspective of integrated water resources management, the economic analysis of water reuse projects should consider the cost of regenerated water (and the benefits), as well as the costs of alternative water supply options, such as drinking water, desalinated water, and/or storm water. Hence, it is possible to determine a ranking of cost-effective solutions for guaranteeing water demand.

Social/cultural barriers: A key challenging area is the acceptance of reused wastewater by people and policy-makers (UN-Water 2020). This acceptance is linked to many aspects including the absence of adequate legislation and the insufficiency of communication on the advantages; the progress in technological performances; and safety for the environment and human health (Ventura et al. 2019). Water reuse is generally well-accepted in corporate industry because it holds no or little risk to human and environmental health (Capodaglio 2020). Wastewater use is promoted as a solution to water scarcity. However, it is very likely to fail if social acceptance is not accounted for. The most common factors to be considered are closely linked to societal development. Involving target groups is an asset and could substantially influence the building of trust in the success of reuse programs (Drechsel et al. 2015).

Environment and health: Recovery of the most beneficial constituents like nutrients and energy is the basis of the systemic approach of the circular economy applied in the context of the water reuse widely promoted in recent years (Voulvoulis 2018). However, health risks incurred by users (of effluents for irrigation) and consumers (of irrigated products) related to the transfer and/or accumulation of the toxic elements and infectious pathogens in the food chain, are of concern among the least developed communities. Environmental risk should be assessed and managed as well, even though it may be overlooked and inadequately addressed in countries with low access to sewerage network (Weber et al. 2006). Risk assessment approaches developed so far are expected to form the basis for regulations. Standards elaboration and putting threshold values are meant to preserve environmental and human health. Consequently, the quality of the treated wastewater must be monitored directly through specific parameters or indicators, or using global parameters.

During the last two decades, the occurrence of emerging contaminants, such as personal-care products, pharmaceutical compounds, disinfectants, and antibiotics in reclaimed water, has raised concern due to the lack of data on their relevance and their impact on long-term human health and ecological systems in many countries, despite the increasing evidence provided by the scientific communities on the transfer of some recalcitrant compounds to edible crops and to consumers (Schapira et al. 2020). Antibiotic-resistant bacteria, found in wastewater and in the vicinity of points of discharge due to multiple factors including the misuse and overuse of antibiotics in humans and industries, reverts in the urban ecosystems and thus via wastewater in the environment. The reuse of wastewater in the agriculture sector to relieve economic pressure on natural resources may lead to soil, crops, and water contamination, and subsequently foster the spread of antimicrobial resistance if not sustainably managed. However, the impact on soil fauna was deemed to be negligible (Negreanu et al. 2012).

More recently, wastewater was recognized as the harbinger of a COVID-19 outbreak caused by SARS-CoV2. The relatively new field of sewage epidemiology (wastewater-based epidemiology: WBE) could play a critical role in forecasting such pandemics (Daughton 2020). Indeed, removing viral contamination from wastewater remains challenging (Aghalari et al. 2020). This kind of emergent risk may increase the barriers for wastewater reuse. It is not yet clear if the SARS-CoV-2 is viable under wastewater and what environmental conditions could facilitate fecal–oral transmission, but there are some reported cases showing the potential of contamination by this pathway.

Finally, education and capacity building will be the key to overcome any barrier in the use of municipal wastewater. This entails generating and sharing knowledge, best practices, and success stories that will trigger the expected change. It is also important for promoting wastewater management related curricula and training to build a network of qualified professionals, as well as investing in data collection, quality control, and analysis.

7.5.2 Response Options

Municipal wastewater reuse for agricultural irrigation is promoted due to the growing water scarcity problem around the world to relieve economic pressure on natural resources. However, if not properly managed, chemical and pathogenic risks that might have detrimental impacts on human health and on environmental systems can occur (Caucci et al. 2016). Given the multiple challenges and barriers that prevent the effective use of municipal wastewater, there is a need to develop a robust and comprehensive response to turn waste into a valuable resource (Table 7.3). Currently, three major organizations: the World Health Organization (WHO), the Food and

Table 7.3 Barriers and responses options to wastewater use for sustainable agricultural irrigation

Barriers	Response options
Technological barrier	• Guidelines, professional curricula, tools; fit-for-purpose use; cooperation (South–South; North–South)
Regulatory barriers	• New regulations promoting a circular economy • Adoptive policy for reuse • Standardization for contaminants of emerging concerns (CECs)
Economic barriers	• Incentives, cost-sharing, adequate tariffs
Social/cultural barriers	• Sensitization; awareness raising; community participation; building trust and acceptance; information sharing
Environmental/health	• Economic evaluation of wastewater: cost of action vs. cost of inaction; sustainability assessment; sanitation safety planning • Highlighting nutrients as a resource • Ecosystem services; nature-based solutions
Educational/knowledge/capacities (cross-cutting) • Knowledge sharing • Changing the yuk factor • Promoting wastewater management-related curricula and training to build up a network of qualified professionals • Investing in data collection, quality control, and analysis	

Agriculture Organization (FAO), and the World Organization for Animal Health (OIE) have created the Tripartite Agreement in joint collaborations under the One Health approach and are providing a platform for cross-disciplinary collaboration between inextricably linked human, animal, and environmental sciences to maintain the health of all (Roberts 2017).

Strengthening the existing framework of One Health approach with Nexus thinking may help to assess the status-quo of resources, break down silos, and balance the needs of humans with shared ecological systems, while preserving the health of natural ecosystems that form the basis of any economic activity (Avellán 2017). The Nexus approach could foster sustainable development by promoting the role of municipal wastewater as a resource that increases the viability of economic activities by alleviating the impacts on water scarcity and reducing the cost of energy and fertilizers. However, the goal is not only to maximize economic profits, or avoid market distortions via increased production and productivity, but to consider the well-being of the ecosystem and prevent potential environmental, sanitary, and nutrition related risks (Caucci and Hettiarachchi 2017).

As a practical and policy-oriented approach, application of Nexus to municipal wastewater use could boost the synergies and make an important contribution towards addressing the technical, institutional, and policy barriers for safe use of municipal wastewater. In the same vein, adopting the circular economy approach that breaks the habit of take–use–dispose and embrace wealth generation from waste is a key solution for safe municipal wastewater use.

7.6 Conclusions

The increased use of reclaimed water has led to the advancement of wastewater treatment technologies that have become more efficient and innovative to not only treat, but to also recover, valuable by-products. Municipal wastewater is now a new water resource that is expected to help in the protection of high-quality freshwater, reducing environmental pollution, alleviating food shortages, and reducing the gap between supply and demand in the future.

An important paradigm shift has been occurring at multiple levels to advance the reuse of municipal wastewater towards a circular economy in which wastewater is considered a valuable resource rather than a burden. Integration of energy production and resource recovery through municipal wastewater is one of the main drivers of circular economy and an opportunity to ensure that growing cities and urban development will not jeopardize sustainable development efforts. Although water and wastewater are key components in transitioning to circular economy, environmental and health risks associated with the use of untreated wastewater hamper the effective use of municipal wastewater.

Systematic and integrated planning and management of wastewater should be economically and ecologically sustainable, and at the same time contribute to preventing negative health impacts and enhancing the health benefits. A Nexus perspective may strengthen the cross-disciplinary 'One Health' approach and create significant synergies for an expanded adoption of municipal wastewater use that protects the human, animal, and environmental health on a global scale and fosters system strengthening and integration by addressing interlinkages between overlapping individual components.

There are still numerous barriers—ranging from technological, regulatory, economic, social, and cultural, as well as environmental and health—to overcome. A wider circular economy perspective has the potential to overcome some of the major barriers to water reuse and increase the feasibility of investing in municipal wastewater treatment and reuse. Overcoming the barriers towards the implementation of municipal water reuse is required to spark the change towards a sustainable society and fill the growing global gaps between low- and high-income countries.

In line with SDG 6.3, adequate treatment and use of municipal wastewater is a prerequisite. A failure to achieve this target will restrict the availability of water for all people for all uses. Adopting circular economy principles, considering the potential of municipal wastewater as a resource and the opportunities and limitations it presents, has the potential to contribute to the alleviation of the worsening water crisis and the realization of Sustainable Development Goals.

- It is much cheaper if various wastewater streams are separated at the user interface (domestic, industrial, agricultural, medical) because this facilitates the use of specific treatment methods and, in the end, produces uniform end products.
- New cities' planning should include smart water management design and an integrated management of its waterways. In this context, a sewerage system that could separate wastewater from households and greywater and industrial water could already be a first step towards a more circular use of urban and peri urban resources.
- Wastewater monitoring should start from its source of production, collection, and treatment to the point of discharge or reuse. It should investigate the whole system in terms of quantity and quality, including source segregation to understand the additional load (volume and pollutants); it is also important to monitor the progress made in water reuse based on the objectives established in the circular economy strategies adopted by governments.
- Successful implementation of sustainable solutions for municipal wastewater reuse requires increasing awareness on its safe use and inclusiveness in the design of oriented solution options. Participatory approaches are thus a fundamental path to be followed when integrated management options at different scales are intended to be applied (local, regional, state level).

References

Adhikari S (2019) Top 10 epidemic diseases that were common in the Ancient World. Ancient History Lists. Retrieved January 1, 2021 from https://www.ancienthistorylists.com/ancient-civilizations/top-10-epidemic-diseases-that-were-common-in-ancient-world/

Aghalari Z, Dahms H, Sillanpää M, Sosa-Hernandez JE, Parra-Saldívar R (2020) Effectiveness of wastewater treatment systems in removing microbial agents: a systematic review. J Globaliz Health 16:13. https://doi.org/10.1186/s12992-020-0546-y

Almuktar SAAAN, Abed SN, Scholz M (2018) Wetlands for wastewater treatment and subsequent recycling of treated effluent: a review. Environ Sci Pollut Res 25(24):23595–23623. https://doi.org/10.1007/s11356-018-2629-3

Andersson K, Rosemarin A, Lamizana B, Kvarnström E, McConville J, Seidu R, Dickin S, Trimmer C (2020) Sanitation, wastewater management and sustainability: from waste disposal to resource recovery. 2nd edition. Nairobi and Stockholm: United Nations Environment Programme and Stockholm Environment Institute

Angelakis AN, Asano T, Bahri A, Jimenez BE, Tchobanoglous G (2018) Water reuse: from ancient to modern times and the future. J Front Environ Sci 6:1–17. https://doi.org/10.3389/fenvs.2018.00026

Antoniou GP (2010) Ancient greek lavatories. Operation with reused water. In Mays LW (eds), Ancient water technologies (pp 67–86). Dordrecht: Springer. https://doi.org/10.1007/978-90-481-8632-7_4

Avellán T, Roidt M, Emmer A, Koerber JV, Schneider P, Raber W (2017) Making the Water–soil–waste nexus work: framing the boundaries of resource flows. J Multidiscip Digital Publish Inst Sustain 9(10):1–17. https://doi.org/10.3390/su9101881

Becerri JE, Jiménez B (2007) Potable water and sanitation in Tenochtitlan: Aztec culture. IWA Publishing, Water Supply 7(1), 147–154. https://doi.org/10.2166/ws.2007.017

Bogler A, Packman A, Furman A, Gross A, Kushmaro A, Ronen A,… Bar-Zeev E (2020) Rethinking wastewater risks and monitoring in light of the COVID-19 pandemic. J Nat Sustain 3:981–990. https://doi.org/10.1038/s41893-020-00605-2

Boretti A, Rosa L (2019) Reassessing the projections of the world water development report. J Npj Clean Water 2(1):1–6. https://doi.org/10.1038/s41545-019-0039-9

Brunet M (2006). La gestion de l'eau en milieu urbain et rural à Délos dans l'Antiquité. Paper presented at Vers une gestion intégrée de l'eau dans l'empire romain, Canada. https://doi.org/10.3406/jds.1999.1619

Capodaglio AG (2020) Taking the water out of "wastewater": an ineluctable oxymoron for urban water cycle sustainability. J Water Environ Res 92(12):2030–2040. https://doi.org/10.1002/wer.1373

Capodaglio A, Olsson G (2020) energy issues in sustainable urban wastewater management: demand reduction and recovery in the urban water cycle. J Multidiscip Digital Publish Inst Sustain 12(1):266. https://doi.org/10.3390/su12010266

Caucci S, Hettiarachchi H (Ed) (2017) The nexus approach and safe use of wastewater in agriculture: a workshop on policy and implementation for Tunisia. (Presented at Dresden, Tunisia, United Nations University Institute for Integrated Management of Material Fluxes and of Resources (UNU-FLORES)

Caucci S, Karkman A, Cacace D, Rybicki M, Timpel P, Voolaid V, Gurke R,… Berendonk TU (2016) Seasonality of antibiotic prescriptions for outpatients and resistance genes in sewers and wastewater treatment plant outflow. J FEMS Microbial Ecology 95(5). https://doi.org/10.1093/femsec/fiw060

Chhipi-Shrestha G, Hewage G, Sadiq R (2017) Fit-for-purpose wastewater treatment: conceptualization to development of decision support tool (I). J Sci Total Environ 607–608, 600–612. https://doi.org/10.1016/j.scitotenv.2017.06.269

Cossio C, Norrman J, McConville J, Mercado A, Rauch S (2020) Indicators for sustainability assessment of small-scale wastewater treatment plants in low and lower-middle income countries. Environ Sustain Indicators 6:100028. https://doi.org/10.1016/j.indic.2020.100028

Crouch DP (1993) Water management in ancient Greek cities. New York: Oxford University Press

D'Este P, Iammarino S, Savona M, von Tunzelmann N (2012) What hampers innovation? Revealed barriers versus deterring barriers. Res Policy 41(2):482–488. https://doi.org/10.1016/j.respol.2011.09.008

Daughton C (2020) The international imperative to rapidly and inexpensively monitor community-wide Covid-19 infection status and trends. J Sci Total Environ 726:138–149. https://doi.org/10.1016/j.scitotenv.2020.138149

De Feo G, De Gisi S (2013) Water and wastewater management in the treatment process of a Roman *fullonica*. J Water Supply 13(3):599–605. https://doi.org/10.2166/ws.2013.003

Drechsel P, Mahjoub O, Keraita B (2015) Social and cultural dimensions in wastewater use. In Drechsel P, Qadir M, Wichelns D (Eds), wastewater: economic asset in an urbanizing world. geographical focus: low- and middle-income countries (pp 75–92). Dordrecht: Springer. https://doi.org/10.1007/978-94-017-9545-6

Earla P (2014) Ancient diseases-microbial impact. J Infectious Diseases Preventive Med 2(1):1–4. https://doi.org/10.4172/2329-8731.r1-001

FAO (Food and Agriculture Organization of the United Nations). (2020). AQUASTAT-FAO's Global Information System on Water and Agriculture. Available at: http://www.fao.org/nr/water/aquastat/data

Goel RK, Flora JRV, Chen JP (2005) Flow equalization and neutralization. In Wang LK, Hung YT, Shammas NK (Eds), Physicochemical treatment processes (pp 21–45). Totowa, NJ: Humana Press.https://doi.org/10.1385/1-59259-820-x:021

Guy H, Mili V (2016) The costs of meeting the 2030 sustainable development goal targets on drinking water, Sanitation, and Hygiene. J Open Knowl Repository- World Bank. Retrieved from https://openknowledge.worldbank.org/handle/10986/23681

Hidenori H, Strande L, Fujii S (2016) Challenges and opportunities of faecal sludge management for global sanitation. In Katsumi T, Hashimoto S (Eds), Towards future earth: challenges and progress of global environmental studies, (pp 81–100). Environmental Science

Hueske AK, Endrikat J, Guenther E (2015) External environment, the innovating organization, and its individuals: a multilevel model for identifying innovation barriers accounting for social uncertainties. J Eng Technol Manag 35:45–70. https://doi.org/10.1016/j.jengtecman.2014.10.001

Huibers F, Redwood M, Raschid-Sally L (2010) Challenging conventional approaches to managing wastewater use in agriculture. In Drechsel P, Scott CA, Raschid-Sally L, Redwood M, Bahri A (Eds), Wastewater irrigation and health: Assessing and mitigating risk in low-income countries (pp. 287–302). London UK: International Development Research Centre (IDRC) & Colombo Sri Lanka: International Water Management Institute (IWMI)

IWA (International Water Association). (2020). Urban Sanitation and Wastewater. Available at: https://iwa-network.org/projects/urban-sanitation-and-wastewater/

Joonaki E, Hassanpouryouzband A, Heldt CL, Areo O (2020) Surface chemistry can unlock drivers of surface stability of SARS-CoV-2 in a variety of environmental conditions. Chem 6 (9):2135–46. https://doi.org/10.1016/j.chempr.2020.08.001

Kehrein P, Loosdrecht MV, Osseweijer P, Garfí M, Dewulf J, Posada J (2020) A critical review of resource recovery from municipal wastewater treatment plants—market supply potentials, technologies, and bottlenecks. J Environ Sci: Water Res Technol 6(4):877–910. https://doi.org/10.1039/C9EW00905A

Khouri N, Kalbermatten JM, Bartone CR (1994) Reuse of wastewater in agriculture: a Guide for planners. Washington, DC: The International Bank for Reconstruction and Development/The World Bank. Available at: https://www.academia.edu/31700393/Reuse_of_Wastewater_in_Agriculture_A_Guide_for_Planners

Kiparsky M, Thompson Jr BH, Binz C, Sedalk DL, Tummers L, Truffer B (2016) Barriers to Innovation in Urban Wastewater Utilities: Attitudes of Managers in California. J Environ Manag 57:1204–1216. https://doi.org/10.1007/s00267-016-0685-3

Korostynska O, Mason A, Al-Shamma'a AI (2013) monitoring pollutants in wastewater: traditional lab based versus modern real-time approaches. In Mukhopadhyay SC, Mason A (Eds), Smart sensors for real-time water quality monitoring, 4, 1–24. https://doi.org/10.1007/978-3-642-370 06-9_1

Lofrano G, Brown J (2010) Wastewater management through the ages: a history of mankind. J Sci Total Environ 408(22):5254–5264. https://doi.org/10.1016/j.scitotenv.2010.07.062

Mahjoub O, Chaibi T (2014) Chapter 13. The sanitary system in ancient Roman civilization: an insight on Tunisia. In Angelakis AN, Rose JB (Eds), Evolution of sanitation and wastewater technologies through the centuries 13, (pp 271–302). London: IWA Publishing. https://doi.org/10.2166/9781780404851

Mahmood Q, Pervez A, Zeb BS, Zaffar H, Yaqoob H, Waseem M, ... Afsheen S (2013) Natural treatment systems as sustainable ecotechnologies for the developing countries. BioMed Research International, 2013. https://doi.org/10.1155/2013/796373

Malissard A (2002) Les Romains et l'eau : fontaines, salles de bain, thèmes, égouts, aqueducs (Romans and water: fountains, baths, thermes, sewers, aqueducts). Realia-Les belles lettres

Markman DW, Antolin MF, Bowen RA, Wheat WH, Woods M, Gonzalez-Juarrero M, Jackson M (2018) Yersinia pestis survival and replication in potential ameba reservoir, Emerging Infectious Diseases, 24(2): 294–302. https://doi.org/10.3201/eid2402.171065

Mays LW (Eds) (2010). A Brief history of water technology during antiquity before the Romans. Ancient Water Technologies, (pp 1–28). Dordrecht: Springer. https://doi.org/10.1007/978-90-481-8632-7_1

Metcalf TG, Melnick JL, Estes MK (1995) Environmental virology: from detection of virus in sewage and water by isolation to identification by molecular biology–a trip of over 50 years. Annual Rev Microbiol (49):461–487. https://doi.org/10.1146/annurev.mi.49.100195.002333

Muralikrishna IV, Manickam V (2017) Wastewater treatment technologies. In Environmental management (pp 249–293). Elsevier. https://doi.org/10.1016/B978-0-12-811989-1.00012-9

Negreanu Y, Pasternak Z, Jurkevitch E, Cytryn E (2012) Impact of treated wastewater irrigation on antibiotic resistance in agricultural soils. J Environ Sci Technol 46(9):4800–4808. https://doi.org/10.1021/es204665b

Perraton SC, Blackwell BD, Fischer A, Gaston TF, Meyers GD (2014) Systemic barriers to wastewater reuse in Australia: some jurisdictional examples. Australasian J Environ Manag 22(3):355–372. https://doi.org/10.1080/14486563.2014.955889

Rizzo L, Gernjak W, Krzeminski P, Malato S, McArdell CS, Perez JA, Schaar H, Fatta-Kassinos D (2020) Best available technologies and treatment trains to address current challenges in urban wastewater reuse for irrigation of crops in EU countries. Sci Total Environ 710. 136312. https://doi.org/10.1016/j.scitotenv.2019.136312

Roberts MC (2017) One health. the role wastewater treatment plants as reservoirs, amplifiers, and transmitters of antibiotic resistance genes and antibiotic resistant bacteria. In Keen PL, Fugère R (Eds), Antimicrobial resistance in wastewater treatment processes 35–53. Wiley. https://doi.org/10.1002/9781119192428.ch3

Rodriguez DJ, Serrano HA, Delgado A, Nolasco D, Saltiel G (2020) From Waste to resource: shifting paradigms for smarter wastewater interventions in Latin America and the Caribbean. J Open Knowl Repository- World Bank. Retrieved from https://openknowledge.worldbank.org/handle/10986/33436

Schapira M, Manor O, Golan N, Kalo D, Mordehay V, Kirshenbaum N, Goldsmith R,... Paltiel O (2020) Involuntary human exposure to carbamazepine: a cross-sectional study of correlates across the lifespan and dietary spectrum. J Environ Int 143:105951. https://doi.org/10.1016/j.envint.2020.105951

Tamburrino A (2010) Water technology in ancient Mesopotamia. In Mays LW (Eds), Ancient water technologies (pp 29–51). Dordrecht: Springer. https://doi.org/10.1007/978-90-481-8632-7_2

Tempelhoff J, Hoag H, Ertsen M (2009) Water history and the modern. J Water History 1:81–82. https://doi.org/10.1007/s12685-009-0014-3

Thebo AL, Drechsel P, Lambin EF, Nelson KL (2017) A global, spacially-explicit assesment of irrigated croplands influenced by urban wastewater flows. J Environ Res Lett. https://doi.org/10.1088/1748-9326/aa75d1

Tzanakakis VE, Koo-Oshima S, Haddad M, Apostolidis N, Angelakis AN (2014) Chapter 24. The history of Land Application and Hydroponic Systems for Wastewater Treatment and Reuse. In Angelakis AN, Rose JB (Eds), Evolution of Sanitation and wastewater management through the centuries (pp 459–482). London: IWA Publishing

UNICEF (United Nations Children's Fund) and WHO (World Health Organization) (2019) Progress on household drinking water, sanitation, and hygiene 2000–2017: Special focus on inequalities. New York: UNICEF and WHO 138. Retrieved from https://www.who.int/water_sanitation_health/publications/jmp-report-2019/en/

UN-Water (2015) UN-Water Analytical Brief on Wastewater Management. UN-Water. Geneva, Switzerland: UN

UN-Water (2020) UN-Water Analytical Brief on Unconventional Water Resources. UN-Water. Geneva, Switzerland, UN

Ventura D, Consoli S, Barbagallo S, Marzo A, Vanella D, Licciardello F, Cirelli GL (2019) How to overcome barriers for wastewater agricultural reuse in sicily (Italy)? J Multidiscip Publishing Inst Water 11(2):335. https://doi.org/10.3390/w11020335

Voulvoulis N (2018) Water reuse from a circular economy perspective and potential risks from an unregulated approach. J Curr Opin Environ Sci Health 2:32–45. https://doi.org/10.1016/j.coesh.2018.01.005

Watts E (2020) What rome learned from the deadly Antonine plague of 165 A.D. Zócalo Public Square. Retrieved on DATE from https://www.smithsonianmag.com/history/what-rome-learned-deadly-antonine-plague-165-d-180974758/

Weber S, Khan S, Hollender J (2006) Human risk assessment of organic contaminants in reclaimed wastewater used for irrigation. J Desalination 187(1–3):53–64. https://doi.org/10.1016/j.desal.2005.04.067

WHO (World Health Organization) & UN-Habitat (United Nations Human Settlements Programme) (2018) Progress on Safe Treatment and Use of Wastewater: Piloting the monitoring methodology and initial findings for SDG indicator 6.3.1.40. Geneva: WHO and UN-Habitat. Retrieved from https://apps.who.int/iris/handle/10665/275967

WHO (World Health Organization) (2019) Critically important antimicrobials for human medicine 6th revision. J World Health Organ 6:45. Retrieved June 30, 2020, from https://www.who.int/foodsafety/publications/antimicrobials-sixth/en/

WRC (Water Research Commission). (2015). Wastewater Treatment Technologies—A Basic Guide. Retrieved from https://www.pseau.org/outils/ouvrages/wrc_wastewater_treatment_technologies_a_basic_guide_2016.pdf

WWAP (United Nations World Water Assessment Programme). (2017). The United Nations World Water Development Report 2017. Wastewater: The Untapped Resource. Paris, UNESCO

Zahari, C (2017). Comparative overview of primary sedimentation-based mechanical stage in some Romanian wastewater treatment systems. IOP conference series: materials science and engineering, 227, 012138. https://doi.org/10.1088/1757-899x/227/1/012138

Chapter 8
Agricultural Subsurface Drainage Water

J. D. Oster, Nigel W. T. Quinn, Aaron L. M. Daigh, and Elia Scudiero

Abstract Drainage waters generated by irrigation are a valuable, unconventional source of irrigation water and efforts to expand their reuse for irrigation are worthwhile, thereby partially mitigating the impacts of increased allocation of freshwater for municipal and industrial use. Because salinity levels in drainage waters are always higher than that of the initial irrigation water, the reuse of drainage water for subsequent irrigation requires more careful management than irrigation with nonsaline water. The first sections of this chapter deal with the basic principles of salinity management, the three irrigation strategies for using saline drainage water (blending, cyclic, and sequential reuse), the results of reuse studies, and farmer experience. The text then examines the utility of transient state models, such as HYDRUS, that simulate changes in soil salinity and crop yields caused by irrigation and rainfall, for designing alternative irrigation management strategies. Advanced methods to monitor soil salinity and crop yields at both the field and regional scales are discussed. The final sections deal with the benefits of managing drainage water reuse at a regional scale with farmers involved in planning regulations. Difficulties and barriers posed by the disposal of unusable drainage water that reuse of saline

J. D. Oster (✉)
Environmental Sciences Department, University of California, Riverside, CA 92521, USA
e-mail: james.oster@ucr.edu

N. W. T. Quinn
Climate & Ecosystem Sciences Division, Berkeley National Laboratory, Berkeley, CA 94720, USA
e-mail: nwquinn@lbl.gov

A. L. M. Daigh
Soil Science Department, North Dakota State University, Fargo, ND 58105, USA
e-mail: aaron.daigh@ndsu.edu

E. Scudiero
U.S. Salinity Laboratory, United States Department of Agriculture—Agricultural Research Service, Riverside, CA 92507, USA; Environmental Sciences Department, University of California, Riverside, CA 92521, USA
e-mail: elia.scudiero@ucr.edu

M. Qadir et al. (eds.), *Unconventional Water Resources*,
https://doi.org/10.1007/978-3-030-90146-2_8

drainage for irrigation can generate are assessed, and a new paradigm in developing regulations where all stakeholders are involved is described.

Keywords Salinity management · Leaching requirement · HYDRUS · Proximal monitoring · Normalized Difference Vegetation Index (NDVI) · Governance

8.1 Introduction

Irrigated agriculture provides nearly 40% of the global food supply but accounts for only 20% of the total cultivated land worldwide.[1] Increases in cropping intensity and in irrigated cropland—from 140 million hectares in 1960 to 275 million in 2020—have augmented yields and food production, which have met the needs of an ever-increasing population. In doing so, water used by irrigated agriculture accounts for 70% of freshwater usage (Tanji and Kielen 2002). Further expansion of irrigated agriculture based on freshwater is in jeopardy due to the increasing water needs of municipalities and industries and the high costs of developing new freshwater resources. Consequently, absent development of new water resources, the productivity of irrigated agriculture needs to increase with the existing water supply: The amount of food produced with the same amount of water needs to increase through conservation of freshwater sources and efficacious reuse of agricultural drainage water generated by irrigation. This unconventional water resource can help to reduce a gap in water demand and supply when reused as irrigation water for salt-tolerant crops and trees.

Irrigation practices invariably generate two types of drainage water: **surface runoff** and **deep percolation**, which can lead to subsurface drainage when deep percolation exceeds natural drainage processes. Surface runoff that occurs during irrigation of a field can be collected and returned to the source of water used to irrigate the field, returned to a nearby irrigation canal, or it can flow into nearby drainage ditches or local streams. For irrigation to be sustainable, water must be able to percolate down and through the root zone. If it does not, the soil will become waterlogged: Water tables and soil salinities will rise, which can reach levels toxic to plant growth (Hilgard 1886; Oster and Wichelns 2014). Waterlogged soils are a major contributor to the 20% of irrigated lands that are salt-affected (Ghassemi et al. 1995). Where waterlogging occurs, the common practice is to install artificial drainage systems, which can include deep drainage ditches spaced at distances of a km or more that surround fields and perforated plastic pipe installed at depths of 1.5–2 m at spacings of 40–100 m. This source of drainage water is commonly referred to as subsurface drainage, with salinity levels that are higher than that of the irrigation water applied originally. Why? Plant roots extract nearly pure water, leaving residual salts behind, concentrating the salt applied in irrigation water. Subsurface drainage will also contain native salts present in the soil.

[1] www.unesco.org/new/en/natural-sciences/environment/water/wwap/facts-and-figures/all-facts-wwdr3/fact-24-irrigated-land/. (Accessed 7/1/2020).

Reuse of subsurface drainage water can add substantial amounts to the irrigation water supply (Qadir et al. 2007; Minhas et al. 2020; Rhoades et al. 1992), thereby partially offsetting the need for freshwater. Reuse can increase economic benefits from additional crop production, and, at the same time, reduce disposal problems (Rhoades 1999; Tanji and Kielen 2002). For reuse to be successful, soil salinity levels—and boron if present—cannot accumulate to levels toxic to crop growth; soil physical conditions conducive to water infiltration must be maintained; and trace element accumulation in crops and forages must remain low enough not to threaten the health of humans or livestock (Grattan et al. 2014). After reuse is started, monitoring soil salinity in and below the root zone and tracking crop yields are crucial because this information is the key to irrigation sustainability.

8.2 Technological Interventions

Reuse and management of agricultural drainage water require the implementation of certain control practices to maintain salt balance throughout the crop root zone. In addition, appropriate crops and crop rotations must be selected; deterioration of soil physical conditions must be avoided; and certain trace elements, such as boron (B), selenium (Se), and molybdenum (Mo), must be controlled (Grattan et al. 2014). Selenium and Mo do not affect crop yields but can limit the ability to discharge drainage to rivers and streams. Accomplishing all of these goals may require changes and improvements to a field's current water and soil management, and, in some cases, adoption of advanced irrigation technologies. The following paragraphs introduce some of the basic design concepts[2] that require attention when reusing drainage waters for irrigation.

8.2.1 Design Basics of Drainage Water Reuse

Salinity Control

Leaching is the key to salinity control. Water from irrigation or rainfall must provide a greater volume of inflow than is needed for the combined outflows due to crop transpiration and evaporation at the soil surface, known as **evapotranspiration (ET)**. This greater volume must exist over the long term, and the excess water must pass through the rootzone. The fraction of excess water leached from the soil is known as the **leaching fraction (LF)**. This excess maintains the balance between the amount of salt supplied to the soil by irrigation water and the amount removed from the

[2] For readers interested in more detail about the basics of irrigation, we recommend Chap. 21, *Irrigation and Water-Use Efficiency* (pp. 407–425), in Daniel Hillel's book, *Introduction to Environmental Soil Physics* (2003, Academic Press, an imprint of Elsevier, ISBN (hardcover): 978-0123486554). Individual chapters, such as Chap. 21, can be downloaded on ScienceDirect.

rootzone in drainage water. If the excess is not sufficient, soil salinity levels will increase and can reach levels toxic to plant growth after several years of irrigation have elapsed. Leaching, a necessity for sustainable irrigation, maintains salt balance in the root zone and results in a discharge of drainage water that is invariably more saline than the applied irrigation water.

Sidebar 8.1 Definitions of Salinity

Management of salinity is always center stage when using drainage water for irrigation because it is more saline than freshwater. Salinity refers to the salts present, not only sodium (Na^+) and chloride (Cl^-), which combine to make table salt (NaCl), but also other ions, such as calcium (Ca^{2+}), magnesium (Mg^{2+}), potassium (K^+), nitrates (NO_3^-), sulfates (SO_4^{2-}), carbonates (HCO_3^-, CO_3^{2-}), and trace elements, such as boron (B) and selenium (Se), which together contribute to total salinity.

How is the salinity of water measured? By measurement of **electrical conductivity (EC)** expressed as **dS/m (decisiemens per meter)**. This is a simple method to quickly measure the amount of salt present in the water. Its use for estimating the soluble salts in water and soil extracts dates to the 19th century (Whitney and Means 1897). Its close relationship with the osmotic potential of waters—of various composition—extracted from soils (Fig. 6, U.S. Salinity Laboratory Staff 1954), is the reason for the long-term acceptance and use of EC to assess the water quality of irrigation waters (Osmotic potential of water and its impact on crop growth are discussed in Sidebar 8.2).

Soil salinity is often measured as the EC of the water extracted from a saturated soil paste (US Salinity Laboratory Staff 1954), and abbreviated as **ECe**. **ECsw** represents the salinity of soil water and is about twice the value of ECe. Finally, **ECiw** represents the EC of irrigation water.

For more information about salinity, see the University of California Division of Agriculture and Natural Resources (ANR) Salinity Management website online.[3]

Typically, where leaching occurs, salinity levels in the soil increase with depth, from values like that of the irrigation water at the soil surface, to levels that severely limit water adsorption by roots at the bottom of the root zone (Shalhevet 1994). Figure 8.1 illustrates three distributions of **soil water salinity, ECsw**, with depth that resulted from irrigation of alfalfa where the **electrical conductivity of the irrigation water (ECiw)** was 2 dS/m, and the LF ranged from 0 to 6.2% (van Schilfgaarde et al. 1974). The ECsw of about 35 dS/m, with a LF of 3.1% at a depth of about 100 cm, was about the maximum level of salinity above which the alfalfa roots were no longer able to absorb water. Reducing the LF to zero had only a small effect on ECsw, but

[3] https://ucanr.edu/sites/Salinity/Salinity_Management/.

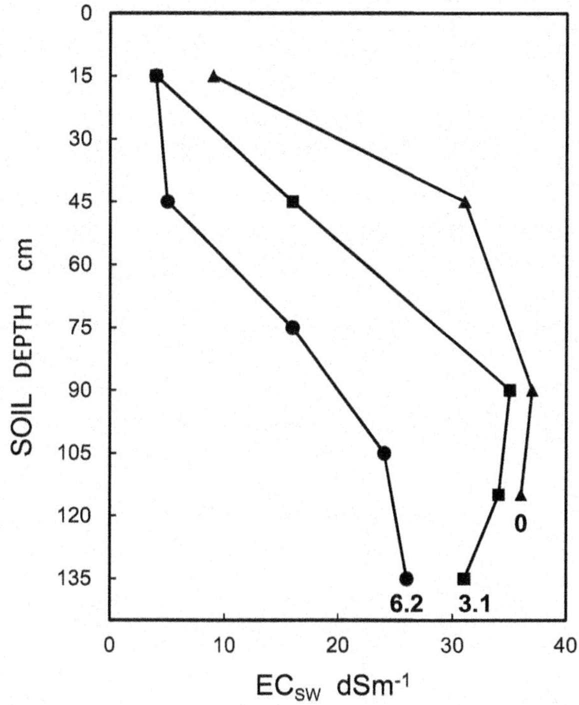

Fig. 8.1 Effects of reducing leaching fraction (LF) for irrigation water salinity of 2 dS/m on the buildup of soil water salinity (ECsw) in the root zone of alfalfa. Numbers below the curves (6.2, 3.1, and 0) indicate the LF in percent[4]

major impact on the salinity at shallow depth, resulting in a decrease in both rooting depth and crop yield (Bernstein and Francois 1973).

Crop Selection

There is a wide range of salt tolerance among common agronomic crops. For instance, salt-sensitive crops include beans, carrots, and muskmelon, where **average salinity** levels (**ECe**) in the root zone >1.0 dS/m can cause yield decline. In contrast, salt-tolerant crops include barley, wheat, cotton, and tall wheatgrass, where yield decline occurs at 6.0 dS/m or higher. As shown by Maas and Hoffman (1977), salinity levels greater than a **threshold salinity** (**ECt**) cause crop yields to decline linearly with increasing average rootzone salinity (Fig. 8.2). The index of soil salinity they chose was the linear-average electrical conductivity of water extracted from saturated soil pastes (ECe) of soil samples obtained within the root zone. Based on these assumptions, the Maas–Hoffman equation relating relative yield (Y) to ECe is

$$Y = 100 - (\mathrm{ECe} - \mathrm{ECt})\ \mathrm{S_L} \tag{8.1}$$

[4] Reprinted from Agricultural Water Management, Vol. 25, Joseph Shalhevet, Using water of marginal quality for crop production: major issues, pp. 233–269, 1994, with permission from Elsevier.

where S_L is the slope of the response function, % decline/(dS/m). Grieve et al. (2012) provide the most recent source of ECt and S_L values for a broad spectrum of crops.

Sidebar 8.2 How Crops Cope with Salinity

Crops differ in their tolerance to salinity. High salt levels can impede crop plants from absorbing water, leading to internal drought within a crop, even if adequate water seems present. Crop salt tolerance is measured by the average ECe of the soil surrounding the roots. Most crops can tolerate an average ECe of 1.4 dS/m, but only a few can tolerate an average ECe of 10 dS/m. The reason that salinity adversely affects plant growth stems from the energy plants must expend (Lauchli and Grattan 2012) to make organic solutes in root cells in order to exclude the salts in the soil during water uptake—the essential component of plant life. These organic solutes counteract the effects of salts, often referred to as the osmotic effect of salts on plants. Their synthesis proceeds at the expense of growth and crop yield. The differences in salt tolerance among crops stems from differences in capabilities to synthesize organic solutes to adjust to the osmotic or water-deficit effects of salts in the soil surrounding the roots. In general, crops are most sensitive to salinity during their early growth stages.

What crops can be grown? The list of crops in Table 8.1 was selected from the many listed in Grieve et al. (2012). Irrigation of sensitive and moderately sensitive crops would not be recommended where ECiw is >2 dS/m unless a considerable

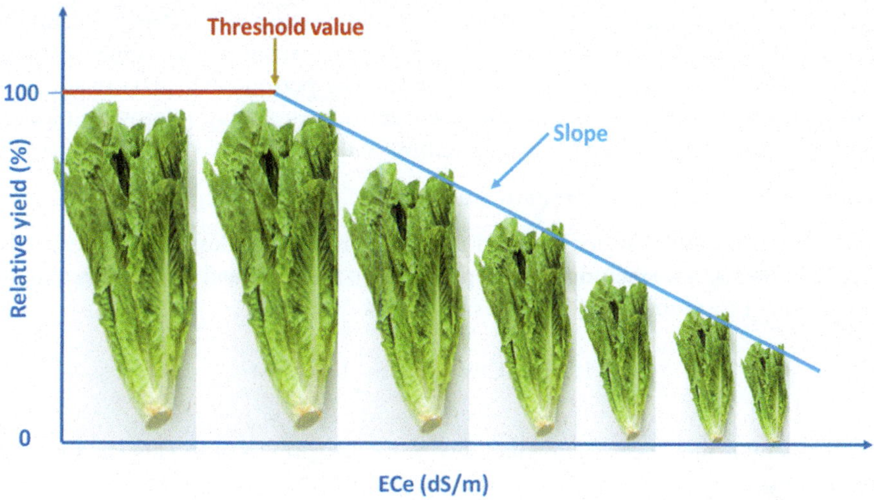

Fig. 8.2 The effect of salinity on romaine lettuce yield. The slope of the line (S_L, Eq. 8.1) shows how growth declines linearly once the salinity in the root zone exceeds the threshold salinity (ECt, Eq. 8.1) (*Source* Dr. Laosheng Wu, Department of Environmental Sciences, University of California, Riverside. Copyright © 2021 Regents of the University of California. Used by permission)

Table 8.1 List of crops classified by tolerance to salinity (Data from Grieve et al. (2012) except for pistachio nuts (Sanden et al. 2004)). ECt represents threshold salinity in dS/m

Sensitive ECt < 1.5	Moderately sensitive 1.5 < ECt < 3	Moderately tolerant 3 < ECt < 6	Tolerant 6 < ECt < 10
Rice	Corn	Sorghum	Barley
Sesame	Peanut	Soybean	Canola
Common bean	Sugarcane	Sunflower	Cotton
Cabbage	Alfalfa	Sweat clover	Kenaf
Carrot	Clover	Safflower	Oats
Onion	Cabbage	Wheat	Sugar beet
Pigeon pea	Cauliflower	Tall fescue	Semidwarf wheat
Strawberry	Cucumber	Artichoke	Wheat, durum
	Lettuce	Lima bean	Salt grass, desert
	Muskmelon	Red beet	Bermudagrass
	Pepper	Broccoli	Tall wheatgrass
	Potato	Celery	Asparagus
	Tomato	Zucchini squash	Swiss chard
	Watermelon		Pistachio nuts

yield reduction were acceptable. Consequently, crops that are moderately tolerant or tolerant of salinity are the most suitable options for irrigation with saline drainage water where ECiw ranges from 3 to 10 dS/m. The performance classifications of crops shown in Table 8.1 are not fixed but have been slowly improving through genetic engineering, traditional crop breeding, and selection processes.

Salinity Control and Water Requirement

The amount of excess water required to maintain a salinity less than the ECt of a crop is known as the **leaching requirement (LR)**. A simple and often used equation to calculate LR, proposed by Rhoades (1974), is

$$LR = ECiw/(5ECt - ECiw) \tag{8.2}$$

where ECiw is the electrical conductivity of the irrigation water. This equation can be used to calculate the LR for different yield objectives, where the ECt of the crop is used to calculate the LR for 100% yields.

How do ECiw and desired yield impact the LR? For moderately sensitive varieties of alfalfa,--ECt of 2 dS/m and S_L of 7.3%/dS/m (Grieve et al. 2012)—the LR for yield objectives of 100 and 80% and four different levels of ECiw, ranging from 1 to 4 dS/m, are given in Table 8.2. Equation 8.1 was used to calculate the ECe for a desired yield of 80%: The resulting ECe was 4.7 dS/m which was then used as the value of ECt in Eq. (8.2) to calculate the corresponding LR values given in Table 8.2. The value of 2 dS/m and the ECt for alfalfa were used to calculate the LR for a yield of 100%. For a relative yield of 100% and an ECiw of 1 and 2 dS/m, the LR is 0.11 and 0.25, respectively, as compared to LR values of 0.43 and 0.67 for ECiw of 3 and 4 dS/m (Table 8.2). The corresponding LR values for a yield (Y) of 80% are much smaller.

Table 8.2 Effects of the electrical conductivity of the irrigation water (ECiw) and relative yield (Y) on leaching requirements (LR) for moderately sensitive varieties of alfalfa

Yield (%)	ECiw, dS/m			
	1	2	3	4
	Leaching Requirement (LR)			
100	0.11	0.25	0.43	0.67
80	0.04	0.09	0.15	0.21

The LR is also used to calculate the amount of **applied water (AW)** to achieve both leaching and adequate water for **crop evapotranspiration (ETc)**. The equation is

$$AW = ETc/(1 - LR) \tag{8.3}$$

where AW and ETc are expressed as a depth, usually in cm, and weather conditions during the crop season are a principal factor affecting AW. The protocol for estimating crop water requirements is beyond the scope of this chapter but covered in detail by Allen et al. (1998) and Minhas et al. (2020).

Whether a LR is achievable will depend on soil texture, the availability of water, and the method of irrigation. The rate of water infiltration into soil and though the crop root zone should be high enough to satisfy crop requirements and supply necessary leaching requirements without waterlogging. When rates of application exceed infiltration rates, runoff occurs at the soil surface. For clay soils that are surface irrigated, a LR > 0.15 likely is not achievable, based on measured LF of cropped fields with such soils in the Imperial Valley (Oster et al. 1986) and on the west side of the San Joaquin Valley (Singh et al. 2020) in California, USA. For sandy soils, a LR of 0.4 probably could be achieved, but the frequency of irrigation would need to be increased because of the soil's low water-holding capacity, but the amount of water required may not be available. The best irrigation methods for controlling the LR are sprinkler and drip irrigation and are especially preferred for a LR < 0.10.

In summary: leaching is the key to irrigation sustainability, and LR is closely linked to the irrigation water salinity, crop salt tolerance, and the amount of water required for irrigation.

Nomogram To Determine Suitability of a Saline Water for Irrigation

Crop tolerance to salinity, LR, and ECiw are the three key factors in determining suitability of an irrigation water. The **nomogram** shown in Fig. 8.3 takes these factors into account and provides a useful resource for a preliminary and quick assessment of LR for a given ECiw and crop. The ordinate in Fig. 8.3 is the linear average ECe at the quarter boundaries of the root zone, assuming water uptake occurs in proportions of 0.4, 0.3, 0.2, and 0.1 in the first, second, third, and fourth quarters of the rootzone, respectively (Rhoades and Merrill 1976; Suarez 2012). To estimate the LR for a crop from the ECiw, if the target yield is 100%, the ECe used would be the crop's ECt. For alfalfa with an ECt of 2 dS/m and an irrigation water with an ECiw of 2 dS/m, the estimated LR is somewhat >0.2—circle labelled 1 in Fig. 8.3. This result is consistent with the LR of 0.25 for alfalfa (Table 8.2) for an ECiw of 2 dS/m calculated using Eq. (8.2). A second example: For irrigation of 'Jose' tall wheatgrass (ECt = 7.5 dS/m) with irrigation water having an ECiw of 4 dS/m, the estimated LR is about 0.1—the circle labelled 2 in Fig. 8.3. Finally, if lower yields (Y < 100%) would be acceptable, the ECe to use would be calculated using Eq. (8.1).

The range of LR in Fig. 8.3, 0.05–0.20, was chosen purposefully, based on its achievability imposed by soil texture and irrigation management. Although a LR > 0.20 can be achieved for a broad range of soil textures, except perhaps for clay soils, excessive irrigation applications can lead to waterlogging and regional, shallow groundwater problems or drainage disposal issues if the excess is intercepted by tile drainage. A LR < 0.05 would be difficult to achieve, even with pressurized irrigation systems, sprinkler, and drip.

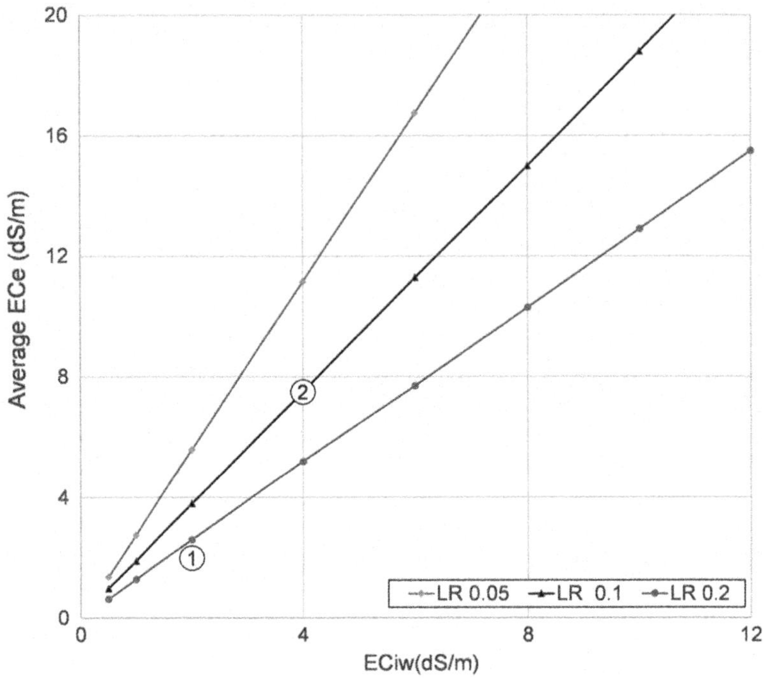

Fig. 8.3 Nomogram. Leaching requirement, LR, as a function of average ECe in the root zone and the electrical conductivity of the irrigation water, ECiw. ECe is the linear average of the calculated ECe in the quarter boundaries of the rootzone, assuming water uptake occurs in proportions of 0.4, 0.3, 0.2, and 0.1 in the first, second, third and fourth quarters of the root zone, respectively

8.2.2 Irrigation Strategies for Reuse of Saline Drainage Water

There are several irrigation strategies that can be applied, depending on where, when, and how saline drainage water is applied. This section will provide a short description of each. Available sources for more detailed information include Ayars and Basinal (2005),[5] Grattan et al. (2014); and Tanji and Kielen (2002).

Blending Water Supplies

A common way of improving the quality of saline water for irrigation is to blend it with water of lower salinity, producing an irrigation water of suitable quality, while, at the same time, expanding the overall water supply volume. Blending does not unconditionally increase the usable water supply (Rhoades et al. 1992). Blending is not an attractive alternative if the saline water does not make up at least 25% of the total irrigation water requirement (Grattan and Oster 2003). The risks of potential

[5] http://www.californiawater.org/californiawater/a-technical-advisors-manual-managing-agricultural-irrigation-drainage-water-a-guide-for-developing-integrated-on-farm-drainage-management-systems/.

crop loss associated with a more saline irrigation supply would likely outweigh the benefits from a modest increase in the available water supply.

Cyclic Use of Saline and Non-saline Water

With a cyclic strategy, the soil salinity is first reduced purposefully by irrigation with nonsaline water, which facilitates germination and permits crops with lower tolerances to salinity to be included in the rotation (Rhoades et al. 1992). Irrigation with saline water then occurs after crops reach a salt-tolerant growth stage. A major disadvantage of this strategy is the storage requirement of the saline drainage water when it cannot be used for irrigation.

Sequential Reuse

The sequential reuse strategy (Fig. 8.4) involves irrigation of salt-sensitive crops with nonsaline water and using the resulting drainage water to irrigate more salt-tolerant crops, including halophytes. The salts contained in the drainage water generated by irrigation of halophytes are harvested in a solar evaporator. A tile drainage system to collect all the drainage water is the key to this strategy.

This reuse system may be employed at the farm or regional scale. Unlike the cyclic strategy, a separate land area is dedicated to either salt-sensitive or salt-tolerant crops (Fig. 8.4). The four main purposes of sequential reuse are the following: 1) Reducing

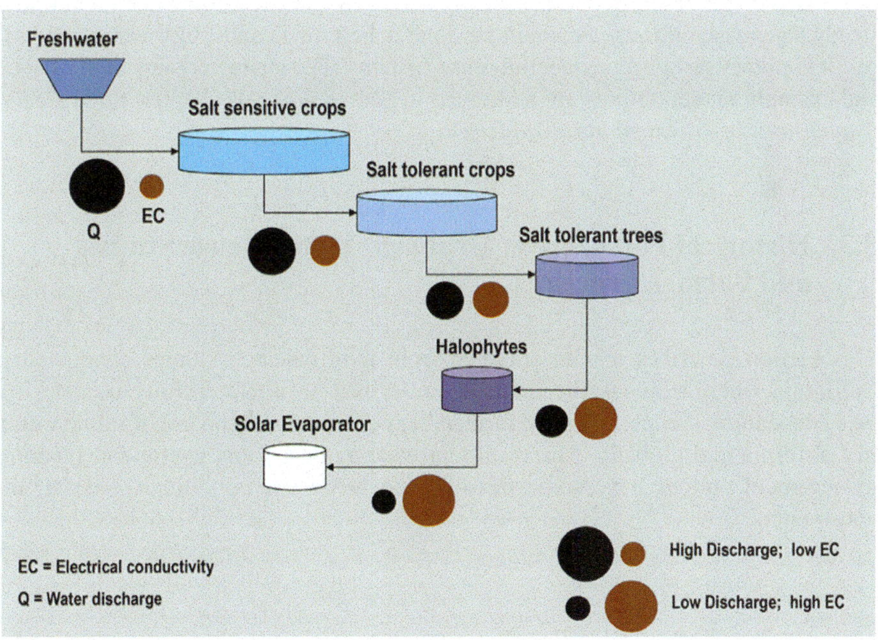

Fig. 8.4 Schematic representation of a sequential drainage-water reuse system (*Source* UN-Water 2020)

the soil salinity in fields irrigated with non-saline water, thereby increasing the area planted to high-value salt-sensitive crops; 2) Obtaining an economic benefit by using drainage water for crop production; 3) Reducing the volume of drainage water that requires disposal; and 4) Harvesting the salt.

On the farm scale, sequential reuse requires a tile drainage system in all the fields to collect the drainage water. On a regional scale, subsurface tile drainage water or drainage intercepted by deep trenches can be conveyed to a reuse area. Some subsurface drainage water is pumped from tile drainage system sumps located at the bottom corner of fields and conveyed into drainage canals. The collected drainage water can be used sequentially in areas dedicated to crops with increasing levels of salt tolerance until the final drainage water is conveyed to a solar evaporator for final disposal (Fig. 8.4).

Although sequential reuse is conceptually attractive, it could take decades or even longer for salts applied at the beginning of the reuse sequence to reach the final stage (Jury et al. 2003), and to establish quasi-steady-state conditions. Travel times will also be affected by regional groundwater extraction and excessive leaching, particularly where the water table depth is periodically below the tile lines. Therefore, caution is advised for those designing sequential reuse systems and estimating the rate of salt movement through the system. During the initial years, the crops and areas irrigated with drainage water will change with time, and the need to grow halophytes in a third stage will take longer. Consequently, the use of transient state salinity models, such as HYDRUS-1D or HYDRUS 2D/3D, to design the system is recommended. Use of steady state assumptions, as would be done if Eqs. (8.1) and (8.2) were used, will result in poorly designed sequential reuse systems. The characteristics of HYDRUS and example model outputs are addressed in Sect. 8.4 of this chapter, titled *Design Capabilities of Transient State Models.*

8.3 Historical Perspective: Drainage Water Reuse Studies and Farmer Practices

This section describes results obtained from field research studies conducted in California where saline–sodic drainage water has been successfully used to irrigate conventional crops, and from farmers' experiences with the use of saline waters in California and globally. The results provide evidence that cyclic and blending management options have some flexibility to fit diverse conditions, criteria, and constraints.

8.3.1 Research Studies Conducted in the San Joaquin Valley of California

Rhoades and coworkers did the first field studies to test the use of saline-sodic drainage water for irrigation in California. Two field research studies were performed: one in San Joaquin Valley (SJV) (Rhoades 1984) and the second in Imperial Valley (Rhoades et al. 1989). In the SJV study, cotton was irrigated with nonsaline water (EC = 0.5 dS/m) during germination and seedling establishment, and thereafter with saline–sodic groundwater (EC = 7.9 dS/m, SAR 11) pumped from wells underlying the cropped area (see Sidebar 8.3 for definition of SAR). Wheat was subsequently irrigated with the nonsaline water, followed by two years of sugar beets with the cyclic strategy used again for irrigation. In the Imperial Valley, Colorado River water (EC = 1.5 dS/m, SAR 4.9) was used to irrigate muskmelon, a moderately salt-sensitive crop (Table 8.1), and for the preplant and early irrigations of wheat and sugar beets. Alamo River water, into which subsurface drainage water from irrigated fields in the Imperial Valley is discharged (EC = 4.6 dS/m, SAR = 9.9), was used for all other irrigations. Sugar beet and wheat yields were not reduced, and irrigation with drainage water often improved crop quality.

Field studies became more numerous following the selenium (Se) crisis at the Kesterson Reservoir in the 1980s, caused by evapoconcentration of Se-laden subsurface drainage water disposed of in the Reservoir (Letey et al. 1986; NRC 1989). Ayars et al. (1990, 1993) used drip irrigation to irrigate cotton, wheat, and sugar beets for three years with saline–sodic drainage water (EC = 7–8 dS/m, SAR = 9), which also contained 5–7 mg/L of B. Cotton was established first with nonsaline water (EC = 0.4–0.5 dS/m) and then irrigated with saline water that supplied 50–59% of the irrigation water requirement. Wheat irrigated with nonsaline water followed cotton, and sugar beets followed wheat, which were irrigated with saline water after stand establishment. Yields under these conditions were the same as from continuous irrigation with good-quality water. Researchers noted a gradual increase in soil B concentrations over time.

Shennan et al. (1995) tested cyclic strategies on processing tomato in rotation with cotton. The consequences in terms of salinity and B (boron) levels in the root zone after three years are shown in Fig. 8.5. Saline drainage water (EC = 7.4 dS/m, SAR = 12, B = 5–7 mg/L) was applied to tomato after first flower to take advantage of salinity's enhancement of fruit quality. For cotton, saline drainage water was applied after seedling establishment and thinning. Nonsaline water (EC = 0.4 dS/m, SAR = 1.6) was used at other times. There were four combinations of cyclic strategies: SSF, FSS, SFF, and FSF where F represents irrigations with only nonsaline ("fresh") water and S represents cyclic use of nonsaline and saline water. Only nonsaline water was applied to one set of plots, which provided a baseline for crop yields and salinity and B levels in the root zone. Where the cyclic strategy was used once in three years, yields of tomatoes were sustained; however, where the cyclic strategy was used twice in three years, tomato yields were reduced.

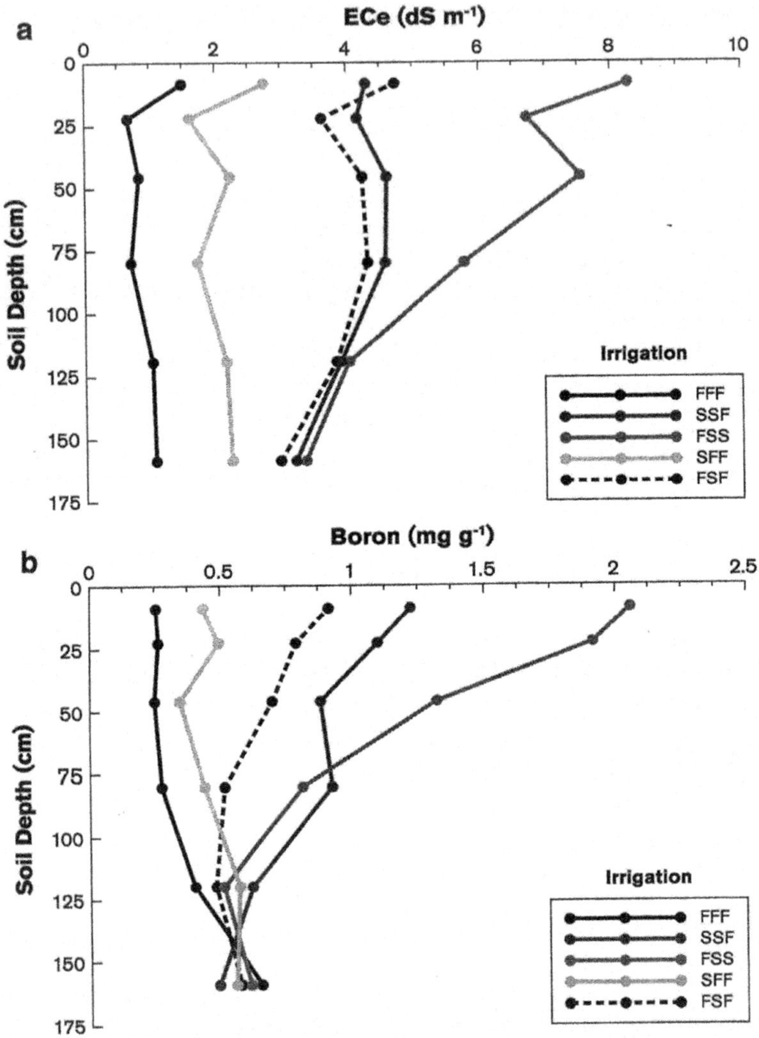

Fig. 8.5 Electrical conductivity, ECe, and boron (B) concentration of soil saturated extracts in a three-year cyclic reuse rotation, where FFF represents irrigation with only nonsaline, freshwater (EC = 0.4 dS/m) and SSF, FSS, SFF, and FSF represent cyclic irrigation with nonsaline and saline water (EC = 7.4 dS/m) (*Source* Grattan et al. 2014, adapted from data in Shennan et al. 1995)

On a relative basis, salts (Fig. 8.5a) were more readily leached than B (Fig. 8.5b). The levels of ECe and B were related both to the fraction of drainage water applied as well as the timing of applications. For example, the FSS sequence had nearly twice the ECe as the SSF sequence. A large reduction in ECe after applying nonsaline, fresh water was also evident in the FSF and SFF sequences (Fig. 8.5a). Similar patterns were observed for B concentration, although most of the changes occurred only in the top 100 cm of the profile, compared to changes in ECe in the entire 160-cm profile. The basic reason for this difference is that B is adsorbed on clay particles; whereas, salts are not adsorbed and remain fully mobile, moving with the soil water. Finally, for the SSF, FSS, and FSF sequences, the ECe levels in the 125–160 depth are similar, but this was not the case for shallower depths, reflecting the damping effects of depth on water movement in the root zone. The rate of downward movement of water decreases with depth because of crop water uptake. Consequently, relative changes in ECe in the root zone occur faster at shallower depths.

Nitrate levels (1.1 mmol/L) in the drainage water used in long-term reuse plots, set up by Shennan et al. (1995) and used by others (Kaffka et al. 1999; Bassil and Kaffka 2002), had significant impacts on crop quality and yields. Shennan et al. (1995) found the high nitrate concentration in saline water prolonged reproductive growth of tomato: The percentage of ‘green’ fruit was double that of the nonsaline control plots at the time of harvest. For sugar beets, use of saline water did not affect root mass, but, because of its high N level, sugar yields were reduced (Kaffka et al. 1999). Nitrogen is often a growth-limiting nutrient so N-fertilization could be adjusted downwards to take advantage of nitrogen applied in the drainage water.

Sidebar 8.3 Salinity and Sodicity Effects on Soil Permeability: How Water Enters and Flows Through Soil

Soil permeability declines when soil clays swell and become dispersed. The phenomenon underlying clay swelling involves the roles of cations adsorbed onto the clay particles and those present as ions in the soil water surrounding them that can prevent their swelling and dispersion (Sposito 2008), a complicated phenomenon beyond the scope of this chapter.

Adsorbed sodium is the primary culprit, but, if salinity is low enough, swelling and dispersion can occur if Ca, Mg, or K are the dominant adsorbed cations (Quirk and Schofield 1955). So, soil permeability decreases with increasing sodicity and increases with increasing salinity, but permeability can be maintained by an optimal combination of salinity and sodicity.

Infiltration rates are more strongly affected than water flow through the soil by a suboptimal combination of salinity and sodicity (Oster and Shainberg 2001; Suarez et al. 2006). Dispersed clay particles at the soil surface can cause hard soil crusts to develop when the soil is dry (Shainberg and Singer 2012; Oster and Jayawardane 1998), and the movement of dispersed clay particles into the soil can plug the pores through which water and air otherwise move (Minhas et al. 2019).

There are two indicators of soil sodicity: the traditional **sodium adsorption ratio (SAR**, Eq. 8.4) **and** the **Cation Ratio of Soil Structural Stability (CROSS**, Eq. 8.5), introduced by Rengasamy and Marchuk (2011),

$$\mathrm{SAR} = C_{\mathrm{Na}} \Big/ \left(\left(C_{\mathrm{ca}} + C_{\mathrm{Mg}} \right) / 2 \right)^{0.5} \tag{8.4}$$

$$CROSS = C_{Na} + \mathbf{a}C_{k}/((C_{ca} + \mathbf{b}C_{Mg})/2)^{0.5} \tag{8.5}$$

where C in both equations is the charge concentration in mmolc/L of the subscripted ion; **a** (Eq. 8.5) is a measure of the dispersing power of K relative to Na; and **b** (Eq. 8.5) is a measure of the aggregating power of Mg relative to Ca. The numerical coefficients posed by Rengasamy and Marchuk (2011) were 0.56 for **a** and 0.60 for **b**. Alternative coefficients were proposed by Smith et al. (2015): 0.335 for **a** and 0.0758 for **b**.

The likelihood that a given combination of salinity and sodicity will have adverse effects on infiltration rates varies greatly among soils (Shainberg and Letey 1984). Consequently, the guidelines in Table 8.3, proposed by Ayers and Westcot (1985) for EC and SAR, delineate when potential problems might occur. Oster et al. (2016) concluded the guidelines can also apply for CROSS, and Qadir et al. (2021) suggested this is the case, regardless of the numbers used for **a** and **b**. During irrigation using a saline drainage water with an ECiw of 2 dS/m, water infiltration problems are unlikely if SARiw (or CROSSiw) ranges from 6–12 $(mmolc/L)^{0.5}$ (Table 8.3). For an ECiw of 5 dS/m, the corresponding range is 20–40 $(mmolc/L)^{0.5}$. If rainfall were to reduce ECe to <0.3 dS/m, the adverse effects are likely where the SARiw (or CROSSiw) ranges from 0 to 3 $(mmolc/L)^{0.5}$.

Table 8.3 Interpretive guidelines for assessing the combined effect of SAR or CROSS in irrigation water on soil infiltration problems. Adapted from Ayers and Westcot (1985)

When SAR or CROSS of the irrigation water or soil water is	Potential water infiltration problems	
	Unlikely if ECe or ECiw is (dS/m)	Likely if ECe or ECiw is (dS/m)
0–3	>0.7	<0.3
3–6	>1.0	<0.4
6–12	>2.0	<0.5
12–20	>3.0	<1.0
20–40	>5.0	<2.0

In a nine-year study by Rains et al. (1987), use of saline drainage water (0.9–11.6 dS/m and 3–30 SAR) on a clay soil, cotton yields began to decline in the fourth year and, by the sixth year, yields declined in all salinity treatments where the salinity and SAR of applied water were >2.5 dS/m and 9 $(mmolc/L)^{0.5}$, respectively. The inability to prepare a seedbed with the tilth necessary for water transfer between the soil and cotton seed contributed to poor stand establishment (Oster 1994) and reduced yields. Application of gypsum to the soil surface in the fall, before the winter rainy season, likely would have prevented this problem. This assertion is supported by results in the western Negev region of Israel (Keren and Shainberg 1978) where cotton was irrigated with a saline–sodic groundwater (EC = 4.6 dS/m, SAR = 26), which prevented deterioration of soil physical properties during the summer because the EC of the irrigation water counteracted the harmful effects of exchangeable sodium; however, deterioration did occur during the rainy season in the winter. Annual application of 5 Mg/ha of phosphogypsum spread on the soil surface, following tillage, prevented the formation of surface seals and crusts and maintained sufficient infiltration of rain to leach salts from the root zone. Coupled with adequate irrigation with saline–sodic water to meet crop needs during the summer, cotton yields were similar to those obtained when only nonsaline water was used for irrigation. See Sidebar 8.3 for discussion of salinity and sodicity effects on soil permeability.

8.3.2 Tree Crops

Pistachio is a very salt-tolerant nut crop. In a nine-year study, Sanden et al. (2004) evaluated *Pistacia vera* 'Kenman' scions grafted on four rootstocks irrigated with 0.5–12 dS/m drainage waters and found no impact on yield with salinities up to 8 dS/m. There were no significant differences in the nut yields among salinity treatments, or among the rootstocks within salinity treatments. The salinity threshold for the tested rootstocks ranged from 9 to 10 dS/m.

A field study testing the effectiveness of eucalyptus (*Eucalyptus camaldulensis*, clones 4543, 4544, and 4573) was conducted in the Tulare Lake Basin, a hydrologically closed basin within the SJV of California (Oster et al. 1999). In 1994, trees were planted and irrigated with nonsaline water to facilitate survival of the young trees, and irrigation with saline–sodic water (EC 8–10 dS/m, SAR 25–30 (mmolc/L$^{0.5}$) began about a year later. The average ECe and SAR in the 0–60 cm depth from 1996 through 1998 was 15 dS/m and 36, respectively. Tree biomass was greatest in those plots that received gypsum applications each fall, which improved soil aeration, infiltration, and drainage during the winter when rain occurred. Substantial rainfall in 1998, between Julian day 31 and 125, resulted in ponding in all treatment plots. Oxygen-diffusion rates remained at 0 $\mu g\ O_2\ cm^{-2}\ min^{-1}$ in the untreated plots from Julian day 80 to 190; whereas, rates increased to 0.3 $\mu g\ O_2\ cm^{-2}\ min^{-1}$ in the gypsum-treated plot after Julian day 140 soon after ponding ended. Hence, gypsum application was shown to increase oxygen-diffusion rates substantially in winter months and to improve tree biomass production. Concurrent with this field study, the salt tolerance characteristics of *E. camaldulensis* clone 4544 was determined to be moderately salt tolerant with a threshold salinity of 6 dS/m (Shannon et al. 1998). Since the soil salinity in the Tulare Lake Study exceeded this threshold salinity, it is likely the lack of aeration compounded the adverse impacts of soil salinity on tree biomass.

8.3.3 Research Results with Salt-Tolerant Forages

Production systems based on salt-tolerant forage crops could provide a year-round supply of feed suitable for grazing and economic weight gains in cattle or sheep, or for sale to dairy farms as ensilage or hay. A 30-ha site in the western SJV was developed in 1999 to study the use of saline drainage and other waste waters (ECw averaged 3.6 dS/m with a range from 1.2 to 12 dS/m) for the production of bermudagrass grazed by cattle (Kaffka et al. 2004). The soil was Lethant clay loam, and, because of high salinity levels (ECe > 20 dS/m), the field had been abandoned. In 1999, the site was laser leveled; tile drains were installed along with instrumentation to monitor the amounts of applied water, subsurface drainage-water flows, and salinity. The initial salinity of the drainage water was >50 dS/m. The methods used to monitor soil salinity during this 10 year study (Corwin et al. 2008) and the changes in ECe that occurred in the 0–120 depth interval between 1999 and 2012 are presented in the Monitoring Methodology section of this chapter. The agronomic and cattle results are summarized here. Bermudagrass (*Cynodon dactylon*, vars. 'Common' and 'Giant') remained productive (1.5–2.5 Mg/ha DW, depending on cultivar) after five years of irrigation with saline drainage and other waste waters during which the average ECe in the 0–30 cm depth averaged about 13 dS/m. In places where ECe exceeded 20 dS/m, stands failed (Kaffka et al. 2004). Livestock studies conducted between 2001 and 2003 confirmed that beef cattle can successfully graze on bermudagrass as a sole source of feed during much of the year (Alonso et al. 2013). Weight gains of 0.7 kg/day were achieved once copper supplementation was administered to offset a deficiency due in part to high S and Mo in the drainage water.

In field studies conducted by Suyama et al. (2007a, b) in the SJV, 'Jose' tall wheatgrass emerged as a top candidate among the forages tested, due to its ability to maintain adequate dry matter yield (7.0 Mg/(ha-yr) and high forage quality (metabolizable energy of 9.3 MJ/kg DW), even when growing in soils having ECe = 19 dS/m, SAR = 37, and B = 24 mg/kg. The concentrations of Se in the soil and drainage water were high, and after multiple years of drainage-water irrigation, the forages accumulated 6–11 mg/kg Se, well above the maximum recommended levels of 2–5 mg/kg DW (NRC 2000). Such levels of Se could presumably cause toxicity in ruminants if used as a sole forage source, but the forage could be used as a Se supplement in Se-deficient areas of the SJV.

8.3.4 Research Results with Drainage Water on Halophytes

The performance of six halophyte species irrigated with saline drainage water (EC = 13.0 dS/m; SAR = 30) was evaluated in terms of water use, biomass production, and animal fodder in a six-year study conducted by Diaz et al. (2013). The species were: *Salicornia bigelovii, Atriplex lentiformis, Distichlis spicata, Spartina gracilis, Allenrolfea accidenialis*, and *Brassia hyssopifolia*. The study, conducted on a commercial farm located in Fresno County, California, had a sequential cropping system, operated from 1995 to 2010.The halophytic species were the terminal-stage crops prior to disposal of the final drainage water to a solar evaporator. The soils were highly saline–sodic (EC = 29 dS/m; SAR = 39) and with high B concentrations ranging from 17 to 70 mg/L. Average biomass ranged from 3.8 to 17.4 Mg dry matter per hectare. Under frequent irrigation, daily evapotranspiration rates ranged from 1 to 1.2 times higher than reference evapotranspiration. The metabolizable energy values were <7 MJ/(kg dry matter) for all halophyte species, the minimum acceptable level for most ruminant animals, and the total ash contents (salt) ranged from 6 to 52%. Consequently, long-term grazing would not be recommended, but halophyte species could be used as a fodder supplement. Benes et al. (2004, 2005) provide more information about plant selection and cultivation practices where saline–sodic drainage water is used to grow halophytic plants.

Others have studied, or are currently studying, halophytes as potential new crops (Boyko and Boyko 1959; Glenn et al. 1999). The International Center for Biosaline Agriculture (ICBA) was established in 1996 in Dubai in the United Arab Emirates to further develop possibilities for the use of saline waters for irrigation, including screening of the most suitable species and varieties, with particular emphasis on forage crops (ICBA 2005). Numerous other salt-tolerant crops, shrubs, and trees have potential for the production of food, fuel, fodder, and fiber when irrigated with saline or saline–sodic water (NRC 1990; Dagar 2018).

8.3.5 Farmer Experiences with Use of Saline Waters Worldwide

Farmers have successfully used waters classified as having moderate-to-severe restrictions for irrigating a broad spectrum of crops in Bahrain, Egypt, Ethiopia, India, Iraq, Israel, Pakistan, Somalia, Tunisia, the United Arab Emirates, and the United States (Ayers and Westcot 1985; Rhoades et al. 1992; Tanji and Kielen 2002). In this section, we will briefly summarize farmer experiences in the United States and India.

In the United States, saline drainage waters up to 8 dS/m have been used for irrigation in several areas of the Southwest, including the Arkansas River Valley of Colorado, the Salt River Valley of Arizona, and the Rio Grande and Pecos River Valleys of New Mexico and west Texas (Dutt et al. 1984; Erickson 1980; Miyamoto et al. 1984). Sustainable use is made possible by several cultural practices: alternate-furrow irrigation during crop establishment to move salts to the dry side of the bed; planting seeds on the edges of flat beds where salt accumulation is minimal, replanting after rainfall, if the resulting crusting limits seedling establishment, and single-row plantings on narrow beds followed by removal of the bed peaks prior to seedling emergence to prevent salt-crust damage to emerging seedlings. Where sprinkler irrigation was used in the Dell City area of Texas to irrigate alfalfa with saline water (Rhoades 1999), leaves frequently exhibited marginal leaf burn but with little impact on yields. Significant reductions in cotton lint yields occurred in west Texas, when cotton was sprinkle irrigated with water of 4 dS/m during the day. However, no significant yield reduction was observed when sprinkler irrigation was applied at night.

In India, canal water supplies are either uncertain or in short supply, so farmers have used saline groundwater, or drainage water, to meet crop water requirements (Minhas 1996; Choudhary et al. 2011). Minhas and Gupta (1992) concluded yields for various patterns of cyclic use were higher than those for blending canal and saline water, based on a large number of multi-locational trials conducted in India with various crops. They recommended that non-saline canal water should be used in the early stages of crop production and saline water should be used later, which is consistent with recommendations made by Rhoades (1999).

Monsoonal rain in India during July through September exceeds evapotranspiration, which induces leaching of the salts added by saline waters used to irrigate winter crops and also provides a source of stored nonsaline water for winter crops. If the growing season for post-monsoon winter crops starts with a surface-leached soil profile, Minhas and Gupta (1992) recommended underirrigation to maximize the use of stored water and to avoid its displacement to depths below the root zone.

Rainfall can have significant negative impacts. Drastic reductions in hydraulic conductivity were observed to occur after rainfall on soils previously irrigated with saline–sodic waters (Minhas et al. 2019). This reduction was not reversible when saline–sodic water was reapplied. Soil clays can become vulnerable to dispersion and movement in soils irrigated with saline–sodic water because electrolyte concentrations in the soil water, after irrigation with a nonsaline water, become too low to counteract the influence of exchangeable sodium on clay swelling and dispersion (Shainberg and Singer 2012). Generally, the sodification process is insidious and the build-up of exchangeable Na is initially gradual. Over the long term, this leads to the formation of a subsoil zone enriched with illuviated clays, which reduce water movement into and through the soil. Minhas et al. (2019) describe the various management and amendment options to mitigate the problem. Because groundwater in India is also characterized by bicarbonate as the predominant anion, mitigation strategies differ somewhat from other places in the world where salinity is dominated by chloride and sulfate (Choudhary et al. 2011). Although decreasing the exchangeable sodium by using gypsum as an amendment is commonly used worldwide, in India for various reasons, including the cost of gypsum, remedial measures (Minhas et al. 2019) commonly include mobilizing native calcite through phytoremediation (Qadir et al. 2001), application of farm-yard manure (Choudhary et al. 2004), growing tolerant crops (Minhas and Sharma 2006), and conservation tillage (Yaduvanshi and Sharma 2008).

8.4 Design Capabilities of Transient State Models

Transient state models that simulate changes in soil salinity in the crop root zone caused by irrigation and rainfall are important tools to assess and design alternative management plans for the use of saline waters. They can be used to relate crop water use and crop yield to the continuous changes of soil salinity (osmotic potential) and soil-water contents (matric potential) that occur in the root zone. When using saline waters, these transient conditions result from changes in irrigation water salinity, amounts of applied water, salination from shallow groundwater levels, rainfall, and climate. They can also determine timelines for when, if ever, steady-state conditions occur for specific designs and weather conditions. Several transient state models are available (Skaggs et al. 2014). We will describe features of only one of these, namely, HYDRUS (Šimůnek et al. 2016) because it is widely used, has an active internet support group, and the author and coworkers have developed online tutorials; both are key to making it simpler for a new user to complete the learning curve, thereby making the model accessible and easy to implement for a new user. Several authors have used HYDRUS to assess the impacts of using saline waters for irrigation (Ramos et al. 2011; Hansen et al. 2009; Kalendhonkar et al. 2012; Lyu et al. 2019; Yang et al. 2019). Many other authors have also used HYDRUS to assess drainage designs (Filipovic et al. 2014), drip-irrigation systems (Roberts et al. 2009; Dabach et al. 2011), and upward salt movement from shallow saline groundwater (van de Craats et al. 2020).

HYDRUS is a transient state model with a graphical user interface (GUI) available for 1D (free to download) and 2D/3D simulations (proprietary; one-time fee to download). HYDRUS-1D and HYDRUS (2D/3D) have more than 50,000 registered users and have been used in thousands of published projects.[6,7] In general, HYDRUS can perform simple-to-advanced simulations of water, solute, and heat movement in soil. Simulations can be performed at resolutions of seconds to days to model processes that occur over minutes, days, years, decades, and centuries. Most model simulations only take a few seconds to run for each year of simulation.

New learners are highly advised to start by downloading the free 1D version[8] and then look through the many step-by-step tutorials[9] and online Library of Projects.[10] For the purposes of this chapter, new users may find of particular value the following tutorials and online projects regarding infiltration and salt/solute transport:

1. '*Water Flow and Solute Transport in a Layered Soil Profile*'[11]

---

[6] http://www.pc-progress.com/en/Default.aspx?h1d-references.

[7] http://www.pc-progress.com/en/Default.aspx?h3d-references.

[8] https://www.pc-progress.com/en/Default.aspx?Downloads.

[9] https://www.pc-progress.com/en/Default.aspx?h1d-tutorials.

[10] https://www.pc-progress.com/en/Default.aspx?h1d-library.

[11] https://www.pc-progress.com/en/Default.aspx?h1d-tutorials#k4.

2. *'Modeling Salinity with the Standard and UnsatChem Modules of HYDRUS-1D'*[12]
3. *'Long-term (3-year) Simulations of Multicomponent Solute Transport (Salinity and Nitrogen Profiles) in Soils Irrigated with Saline Water'*[13]

Additionally, a free tutorial eBook (Rassam et al. 2018) can be downloaded.[14] The examples for the eBook can also be downloaded.[15]

New users often try to build something rather complex in their first set of simulations. Undoubtedly, they run into the issues of long run-times and a model that does not converge numerically to yield a final output. New users are advised to get each module set up and working correctly, sequentially one at a time (water flow, heat, and solute transport). This approach makes run-time problems with the HYDRUS model easier to diagnose and remedy. Users should build complexity into the simulations systematically. For instance, create a soil and then get water flowing through it. If the model converges and gives realistic outputs, then move on to adding in more complexity by adding chemical transport modules, crop and root growth modules, and then plant water and salt stress modules, while checking for model convergence at each phase of model development.

While the tutorials and example projects show example input data, finding input data for other soils may often elude new users. Soil maps and databases such as the USDA Web Soil Survey[16] and the KSSL database[17] can be used to locate representative soil physical properties and saturated hydraulic conductivity (Ksat) values for soil series of interest. All HYDRUS projects need water retention-curve parameters and Ksat values before any simulation can be performed. Almost always, practitioners are unlikely to have specific water retention-curve parameters for their fields, but they might have data that provide a good estimate of Ksat. Fortunately, these values can be estimated by selecting tabular combinations of texture, bulk density, field capacity, and permanent wilting point from the KSSL database for use in formulating HYDRUS' built-in pedotransfer functions (i.e., a hieratical neural network analyzer of the Rosetta dataset).

When users need support, the help command is quite useful and available at each step of any project. Moreover, the Hydrus discussion forum is a place to get feedback for specific questions.[18] If the help command is not sufficient for their needs, HYDRUS users should go to the forum and post their questions. From experience, most questions get a response within hours to a day or two from the creators and support staff at PC-Progress (the company developing the HYDRUS GUI and distributing the HYDRUS software). Users can also acquire paid services,

[12] https://www.pc-progress.com/en/Default.aspx?h1d-tutorials#k9.

[13] https://www.pc-progress.com/en/Default.aspx?h1d-lib-Portugal.

[14] https://www.pc-progress.com/Downloads/Public_Lib_H1D/HYDRUS-1D_Tutorial_V1.00_2018.pdf.

[15] https://www.pc-progress.com/en/Default.aspx?h1d-tut-TutorialBook.

[16] https://www.pc-progress.com/en/Default.aspx?h1d-tut-TutorialBook.

[17] https://ncsslabdatamart.sc.egov.usda.gov/.

[18] https://www.pc-progress.com/forum/.

including project setup, calibrations of the standard HYDRUS models, and custom modifications to its code and GUI through PC-Progress services.[19]

As with all physics-based numerical simulations, three key components are needed: (1) soil hydraulic and physical properties that adequately characterize the system, (2) initial conditions (i.e., the initial state of the system), and (3) boundary conditions over the duration of the simulation. The following is a brief summary of the inputs that would be required to use HYDRUS to assess effects of irrigation water salinity on crop yields (Ramos et al. 2011; Oster et al. 2011).

1. Soil hydraulic properties for water flow.
2. Initial soil water salinity and salinity of applied irrigation water.
3. Soil hydraulic properties.
4. Matric stress effects on crop water use. HYDRUS has an internal database for more than 30 types of crops (Feddes et al. 1978).
5. Osmotic and matric stress effects on crop water use [options for additive (i.e., lower transpiration) or multiplicative (higher transpiration) effects]. The crop's threshold salinity and slope to describe crop salt tolerance.
6. Initial and final root depths with options to control initial growth and harvest dates, the distribution of root growth during the season, or changes in root growth based on atmospheric conditions.
7. The irrigation schedule (irrigation date and depth of water applied).
8. The crop water requirement during the crop season (the potential transpiration rate). This changes from the seedling stage to full maturity and is calculated from the reference crop evapotranspiration (ET) (measured by a climate weather station) using procedures recommended by Allen et al. (1998).

Changes in soil salinity can be simulated by one of three modules: (1) General Solute Transport, (2) Major Ion Chemistry, and (3) HYDRUS coupled with PHREEQC (HP1). The General Solute Transport module is a general package to simulate basic transport of solutes. The Major Ion Chemistry module, also known as the UnsatChem code, considers cation exchange and precipitation and dissolution of such minerals as calcite and gypsum. The HP1 module is for advanced simulations by coupling HYDRUS with the geochemistry database and model PHREEQC. For salinity simulations, users are advised to use the General Solute Transport or Major Ion Chemistry modules.

HYDRUS does not simulate plant growth and yield per se, but rather simulates actual transpiration, in other words, crop water uptake. Users can calculate the relative yield as the ratio of simulated seasonal transpiration to the seasonal potential transpiration. Relative transpiration is assumed to equal relative crop yield. For example, Fig. 8.6 shows the results of four irrigation scenarios using HYDRUS-1D to calculate the change in soil water salinity (ECsw) with soil depth and changes in relative alfalfa yield. Each irrigation occurred every 15 days, and the calculation involved 98 irrigations (1,470 days) for each scenario. The total computer run time was 14 min to generate the data for the four scenarios shown in Fig. 8.6.

[19] https://www.pc-progress.com/en/Default.aspx?services.

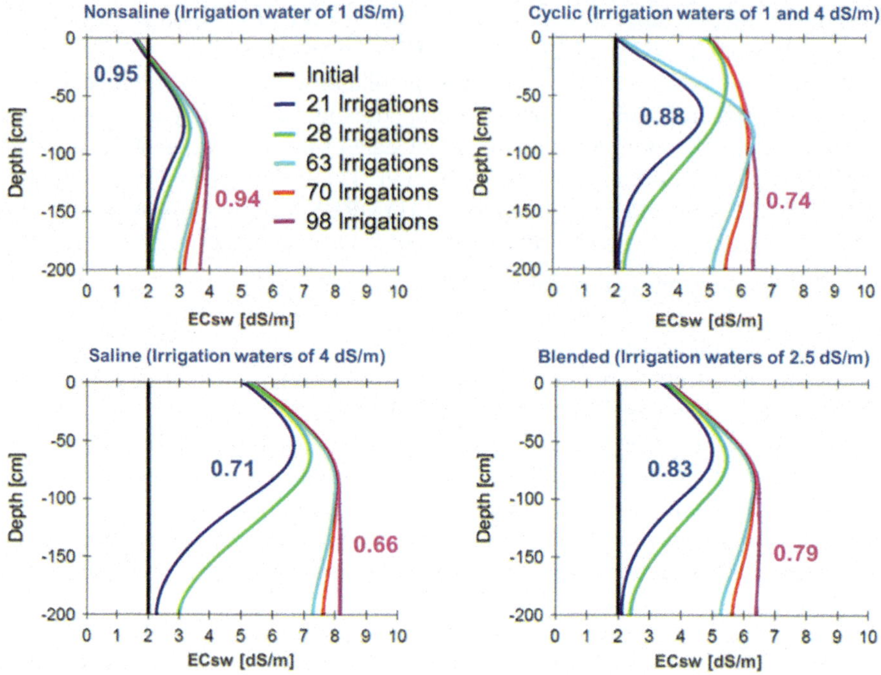

Fig. 8.6 Soil salinity [electrical conductivity of soil water (ECsw)] with depth after 21, 28, 63, 70 and 98 irrigations for a simulated alfalfa crop irrigated with saline and nonsaline waters every 15 days. The blue and pink numbers are relative alfalfa yields after 21 and 98 irrigations. Cyclic irrigation cycles consisted of low salinity water for 7 irrigations and high salinity water for 7 irrigations; then, the sequence repeated. Alfalfa harvest was every 30 days. Soil was a Panoche clay loam with initial soil water at −500 cm H_2O matric potential (i.e., 0.159 m^3/m^3), and ECsw of 2 dS/m was uniform throughout 200 cm depth. Total rooting depth was 100 cm and free drainage occurred at 200 cm. The van Genuchten S-shape model (van Genuchten 1980) for water stress with a multiplicative Maas threshold model for osmotic stress was used for all scenarios

The soil was Panoche clay loam, and the initial ECsw was 2 dS/m. The water applied for each irrigation was 20% more than potential crop ET, where potential crop ET was 1 cm/day and held constant for all irrigation cycles. Crop coefficients for initial, mid, and final crop growth stages, during a 30-day crop cycle (two irrigation cycles) beginning after harvest and ending the day of the irrigation, were calculated using the methods described in Allen et al. (1998). The soil depth was 200 cm; alfalfa rooting depth was 100 cm; and free drainage was assumed for the lower boundary condition at 200 cm.

Crop water uptake causes ECsw to increase with depth. The greater the water salinity (ECiw) of the irrigation applied, the greater the increase in salinity within the soil profile (Fig. 8.6). Over time, changes in ECsw were small after 70 irrigations (980 days) for all four scenarios shown in Fig. 8.6, suggesting that 98 irrigations were sufficient to calculate the ECsw depth distribution for steady-state conditions to a root zone depth of 100 cm. This was confirmed by running the nonsaline scenario for 392 irrigations, where the increase in ECsw at a depth of 100 cm was about 5%, and ECsw did not change in the 0–60 cm depth interval.

For the nonsaline scenario, where ECiw was 1 dS/m, the relative yield of 0.94 for the 98th irrigation was reasonable, although <1.0. It is likely that transient changes in salinity within the root zone exceeded 4 dS/m, the threshold soil-water salinity for alfalfa. The cyclic scenario consisted of using nonsaline water (ECiw of 1 dS/m) for seven irrigations, followed by seven irrigations with saline water (ECiw of 4 dS/m), with this sequence repeated fourteen times. Between the 21st and 28th irrigation, the cyclic scenario resulted in a large increase in ECsw in the 0–50 cm depth interval because 4 dS/m water was applied in irrigations 21–28. This increase in ECsw reduced the relative yield: for the 28th irrigation, it was 0.77, compared to 0.88 for the 21st irrigation. The relative crop yields for the cyclic strategy were very similar to those obtained for irrigation with (blended) 2.5 dS/m water, which simulated the option of blending 1 and 4 dS/m water. Finally, as expected, irrigation with only the saline water (ECiw of 4 dS/m) resulted in the lowest relative yields.

8.5 Monitoring Methodology for Use of Saline Drainage Waters

Soil Sampling Scheme Delineation

When using saline drainage waters for irrigation over entire fields, one should monitor their effects on soil salinity, sodicity levels, and crop yield. Because soils are inherently spatially variable, measurements should be carried out at multiple locations in an attempt to represent the average salinity and variability (e.g., standard deviation) of soil salinity across the entire field. If no accurate information is available on the spatial variability of soil physical properties and soil salinity, a fairly intense soil sampling scheme may be required to reliably assess field average values of salinity, sodicity, or crop yield. Sampling on a grid with size equal to 100 × 100 to 150 × 150 m is often recommended (Burt and Soil Survey Staff 2014). Brus (2014) discusses options (alternatives to grid sampling) for reliably quantifying average and variability of spatial variables over time. Of practical relevance is that, without auxiliary spatial information, many locations need to be sampled to obtain the averages of the target soil and plant variables (Reyes et al. 2018).

Fortunately, inexpensive auxiliary information from near-ground sensors and satellite imagery can be used as a proxy for the spatial variability of target soil and plant variables. Sensor measurements can be acquired over an entire field (e.g., hundreds of measurements) in a timely manner (Corwin and Scudiero 2016). Using information from such proxies, it is easier to estimate field-average and variability of the target soil and plant properties with fewer samples. In most cases, instead of hundreds of sampling locations, as few as 5–20 locations may be needed (Lesch 2005). Figure 8.7 provides an example of how proximal or remote sensing data can be used to direct sampling scheme delineation.

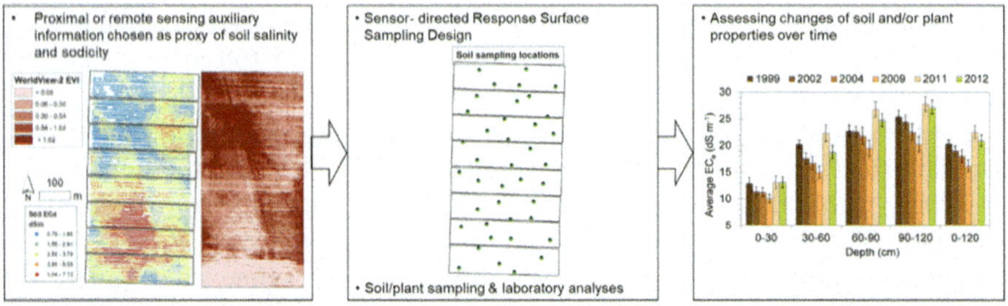

Fig. 8.7 Example of workflow for sensor-directed sampling scheme delineation and assessment of changes over time of target plant and/or soil properties. **a** Left Panel: Selection of data from proximal and remote sensing as proxy for soil salinity and sodicity. In the panel, soil apparent electrical conductivity and multi-year maximum enhanced vegetation index (EVI) from the WorldView-2 satellite are compared at the same site (Data from Scudiero et al. 2017a, *California Agriculture 71*(4), pp. 231–238. University of California Copyright © 2017 Regents of the University of California (CC-BY-NC-ND 4.0)); **b** Middle Panel: Sensor-directed sampling scheme for soil and/or plant sampling for laboratory analyses; and **c** Right Panel: Histograms showing changing salinity (ECe) over time at various depths (Scudiero et al. 2017b)

Proximal Sensing

In saline and sodic soils, near-ground (e.g., mounted on a field vehicle) sensor measurements, such as apparent electrical conductivity (ECa, i.e., current conducted by a volume of soil) and visible and near-infrared spectrometry can be used as proxy for soil properties of interest (e.g., soil salinity or ECe, SAR) (Corwin and Scudiero 2019; Rossel et al. 2011). Soil ECa is arguably the most widely used and studied proximal sensor measurement for soil salinity and sodicity mapping and monitoring applications (Corwin and Scudiero 2019). On-the-go measurements of ECa can be acquired using electromagnetic induction (EMI) or electrical resistivity (ER) sensors connected with a global navigation satellite systems (GNSS) receiver and datalogger. Several EMI (e.g., Geonics Ltd., Mississauga, Ontario, Canada Geonics; GF Instruments, S.R.O.; Brno, Czech Republic) and ER sensors (e.g., Veris Technologies, Inc., Salina, KS, USA) are available in the market and are currently widely used to characterize the spatial variability of salinity and other soil properties. Commercially available sensors provide ECa information over entire soil layers, from the soil surface to a nominal depth, e.g., 0–0.75 m, 0–1.5 m. On-the-go EMI and ER sensors generally measure ECa over small footprints (e.g., 2 × 2 m); therefore, when thousands of sensor measurements are taken at a site, high-resolution ECa maps can be generated (e.g., via Inverse Distance Weighing or other spatial interpolation techniques). Systems, including EMI sensor, datalogger, and GNSS receiver with sub-meter accuracy, can be purchased for < US$10,000 (measuring a single soil layer) to US$60,000 (measuring multiple soil layers simultaneously). Maps of ECa can be complemented with maps of soil sensors that are proxies for soil pH (Schirrmann et al. 2011) to better characterize spatial variability of sodic soils.

Remote Sensing

Satellite imagery can be also used to characterize spatial and temporal variability of saline and sodic soils. Bare-soil (or scarcely vegetated soil) imagery (e.g., visible and near-infrared) can be used to map and monitor soil surfaces (Aldabaa et al. 2015). A single satellite image cannot be used reliably as a predictor of root zone (e.g., 0–1 m) salinity and sodicity because of the possible presence of other limiting factors affecting crop growth in any given season and/or crop growing phase. However, in irrigated agricultural soils, salinity and sodicity indicators tend to remain fairly stable across the entire root zone over a short number of years (Shouse et al. 2010), permitting multiple-year crop-imagery time series to be used to predict root zone soil salinity and sodicity levels (Scudiero et al. 2017a).

Several vegetation indices can be used in the time-series analyses, including the Normalized Difference Vegetation Index (NDVI), the Enhanced Vegetation Index (EVI), among others. The Canopy Response Salinity Index (CRSI), originally proposed by Scudiero et al. (2014), is reported by multiple authors (e.g., Ramos et al. 2020) to be one of the best proxies of root zone soil salinity. The CRSI is formulated as follows:

$$\mathrm{CRSI} = \sqrt{\frac{(\mathrm{NIR} \times \mathrm{R}) - (\mathrm{G} \times \mathrm{B})}{(\mathrm{NIR} \times \mathrm{R}) + (\mathrm{G} \times \mathrm{B})^2}}$$

where NIR, R, G, and B are the surface reflectance values near-infrared, red, green, and blue spectral bands, respectively. These spectral band data are available from most satellites commonly used for vegetation monitoring, including MODIS, Landsat 8, and Sentinel 2. High plant vigor corresponds to a larger CRSI value.

Using Auxiliary Information To Direct Soil Sampling

The response surface-sampling design (RSSD) method (Lesch 2005) can be used to identify a handful (5–20) representative locations when auxiliary information is available. Sensor-directed soil-sampling scheme delineation with the RSSD method can be carried out with the free software ESAP (Lesch et al. 2000). Corwin and Scudiero (2016) describe in detail the use of RSSD methods in ESAP. Briefly, sensor-directed RSSD can be used to select representative locations in a field sufficiently far apart to avoid spatial autocorrelation bias, which will provide a good estimation of field average and variability for a target soil property (e.g., ECe, SAR, CROSS). The laboratory results obtained from the soil samples can be used to generate maps with a regression and interpolation approach (e.g., kriging or inverse distance weighing): The laboratory results for each soil sample at a specific location are correlated, using a linear regression model (Lesch and Corwin 2008), with proximal data obtained at the same location. Then, the regression is applied to all (hundreds) of locations where the auxiliary proximal data are available. Finally, kriging or another interpolation method is used to produce a map for the entire plot/field. Detailed protocols and suggested procedures for this method are described by Corwin and Lesch (2005).

8.6 Drainage Water Reuse and Disposal Issues: Major Barriers and Responses by Stakeholders

Implementation of a strategy to reuse drainage water for irrigation often requires the acquisition of new skills and a change in mindset. In regions where subsurface tile drainage is a necessity to sustain agricultural production, drainage service is typically considered a normal cost of production, not a potential source of irrigation water. That's why reuse of agricultural drainage water for irrigation is still considered an unconventional water resource. Subsurface drainage reuse can have the dual benefit of maximizing the water supply while minimizing the volume of drainage water that ultimately will require disposal. After sequential reuse, the salinity of drainage water can reach levels of 20–40 dS/m. Once achieved, such drainage water is unusable for irrigation of salt-tolerant crops but could be used to irrigate halophilic crops (Fig. 8.4). Disposal of unusable drainage water into irrigation canals where it may be comingled with irrigation water supply is counterproductive because it reduces the volume of useable water (Rhoades 1999). Keeping irrigation water supply and drainage return flows separate to the extent possible is good resource management.

Disposal of unusable subsurface drainage water poses a major environmental problem. Ocean disposal via dedicated drainage canals is usually regarded as the safest option. However, with increasing social concerns about adverse environment impacts, there now is an awareness of potential harm to fragile coastal ecosystems, such as mangroves and coral reefs, due to toxic effects of plant nutrients (N, P), herbicides, and pesticides contained in the agricultural drainage water. In inland basins without access to ocean disposal, storage and disposal of agricultural drainage in evaporation ponds has been used by irrigation districts in Australia (Leaney et al. 2000) and in the closed Tulare Basin in California's San Joaquin Valley (Quinn 2014). In the Tulare Basin, the management of evaporation ponds includes construction and maintenance of wetland habitats to compensate for the loss of habitat and potential harm to migratory waterfowl and other wildlife due to Se ecotoxicity of the ponded water. The salt concentration in the Salton Sea in southern California has increased beyond the tolerance threshold for ocean fish, and Se levels in the discharged drainage water exceed safe levels for sustainable waterfowl habitat. Wind-blown, salt-laden dust downwind from the beaches surrounding the Salton Sea has led to health impacts in the local population and a rise in respiratory ailments directly linked to air pollution.

Another disposal option is deep well injection into deep geologic formations near agricultural areas. Such formations would need to have high permeability and sufficient storage capacity to accommodate the volume of agricultural subsurface drainage requiring disposal. In the 1990s, efforts to develop deep well injection by the Westlands Water District, a district in the San Joaquin Valley, California, without access to ocean disposal, were terminated because of high operation costs for pretreatment and power, difficulties with screen fouling, and obtaining the necessary permits (Linneman et al., private communication 2020).

The sequential reuse strategy (Fig. 8.4) was investigated by the owner of a salinity-impacted ranch in the Westlands Water District in conjunction with academic and water agency researchers. Three quarters of a 260 ha tract of irrigated land was dedicated to high-value salt sensitive crops and the remaining one quarter planted to salt tolerant crops, such as cotton, and agroforestry crops, with a small area dedicated to irrigation of halophilic crops. The halophyte crops had threshold salinities of about 40 dS/m (Diaz et al. 2013). A solar evaporation facility (Benes et al. 2004) converted the drainage water from the halophytes to a solid that could be transported to a local landfill. The project lasted almost ten years but was discontinued by the grower because operating costs exceeded income from the harvested crops. Another key problem was that the harvested salts were not salable, and their disposal posed serious environmental issues because of their Se content, whether disposed in a landfill or the ocean.

Farmers within the 44,000 ha Se-impacted Grasslands subarea, on the westside of the San Joaquin Valley, use a combination of techniques to manage and dispose of drainage water. Prior to 1998, the drainage water intercepted by tile drainage was routed to the San Joaquin River via a 160 km network of shallow ditches that passed through a seasonal wetlands complex. This complex comprises numerous private duck clubs, and State and Federal wildlife refuges that provide hunting and birding opportunities for outdoor enthusiasts and a seasonal habitat for migratory waterfowl. Toxic concentrations of Se in the drainage water endangered the waterfowl that foraged in these canals while they carried drainage water. They also needed to be flushed with freshwater before they could be used to convey wetland supply water—a loss of water that could have been used for irrigation. After six years of negotiations with State and Federal water and resource management agencies and environmental groups (Entrix 2009; USBR 2009), the Grassland Bypass Project Use Agreement plan was adopted in 1996 to re-route this drainage through the federal San Luis Drain (Quinn 2020). This Agreement included a commitment by the Grassland Farmers to permanently reduce salt and Se-laden drainage flows discharged into the San Joaquin River. In the 24 years to date that the agreement has been in force, it has resulted in positive impacts on salt loading to the San Joaquin River that have benefited management of the wetlands and farmers, who use the river water downstream to irrigate vegetables. Prior to Water Year 1996, more than 157,000 Mg of salts (Fig. 8.8) were discharged annually to the San Joaquin River. Since 2015, the salt loads have decreased to approximately 18,000 Mg of salts, the consequence of extreme flood events, due to rainfall, which required waivers issued by the Regional Water Quality Control Board.

These major changes in salt loads resulted from the reuse of drainage water for irrigation in a separate area, known as the San Joaquin River Improvement Project (SJRIP), developed by the Grassland Farmers with the support of the U.S. Bureau of Reclamation. During Water Year 2020, approximately 236,700 Mg of salts (Fig. 8.8) were displaced to the SJRIP.

The Grasslands Area Farmers have also made major investments in improved methods of irrigation and drainage management technologies to reduce the volume of drainage water used in the SJRIP. This included a moratorium on runoff from irrigated

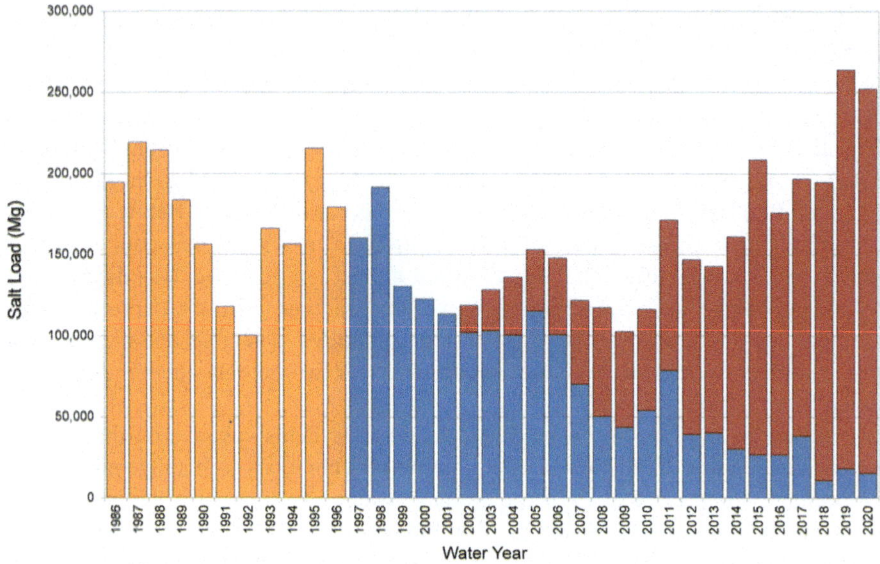

Fig. 8.8 Salts (Mg) discharged from the Grasslands Drainage Area, Water Years 1986–2020: Orange Bars—pre-San Joaquin River Improvement Project (SJRIP), 1986–1996; Blue Bars—salts discharged to the San Luis Drain, 1997–2020; Red Bars—salts displaced to the SJRIP, 2002–2020 (Adapted from Fig. 3, USBR Management Agency Agreement 2020 Annual Report)

fields. As a result, all return flows are now captured by tailwater sumps and returned to the head ditches, ending in its blending with subsurface drainage. In 2004, farmers began to convert surface-irrigation methods to high-efficiency irrigation systems, such as buried drip and micro-sprinklers. By 2014, these irrigation systems were in use on more than 18,000 ha (Linneman et al. 2014). Subsurface drainage water from certain sumps and groundwater from shallow wells are blended deliberately with canal water so as not to exceed a target salt concentration of around 1.4 dS/m. The remaining surface drainage volume is conveyed to the 2,400 ha SJRIP reuse facility where it is used to irrigate 'Jose' tall wheatgrass, alfalfa, and salt-tolerant trees, such as pistachios. This operation has been fully sustainable for 20 years (Singh et al. 2020). Only a small portion of the SJRIP, with the poorest natural drainage, has tile drains, and this drainage water is reused for irrigation within the SJRIP. Consequently, the underlying soil strata is the final disposal site for salt loads generated by Grasslands Area Farmers. While the SJRIP reduces the spatial scale of the environmental impacts of agricultural subsurface drainage from an irrigated area of 44,000 ha to a localized area of 2400 ha, there are environmental impacts on groundwater beneath the SJRIP that have yet to be resolved.

Impacts on groundwater pumped from depths of 100–200 m will likely take decades to occur and even longer for groundwater pumped from below the Corcoran Clay layers at depths >200 m. There are many unresolved issues involved with this method of disposal (Quinn 2014), largely associated with the complexity of the hydrogeology in this region. The outstanding issues include:

1. Lateral extent and depth to which the underlying groundwater aquifers will be contaminated over time;
2. Relationship between groundwater pumpage from aquifers above and below the Corcoran Clay and the potential for induced contamination of the deeper aquifers across the Corcoran Clay;
3. Development and adoption of methods to monitor the long-term impacts on the land and groundwater system.

As was the case with the six years of negotiations that resulted in the Grassland Bypass Project Use Agreement, it is reasonable to expect these outstanding issues will require considerable research and years of negotiation to develop equitable and cost-effective strategies that include the reuse of saline drainage water for irrigation.

Disposal of unusable drainage water into soil strata and underlying groundwater will result in adverse environmental impacts on groundwater quality. Regulations of how and where such waters can be disposed need to be consistent with what is known about environmental impacts as set out by local, regional, state, and federal governing agencies. Governance associated with disposal of unusable drainage water is a complicated process. Five governance components are related to priorities, objectives, monitoring, and control:

1. An agreed understanding of the hydrological system is necessary, such that interventions can be evaluated in a way that all stakeholders accept;
2. The priorities, laws and institutions must be coherent;
3. Decisions must be transferable into enforceable regulations;
4. Institutions responsible for monitoring and management must be established and equipped;
5. Ever-shifting societal attitudes will impact governance.

The environmental crisis caused by Se toxicosis at Kesterson Reservoir in California forever changed societal attitudes related to the potential adverse environmental impacts of irrigation.

Stakeholders' involvement in the decision process to generate the Grassland Bypass Agreement was a new paradigm in governance of water resources in California. What was new? The inclusion in the planning process of farmers directly impacted by the final decisions that were made. It was a bottom-up process as compared to the traditional top-down planning process where those directly impacted are not involved in developing new rules and regulations. An earlier example (1975–present) of this new paradigm was its use by the Santa Ana Watershed Planning Authority to establish resource plans for all the waters used within the Santa Ana Watershed in southern California. The goal of this Authority's planning process, entitled "One Water, One Watershed," simply captures what is involved in a bottom-up planning process: "all sectors of communities in the watershed adopt a water ethic that focuses on understanding where their water comes from, how much they use of it, what they put into water, and where it goes after they finish using it."[20]

Another recent stakeholder-led policy initiative that may have implications for salinity drainage in California is the passage of the Sustainable Groundwater Management Act (SGMA) in September 2014. Although the main objective of this policy is directed at groundwater resource sustainability, water quality issues are being considered in a coequal decision framework as water resource management issues for the first time. Management zones for groundwater configured as part of requisite SGMA Groundwater Sustainability Plans are being considered by the regional association of stakeholders known as CVSALTS as the basis for salinity management zones for long-term regional salinity management. The goal of managing root zone salinity and groundwater quality issues from a whole system perspective may yet be realized.

8.7 Conclusions

Subsurface drainage waters generated by irrigation are a valuable, unconventional source of irrigation water, and efforts to expand their reuse for irrigation are worthwhile, partially mitigating the impacts of the increased allocation of freshwater for municipal and industrial use. Such waters are always more saline than the water used for initial irrigation, which means their use requires an extra degree of care and management skills. Trace elements, such as B, Se, and Mo, if present, can affect the extent to which saline drainage water can be used to irrigate certain crops. Much has been learned in the past three decades about the use and reuse of subsurface drainage water, including the impacts of trace elements, as documented in this chapter.

Although the same set of scientific principles apply in all irrigation and reuse cases, there is not one management practice that uses saline surface drainage water that will be appropriate in all areas and appropriate to every farmer, or farming operation. Rather, use and reuse of saline drainage water must be customized to site-specific conditions to be sustainable. There are several efficacious strategies for reuse:

- Blending the saline subsurface drainage water with nonsaline water;
- Cyclic use of saline drainage and nonsaline water for a portion of the irrigation season;
- Sequential reuse where the saline drainage water from one field is used to irrigate appropriately salt-tolerant crops on another. The process is repeated in separate fields until the drainage water becomes so saline that it can only be used to irrigate halophilic crops. The salt in the resulting super-saline drainage water is harvested using solar evaporation.

The blending, cyclic, and sequential strategies for reuse of drainage water for irrigation are best handled on a regional basis or a national basis as occurs in the

[20] https://sawpa.maps.arcgis.com/home/index.html.

blending strategy adopted by Egypt (Dayem et al. 2007) in the northern portion of the Nile River. Having an infrastructure in place to separate, collect, store, and distribute surface and subsurface drainage water can help to optimize its use at an appropriate time and place. This optimization can be facilitated using transient computer-based simulation models, such as HYDRUS, to assess alternative irrigation strategies for using drainage water on selected crops appropriate for the region. Once a reuse scheme is adopted and used, monitoring soil salinity and tracking crop yields are necessary for sustainability and can be accomplished using remote -sensing techniques.

Continued funding for scientific research on drainage water reuse in arid and semi-arid regions will lead to new technologies that can expand the use of this important unconventional water resource, whose value for conservation of water resources has been established by the research completed to date. The challenge is to design and operate reuse strategies that maximize multiple agronomic, environmental, and societal benefits, while minimizing the risk of adverse impacts related to the disposal of saline drainage waters that can result from reusing drainage water for irrigation. Once a reuse strategy is designed and started, monitoring provides the information needed to make whatever changes in water and crop management are needed to assure long term sustainability. Based on continued research and documentation of user experience with reuse strategies, improvements in both the methods available to design a reuse system and to monitor the results can be expected to occur. Finally, bringing all the stakeholders into the decision process is a promising new paradigm that may be at hand to establish the rules and regulations for disposal that are acceptable to all.

Acknowledgements We are thankful for the helpful comments we have received from the following colleagues during our development of this chapter: S. Abdel-Dayem, J. Ayars, S. Benes, T. DeSutter, S. Grattan, C. Linneman, J. N. Lu, P.S. Minhas, M. Qadir, A.S. Qureshi, P.A.C. Raats, J. Šimůnek, G. Sposito, and L. Wu. Deborah Brawer Silva edited the chapter, for which we are particularly thankful.

References

Aldabaa AAA, Weindorf DC, Chakraborty S, Sharma A, Li B (2015) Combination of proximal and remote sensing methods for rapid soil salinity quantification. Geoderma 239:34–46. https://doi.org/10.1016/j.geoderma.2014.09.011

Allen RG, Pereira LS, Raes D, Smith M (1998) Crop evapotranspiration: guidelines for computing crop water requirements. FAO Irrigation and Drainage Paper No. 56. FAO, Rome, Italy

Alonso MF, Corwin DL, Oster JD, Maas J, Kaffka SR (2013) Modeling a sustainable salt tolerant grass-livestock production system under saline conditions in the western San Joaquin Valley of California. Sustainability 5:3539–3857. https://doi.org/10.3390/su5093839

Ayars JE, Hutmacher RB, Hoffman GJ, Ben-Asher J, Pilaum T (1990) Response of sugar beet to non-uniform irrigation. Irrig Sci 11:101–109

Ayars JE, Hutmacher RB, Schonerman RA, Vail SS, Pilaum T (1993) Long-term use of saline water for irrigation. Irrig Sci 14:27–34

Ayars, JE, Basinal, L (2005) A technical advisor's manual. Managing agricultural irrigation drainage water: a guide for developing integrated on-farm drainage management systems. Westside resources conservation district and center for irrigation technology, California State University, Fresno, Calif

Ayers RS, Westcot DW (1985) Water quality for agriculture. Irrigation and Drainage Paper No. 29, Rev. 1. FAO, Rome, Italy

Bassil E, Kaffka SR (2002) Response of safflower (Carthamus tinctorius L.) to saline soils and irrigation. II. Crop growth and yield. Agric Water Manag 54:81–92

Benes SE, Grattan S, Finch C, Basinal L (2004) Plant selection for IFDM. In: Jacobsen T, Basinal L (eds) A landowner's manual: managing agricultural irrigation drainage water. A guide for developing integrated on-farm drainage management systems. California State Water Resources Control Board. Hudson Orth Communication, Fresno, CA, pp 1–21

Benes SE, Grattan SR, Robinson PH (2005) Cultivation of halophytes to reduce drainage volumes on the westside San Joaquin Valley of California. In: Final Report to the California State University Agricultural Research Institute (ARI), Project No. 00-1-003, October 18, 2005, California State University, Fresno, Calif

Bernstein L, Francois LE (1973) Leaching requirement studies: sensitivity of alfalfa to salinity of irrigation and drainage waters. Soil Sci Soc Am J 37(6):931–943

Boyko H, Boyko E (1959) Seawater irrigation a new line of research on a bioclimatic plant-soil complex. Int J Bioclimatol Biometeorol 3:33–61

Brus DJ (2014) Statistical sampling approaches for soil monitoring. Eur J Soil Sci 65:779–791. https://doi.org/10.1111/ejss.12176

Burt R, Soil Survey Staff (eds) (2014) Soil survey field and laboratory methods manual. Soil survey investigations report No. 51, version 2. USDA, NRCS (Natural Resources Conservation Service), Washington, DC

Choudhary OP, Josan AS, Bajwa MS, Kapur ML (2004) Effect of sustained sodic and saline-sodic irrigation and application of gypsum and farmyard manure on yield and quality of sugarcane under semi-arid conditions. Field Crop Res 87(2–3):103–116

Choudhary OP, Grattan SR, Minhas PS (2011) Sustainable crop production using saline and sodic irrigation waters. In: Lichtfouse E (ed) Alternative farming systems, biotechnology, drought stress and ecological fertilisation. Sustainable Agriculture Reviews, vol 6. Springer, Dordrecht, Netherlands, pp 293–318

Corwin DL, Lesch SM, Oster JD, Kaffka SR (2008) Short-term sustainability of drainage water reuse: spatio-temporal impacts on soil chemical properties. J Environ Qual 37(5 Suppl.):S8–S24. https://doi.org/10.2134/jeq2007.0140

Corwin DL, Lesch SM (2005) Characterizing soil spatial variability with apparent soil electrical conductivity: I. Survey protocols. Comput Electron Agricult 46(1–3):103–133. https://doi.org/10.1016/j.compag.2004.11.002

Corwin DL, Scudiero E (2016) Field-scale apparent soil electrical conductivity. In: Logsdon S (ed) Methods of soil analysis, vol 1. Soil Science Society of America, Madison, WI, pp 1–29. https://doi.org/10.2136/methods-soil.2015.0038 (reprinted online 18 Nov 2020 in Soil Science Society of America Journal)

Corwin DL, Scudiero E (2019) Review of soil salinity assessment for agriculture across multiple scales using proximal and/or remote sensors. In: Sparks DL (ed) Advances in agronomy, vol 158. Academic Press, Cambridge, MA, pp 1–130. https://doi.org/10.1016/bs.agron.2019.07.001

Dabach S, Lazarovitch N, Šimůnek J, Shani U (2011) Numerical investigation of irrigation scheduling based on soil water status. Irrig Sci 31:27–36. https://doi.org/10.1007/s00271-011-0289-x

Dagar JC (2018) Utilization of degraded saline habitats and poor-quality waters for livelihood security. Scholarly J Food Nutr 1(3). https://doi.org/10.32474/sjfn.2018.01.00015

Dayem SA, Abdel-Gawad S, Fahmy H (2007) Drainage in Egypt: a story of determination, continuity, and success. Irrig Drain 56(S1):S101–S111. https://doi.org/10.1002/ird.335

Diaz FJ, Benes SE, Grattan SR (2013) Field performance of halophytic species under irrigation with saline drainage water in the San Joaquin Valley of California. Agric Water Manag 118:59–69. https://doi.org/10.1016/j.agwat.2012.11.017

Dutt GR, Pennington DA, Turner F (1984) Irrigation as a solution to salinity problems in river basins. In: Frenc RH (ed) Salinity in watercourses and reservoirs. Ann Arbor Science Publishers, Ann Arbor, MI, pp 465–472

Entrix (2009) Grassland Bypass Project (2010–2019) Environmental impact statement and final environmental impact report, 391 pp. Prepared by Entrix, Concord, CA for US Bureau of Reclamation, (USBR) Mid-Pacific Region (Sacramento, CA), USBR South Central California Area Office (Fresno, CA) and San Luis & Delta-Mendota Water Authority (Los Banos, CA)

Erickson JR (1980) Using high salinity waters in the Southwest (1980). In: Eggleston J (ed) Today's challenges. Proceedings 1980 specialty conference on irrigation and drainage, Boise, ID, 23–25 July 1980. ASCE (American Society of Civil Engineers), New York, NY, pp 198–204

Feddes RA, Kowalik PJ, Zaradny H (1978) Simulation of field water use and crop yield. Simulation monographs. Centre for Agricultural Publishing and Documentation (pudoc), Wageningen, The Netherlands, 189 pp

Filipovic V, Mallman FJK, Coquet Y, Šimůnek J (2014) Numerical simulation of water flow in tile and mole drainage systems. Agric Water Manag 146:105–114

Ghassemi F, Jakeman AJ, Nix HA (1995) Salinisation of land and water resources: human causes, extent, management & case studies. CABI Publishing, Wallingford, UK, p 540

Glenn EP, Brown JJ, Blumwald E (1999) Salt tolerance and crop potential as halophytes. Crit Rev Plant Sci 18(2):227–255

Grattan SR, Oster JD (2003) Use and reuse of saline-sodic waters for irrigation of crops. In: Goyal SS, Sharma SK, Rains DW (eds) Crop production in saline environments: global and integrative perspectives. Haworth Press, New York, pp 131–162

Grattan SR, Oster JD, Letey J, Kaffka SR (2014) Drainage water reuse: concepts, practices and potential crops. In: Chang AC, Brawer Silva D (eds) Salinity and drainage in San Joaquin Valley, California: Science, technology, and policy. Springer, Dordrecht, pp 277–302

Grieve CM, Grattan SR, Maas EV (2012) Plant salt tolerance. In: Wallender W, Tanji KK (eds) Agricultural salinity assessment and management, 2nd ed. ASCE Manuals and Reports on Engineering Practice No. 71, Reston, VA, pp 405–459

Hansen BR, May DE, Šimůnek J, Hopmans JW, Hutmacher RB (2009) Drip irrigation provides the salinity control needed for profitable irrigation of tomatoes in the San Joaquin Valley. Calif Agric 63(3):131–136

Hilgard EW (1886) Irrigation and alkali in India. College of Agriculture, University of California Press, Berkeley, Bulletin 86, 34 pp

ICBA (International Center for Biosaline Agriculture). Annual report 2005 (1425–26H, ICBA, Dubai, UAE

Jury WA, Tuli A, Letey J (2003) Effect of travel time on management of a sequential reuse drainage operation. Soil Sci Soc Am J 67:1122–1126

Kaffka SR, Daxue D, Peterson G (1999) Saline water can be used to irrigate sugarbeets, but sugar may be low. Calif Agric 53(1):11–15

Kaffka S, Oster J, Maas J, Corwin D (2004) Forage production and soil reclamation using saline drainage water. In: Proceedings, 2004 National Alfalfa symposium, 13–15 December 2004, San Diego, CA, pp 247–253

Kaledhonkar MJ, Sharma DR, Tyagi NK Kumar A, van Der Zee SEATM (2012) Modeling for conjunctive use irrigation planning in sodic groundwater areas. Agricult Water Manag 107(C):14–22

Keren R, Shainberg I (1978) Irrigation with sodic and brackish water and its effect on the soil and on cotton fields. Harrade 58:963–976

Lauchli A, Grattan SR (2012) Plant responses to saline and sodic conditions. In: Wallender WW, Tanji KK (eds) Agricultural salinity assessment and management, 2nd ed. ASCE Manuals and Reports on Engineering Practice No. 71. Reston, VA, pp 169–205

Leaney F, Christen E, Jolly I, Walker G (2000) On-farm and community-scale salt disposal basins on the Riverine Plain: Guidelines summary. CRC for Catchment Hydrology Report 00/17 and CSIRO Land and Water Technical Report No. 24/00. Adelaide, Australia

Lesch SM (2005) Sensor-directed response surface sampling designs for characterizing spatial variation in soil properties. Comput Electron Agric 46:153–179. https://doi.org/10.1016/j.compag.2004.11.004

Lesch SM, Corwin DL (2008) Prediction of spatial soil property information from ancillary sensor data using ordinary linear regression: model derivations, residual assumptions and model validation tests. Geoderma 148:130–140. https://doi.org/10.1016/j.geoderma.2008.09.014

Lesch SM, Rhoades JD, Corwin DL (2000) ESAP-95 version 2.01R: User manual and tutorial guide, Research Report No. 146, USDA, George E. Brown Jr. Salinity Laboratory, Riverside, CA

Letey J, Roberts C, Penberth M, Vasek C (1986) An agricultural dilemma: Drainage water and toxics disposal in the San Joaquin Valley. Agricultural Experiment Station Special Publication 3319. University of California Press, Oakland, CA

Linneman C, Falaschi A, Oster JD, Kaffka S, Benes SE (2014) Drainage reuse by Grassland area farmers: the road to zero discharge. In: Macauley S, Thoreson BP, Anderson SS (eds) Groundwater issues and water management—strategies addressing the challenges of sustainability, Proceedings of the USCID Water Management Conference Sacramento, CA, March 4–7, 2014. The U.S. Committee on Irrigation and Drainage (USCID), Denver, CO, pp 65–78

Lyu S, Chen W, Wen X, Chang AC (2019) Integration of HYDRUS-1D and MODFLOW for evaluating the dynamics of salts and nitrogen in groundwater under long-term reclaimed water irrigation. Irrig Sci 37:35–47. https://doi.org/10.1007/s00271-018-0600-1

Maas EV, Hoffman GJ (1977) Crop salt tolerance: Current assessment. J Irrigation Drainage Div (ASCE) 103(2):115–134

Minhas PS (1996) Saline water management for irrigation in India. Agric Water Manag 30(1):1–24

Minhas PS, Gupta RK (1992) Quality of irrigation water: assessment and management. Information and Publication Section, Indian Council of Agricultural Research, New Delhi

Minhas PS, Qadir M, Yadav RK (2019) Groundwater irrigation induced soil sodification and response options. Agric Water Manag 215:74–85

Minhas PS, Ramos TB, Ben-Gal A, Pereira LS (2020) Coping with salinity in irrigated agriculture: crop evapotranspiration and water management issues. Agric Water Manag 227:105832. https://doi.org/10.1016/j.agwat.2019.105832

Minhas PS, Sharma DR (2006) Predictability of existing indices and an alternative coefficient for estimating sodicity build-up using adj.R_{Na} and permissible limits for crops grown on soils irrigated with water having residual alkalinity. J Indian Soc Soil Sci 54(3):331–338

Miyamoto S, Moore J, Stichler C (1984) Overview of saline water irrigation in far west Texas. In: Replogle JA, Renard KG (eds) Water today and tomorrow. Proceedings ASCE irrigation and drainage specialty conference, Flagstaff, AZ, 24–26 July 1984. American Society of Civil Engineers, New York

National Research Council (NRC) (1989) Irrigation-induced water quality problems. What can be learned from the San Joaquin Valley experience? The National Academies Press, Washington, DC. https://doi.org/10.17226/19051

National Research Council (NRC) (1990) Saline agriculture: salt-tolerant plants for developing countries. The National Academies Press, Washington, DC. https://doi.org/10.17226/1489

National Research Council (NRC) (2000) Nutrient requirements of beef cattle, 7th Revised ed. The National Academies Press, Washington, DC

Oster JD (1994) Irrigation with poor quality water. Agric Water Manag 25(3):271–297. https://doi.org/10.1016/0378-3774(94)90064-7

Oster JD, Jayawardane NS (1998) Agricultural management of sodic soils. In: Sumner ME, Naidu R (eds) Sodic soils: Distribution, processes, management and environmental consequences. Oxford University Press, New York, pp 125–147

Oster JD, Shainberg I (2001) Soil responses to sodicity and salinity: challenges and opportunities. Aust J Soil Res 39(6):1219–1224

Oster JD, Meyer JL, Hermsmeier L, Kaddah M (1986) Field studies of irrigation efficiency in the Imperial Valley. Hilgardia 54(7):1–15

Oster JD, Letey J, Vaughan P, Wu L, Qadir M (2011) Comparison of transient state models that include salinity and matric stress effects on plant yield. Agric Water Manag 103:167–175. https://doi.org/10.1016/j.agwat.2011.11.011

Oster JD, Sposito G, Smith CJ (2016) Accounting for potassium and magnesium in irrigation water quality assessment. Calif Agric 70(2):71–78. https://doi.org/10.3733/ca.v070n02p71

Oster JD, Wichelns D (2014) E.W. Hilgard and the history of irrigation in the San Joaquin Valley: stunning productivity, slowly undone by inadequate drainage. In: Chang AC, Brawer Silva D (eds) Salinity and drainage in San Joaquin Valley, California: Science, technology, and policy. Springer, Dordrecht, pp 7–46

Oster JD, Fernandes-Macedo T, Davis D, Fulton A (1999) Developing sustainable reuse and disposal of saline drainwater on Eucalyptus *(E. camaldulensis)*. Final Report. United States Bureau of Reclamation, Tulare Lake Drainage District, Corcoran, CA, 56 pp

Qadir M, Ghafoor A, Murtaza G (2001) Use of saline-sodic waters through phytoremediation of calcareous saline-sodic soils. Agric Water Manag 50(3):197–210

Qadir M, Sharma BR, Bruggeman A, Choukr-Allah R, Karajeh F (2007) Non-conventional water resources and opportunities for water augmentation to achieve food security in water scarce countries. Agric Water Manag 87(1):2–22

Qadir M, Sposito G, Smith CJ, Oster JD (2021) Reassessing water quality guidelines for sodicity hazard. Agricultural Water Management (submitted)

Quinn NWT (2014) The San Joaquin Valley: Salinity and drainage problems and the framework for a response. In: Chang AC, Brawer Silva D (eds) Salinity and drainage in San Joaquin Valley, California: Science, technology, and policy. Springer, Dordrecht, pp 47–97

Quinn NWT (2020) Policy Innovation and Governance for Irrigation Sustainability in the Arid Saline San Joaquin River Basin. Sustainability 12(11):4733. https://doi.org/10.3990/su12114733

Quirk JP, Schofield RK (1955) The effect of electrolyte concentration on soil permeability. J Soil Sci 6:163–178

Rains DW, Goyal S, Weyrauch R, Läuchli A (1987) Saline drainage water reuse in a cotton rotation system. Calif Agric 41(9):24–26

Ramos TB, Šimůnek J, Goncalves JD, Martins JC, Prazeres A, Castanheira NL, Pereira LS (2011) Field evaluation of a multicomponent solute transport model in soils irrigated with saline waters. J Hydrol 407(1–4):129–144. https://doi.org/10.1016/j.hydrol.2011.07.016

Ramos TB, Castanheira N, Oliveira AR, Paz AM, Darouich H, Simionesei L, Farzamian M, Goncalves MC (2020) Soil salinity assessment using vegetation indices derived from Sentinel-2 multispectral data. application to Lezíria Grande, Portugal. Agricult Water Manag 241(C). https://doi.org/10.1016/j.agwat.2020.106387

Rassam D, Šimůnek J, Mallants D, van Genuchten MT (2018) The HYDRUS-1D software package for simulating the one-dimensional movement of water, heat, and multiple solutes in variably-saturated media: tutorial, Version 1.00, July 2018. CSIRO (Commonwealth Scientific and Industrial Research Organisation) Land and Water, Adelaide, Australia, 183 pp. ISBN 978-1-4863-1001-2

Rengasamy P, Marchuk A (2011) Cation ratio of soil structural stability (CROSS). Soil Res 49:280–285

Reyes J, Wendroth O, Matocha C, Zhu J, Ren W, Karathanasis AD (2018) Reliably mapping clay content coregionalized with electrical conductivity. Soil Sci Soc Am J 82(3):578–592. https://doi.org/10.2136/sssaj2017.09.0327

Rhoades JD (1984) Use of saline water for irrigation. Calif Agric 38(10):42–43

Rhoades JD, Bingham FT, Letey J, Hoffman GD, Pinter AR, Alves WJ, Swain R, Pacheco P, Lemert R, Replogle JA (1989) Use of saline drainage water for irrigation: imperial Valley study. Agric Water Manag 16(1–2):25–36

Rhoades J, Merrill SD (1976) Assessing the suitability of water for irrigation: theoretical and empirical approaches. In: Prognosis of salinity and alkalinity. FAO Soils Bulletin 31. FAO, Rome, Italy, pp 68–108
Rhoades JD, Kandiah A, Mashali AM (1992) The use of saline waters for crop production. FAO Irrigation and Drainage Paper No. 48. FAO, Rome
Rhoades JD (1974) Drainage for salinity control. In: van Schilfgaarde J (ed) Drainage for agriculture. Agronomy Monographs No. 17, vol 17. American Society of Agronomy, Inc. (ASA), Madison, WI, pp 433–461
Rhoades JD (1999) Use of saline drainage water for irrigation. In: Skaggs RW, van Schilfgaarde J (eds) Agricultural drainage. Agronomy Monographs No. 38, vol 38. American Society of Agronomy (ASA), Madison, WI, pp 615–657
Roberts T, Lazarovitch N, Warrick AW, Thompson TL (2009) Modeling salt accumulation with subsurface drip irrigation using HYDRUS-2D. Soil Sci Soc Am J 73:233–240
Rossel RAV, Adamchuk VI, Sudduth KA, McKenzie NJ, Lobsey C (2011) Proximal soil sensing: an effective approach for soil measurements in space and time. Adv Agron 113:237–282. https://doi.org/10.1016/B978-0-12-386473-4.00010-5
Sanden BL, Ferguson L, Reyes HC, Grattan SR (2004) Effect of salinity on evapotranspiration and yield of San Joaquin Valley pistachios. Acta Hort 664:583–589
Schirrmann M, Gebbers R, Kramer E, Seidel J (2011) Soil pH mapping with an on-the-go sensor. Sensors 11:573–598
Scudiero E, Skaggs TH, Corwin DL (2014) Regional scale soil salinity evaluation using Landsat 7, western San Joaquin Valley, California, USA. Geoderma Reg 2–3:82–90. https://doi.org/10.1016/j.geodrs.2014.10.004
Scudiero E, Corwin DL, Anderson RG, Yemoto K, Clary W, Wang ZL (2017a) Remote sensing is a viable tool for mapping soil salinity in agricultural lands. California Agricult 71(4):231–238. https://doi.org/10.3733/ca.2017a0009
Scudiero E, Skaggs TH, Corwin DL (2017b) Simplifying field-scale assessment of spatiotemporal changes of soil salinity. Sci Total Environ 587:273–281. https://doi.org/10.1016/j.scitotenv.2017.02.136
Shainberg I, Singer MJ (2012) Soil response to saline and sodic conditions. In: Wallender WW, Tanji KK (eds) ASCE manuals and reports on engineering practice No. 71 Agricultural salinity assessment and management, 2nd ed. ASCE, Reston, VA, pp 139–167
Shainberg I, Letey J (1984) Response of soil to sodic and saline conditions. Hilgardia 52(2):1–57
Shalhevet J (1994) Using water of marginal quality for crop production: major issues. Agric Water Manag 25(3):233–269
Shannon MC, Grieve CM, Wilson C, Poss J, Suarez DL, Lesch S, Rhoades JD (1998) Growth and water relations of plant species suitable for saline drainage water reuse systems. Final report to California Dept. of Water Resources, Project DWR B-59922. California Department of Water Resources, Sacramento, CA
Shennan C, Grattan SR, May DM, Hillhouse CJ, Schactman DP, Wander M, Roberts B, Gurau RG, McNeish C, Zelinski L (1995) Feasibility of cyclic reuse of saline drainage in a tomato-cotton rotation. J Environ Qual 24(3):476–486
Shouse PJ, Goldberg S, Skaggs TH, Soppe RWO, Ayars JE (2010) Changes in spatial and temporal variability of SAR affected by shallow groundwater management of an irrigated field. California. Agricult Water Manag 97(5):673–680
Šimůnek J, van Genuchten MT, Sejna M (2016) Recent developments and applicatioins of the HYDRUS computer software packages. Vadose Zone J. https://doi.org/10.2136/vzj2016.04.0033
Singh A, Quinn NWT, Benes SE, Cassel F (2020) Policy-driven sustainable saline drainage disposal and forage production in the western San Joaquin Valley of California. Sustainability 12(16):6362. https://doi.org/10.3390/su12166362
Skaggs TH, Anderson RG, Corwin DL, Suarez DL (2014) Analytical steady-state solutions for water-limited cropping systems using saline irrigation water. Water Resour Res 50:9656–9674

Smith DJ, Oster JD, Sposito G (2015) Potassium and magnesium in irrigation water quality assessment. Agric Water Manag 157:59–64

Sposito G (2008) The chemistry of soils, 2nd edn. Oxford University Press, New York

Suarez DL, Wood JD, Lesch SM (2006) Effect of SAR on water infiltration under a sequential rain-irrigation management system. Agric Water Manag 86(1–2):150–164. https://doi.org/10.1016/j.agwat.2006.07.010

Suarez DL (2012) Irrigation water quality assessments. In: Wallender WW, Tanji KK (eds) ASCE manuals and reports on engineering practice No. 71 Agricultural salinity assessment and management, 2nd ed. ASCE, Reston, VA, pp 343–370

Suyama H, Benes SE, Robinson PH, Getachew G, Grattan SR, Grieve CM (2007a) Biomass yield and nutritional quality of forage species under long-term irrigation with saline-sodic drainage water: field evaluation. Animal Feed Sci Technol 135(3–4):329–345

Suyama H, Benes SE, Robinson PH, Grattan SR, Grieve CM, Getachew G (2007b) Forage yield and quality under irrigation with saline-sodic drainage water: greenhouse evaluation. Agricult Water Manag 88(1–3):159–172

Tanji KK, Kielen NC (2002) Agricultural drainage water management in arid and semi-arid areas: Issue 61 of FAO Irrigation and Drainage Paper. Food and Agriculture Organization of the United Nations (FAO), Rome, Italy

U.S. Salinity Laboratory Staff (1954) Diagnosis and improvement of saline and alkali soils. USDA Agriculture Handbook 60. Washington, DC

UN-Water (2020) UN-water analytical brief on unconventional water resources. Geneva, Switzerland

USBR (U.S. Bureau of Reclamation) (2009) Record of Decision Grassland Bypass Project, 2010–2019 ROD-07–141. U.S. Department of the Interior, Bureau of Reclamation, South-Central California Area Office, Fresno, CA, 15 pp

USBR (U.S. Bureau of Reclamation) (2020) Management Agency Agreement 2020 Annual Report, October 1, 2019–September 30, 2020

van de Craats D, van der Zee SEATM, Sui C, van Asten PJA, Cornelissen P, Leijnse A (2020) Soil sodicity originating from marginal groundwater. Vadose Zone J 19:e20010. https://doi.org/10.1002/vzj2.20010

van Genuchten MT (1980) A closed-form equation for predicting the hydraulic conductivity of unsaturated soils. Soil Sci Soc America J 44:892–898

van Schilfgaarde J, Bernstein L, Rhoades JD, Rawlins SL (1974) Irrigation management for salt control. J Irrigation Drainage Div, American Soc Civil Eng (ASCE) 100(3):321–338

Whitney M, Means TH (1897) An electrical method of determining the soluble salt content of soils. (USDA Soils Division Bulletin 8). US Government Printing Office, Washington, DC, 30 pp

Yaduvanshi NPS, Sharma DR (2008) Tillage and residual organic manures/chemical amendment effects on soil organic matter and yield of wheat under sodic water irrigation. Soil Tillage Res 98(1):11–16. https://doi.org/10.1016/j.still.2007.09.010

Yang T, Šimůnek J, Mo M, Mccullough-Sanden B, Shahrokhnia H, Cherchian S, Wu L (2019) Assessing salinity leaching efficiency in three soils by the HYDRUS-1D and -2D simulations. Soil Tillage Res 194:104342. https://doi.org/10.1016/j.still.2019.104342

Part V
Moving Water Physically

Chapter 9
Water Transportation via Icebergs Towing

Manzoor Qadir and Nisal Siriwardana

Abstract Climate change has fast-tracked the breaking of huge chunks of ice—icebergs—in the polar regions and subsequent drifting of the icebergs across the ocean. Despite being the world's largest untapped freshwater source, icebergs continue to decay in the ocean over time in an era when freshwater shortages intensify in dry areas of the world, which desperately look for every option to augment water resources. Thus, an environmental concern—increasing iceberg calving—may just offer relief to a troubling reality—intensifying water scarcity. However, the idea of harnessing icebergs to produce freshwater is not a new one, although no one has yet towed icebergs from the Arctic and Antarctic oceans to provide freshwater to water-scarce areas. Frequent droughts and growing water scarcity in recent years have led to renewed interest in towing icebergs from polar ice caps to dry areas in Africa and Middle East. The timing is pertinent due to the increasing need for freshwater, the continued abundance of icebergs, and advancements in the science and technology to make iceberg harvesting possible despite skepticism over financial and technological challenges and the lack of legal instruments. This chapter addresses the history, technological interventions, research status, and major tradeoffs of water transportation related to icebergs' towing, while highlighting the importance of icebergs as an unconventional water resource with massive potential to address growing water scarcity across the world.

Keywords Water scarcity · Iceberg calving · Global warming · Antarctic ice · Polar ice caps · Water resources

9.1 Introduction

Climate change-driven rise in temperatures is common throughout the world, thereby impacting the face of our planet. Huge chunks of ice in the polar regions have been observed breaking off and subsequently drifting in slow decay across the ocean

---

M. Qadir (✉) · N. Siriwardana
United Nations University Institute for Water, Environment and Health (UNU-INWEH), 204-175 Longwood Road South, Hamilton, Ontario L8P 0A1, Canada
e-mail: Manzoor.Qadir@unu.edu

M. Qadir et al. (eds.), *Unconventional Water Resources*,
https://doi.org/10.1007/978-3-030-90146-2_9

(UN-Water 2020). Meanwhile, around 60% of the global population lives in areas of water stress where available water resources cannot sustainably meet the demand for at least part of the year (Damania et al. 2017). While global warming may potentially alter global hydrological cycle patterns, the two troubling realities—huge icebergs breaking and water scarcity—owe their fate to the same phenomenon as one may just offer relief to the other (UN-Water 2020).

Approximately 75% of the world's freshwater is held in ice, and of that volume, approximately 90% sits in the Antarctic. The total volume of Antarctic ice contains 27 million km^3 of water (Lewis 2015). Of the total volume of ice, Antarctica annually calves approximately 93% of the world's total iceberg mass. As massive chunks of ice, icebergs can provide a substantial, constantly renewable, and potentially environmentally neutral untapped freshwater source. More than 100,000 Antarctic icebergs melt into the ocean each year. They range from merely large to country-size (the biggest seen recently was the size of Jamaica), and by some calculations they contain more than the annual global consumption of freshwater.

Towing an iceberg from one of the polar ice caps to a water-scarce country may not seem like a practical solution to water shortages, but scientists, scholars, and politicians have been considering iceberg harvesting as a potentially viable freshwater source for several decades (Spandonide 2009; Lewis 2015). Furthermore, iceberg-towing technology is available as the Canadian oil and gas industry regularly tows icebergs away from offshore platforms when there is a risk of collision. Although it has not yet been carried out on a large scale, the increasing need for freshwater around the world, the continued abundance of icebergs, and the advancements in technology and science for iceberg harvesting might soon expedite the efforts to make it possible to launch iceberg towing in practice.

As water scarcity is expected to continue and intensify in dry and overpopulated areas, the water-scarce areas must sustainably access and utilize every available option for enhancing water resources to minimize the pressure that continues to grow. Iceberg towing to dry areas can provide critical support to the associated communities for addressing local water shortages. Moreover, the fate of towed icebergs for reducing water scarcity is difficult to evaluate due to lack of experience, and the development of future scenarios and projections utilizing icebergs for water supply to dry areas is likewise difficult.

In addition to water supply, it is interesting to note that icebergs have the potential to produce energy (Cohen et al. 1982). A large amount of energy can be obtained through the thermal gradient if the icebergs are transported to lower latitudes. In the case of conventional stations operating with fossil fuels, ice can lower the condensation temperature.

This chapter addresses various aspects of water transportation regarding iceberg towing and highlights its importance as an unconventional water resource.

9.2 Technological Interventions

Based on modern science and technology, it appears that towing an iceberg from one of the polar regions to a warmer climate across the ocean is possible. The technical feasibility of iceberg towing can be broken down into four parts: (1) locating a suitable source and supply; (2) calculating the necessary towing power requirements; (3) accurately predicting and accounting for in-transit melt; and (4) estimating the economic feasibility of the entire endeavor (Lewis 2015).

The Canadian oil and gas industry regularly tows icebergs away from offshore platforms when there is a risk of collision. But by Antarctic standards, these icebergs are small and are not towed to distant locations. On average, these icebergs are 60–80 m wide at the waterline, weighing around 0.1 million tons. Larger icebergs weighing up to 10 times that (1 million tons) have reportedly been towed with a rope slung between two vessels. Typically, icebergs are towed for a few dozen km to get them out the way of oil and gas platforms sitting offshore (Spandonide 2009).

Because the iceberg towing process needs to move large icebergs from polar ice caps to water-scarce countries in Africa and Middle East, the key challenge would be towing a large mass of ice, for example a 100-million-ton iceberg, through the notoriously rough Antarctic Ocean, where swells regularly reach 15 m and winds can be up to 130 km/h. With the net in place, the iceberg needs to be attached to two supertankers at about 1.6 km, while the tankers remain about 300 m from one another and travel at about 1.6 km/h. Because the tankers have little ability to steer at such low speeds, each tanker needs to be led by a tugboat (Fig. 9.1).

Not any iceberg can be towed and harvested. Due to the massive size, weight, and density of an iceberg, there is a great risk of rolling over while being towed. As such, rectangular icebergs with tabular shapes and horizontal dimensions much larger than their thicknesses are the most desirable (Lewis 2015).

Since transportable icebergs are abundant and exist in various sizes and shapes, the selection of icebergs for towing long distances is feasible via remote-sensing techniques. Primarily found in Antarctica, the tabular icebergs are the biggest in size and the most suitable specimens for iceberg towing (Fig. 9.2).

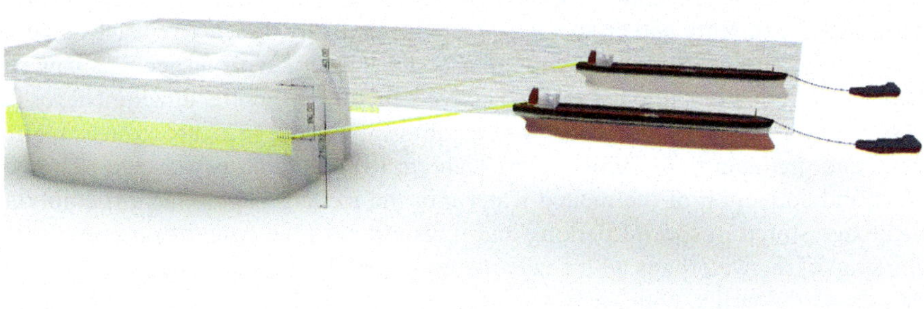

Fig. 9.1 Schematic presentation of iceberg towing technology (*Credit* Nicholas Sloane, Southern Ice Forum)

Fig. 9.2 Antarctic ice has the potential to provide adequate supplies of water to water-scarce areas (*Credit*: Nicholas Sloane, Southern Ice Forum)

There are a range of technological approaches and measures that may play an important role in moving icebergs from Antarctica to countries in dry areas of Africa and Middle East (Spandonide 2009). Such measures consist of: (1) selection of a representative iceberg by detection and selection of a tabular iceberg, elliptical section, without cracks; (2) modelling the physical properties of the selected iceberg by considering the geometry of a variable volume, center of gravity, and axis; (3) assessment of the permanent sensing of the center of gravity of the iceberg and possible cracking, particularly for security reasons for humans and equipment; (4) establishment of a supporting structure for peripheral circulation for humans, maintenance, materials, and equipment; (5) possible wrapping of the iceberg with a filet to enhance its stability and protect it against mechanically destructive waves' action, degradation, and cracking; (6) wrapping of the iceberg with an iceberg bag and emplacement of a metallic net and protective structure; (7) collar attachment for the iceberg's bag, on the floating line, with an auto supporting structure to store melted water and maintain the stability of the iceberg; (8) reduction of wastage of melted water and collection of the melted water into the iceberg bag collar; (9) emptying waterbags stored in special parking and transported to the connection stands; (10) filling-up of the waterbags with newly melted water; (11) disconnection of the water bags from the collar with each waterbag becoming an independent link and the links assembled by three bags; (12) formation of a train of waterbags with the train composed from 10 rows of three links.; and (13) sustainable transportation of the train with maritime currents (Spandonide 2009).

9.3 History

Harvesting and redirecting icebergs is not a new idea. In the mid-1800s, breweries in Chile towed small icebergs from Laguna San Rafael to Valparaiso, where they were used for refrigeration (Winter 2019). In the late 1940s, John Isaacs of the Scripps Institution of Oceanography began exploring an ambitious plan for the possibility of transporting a massive iceberg to San Diego to mitigate California droughts. Such extensive icebergs—approximately 30-km long, 900-m wide, and 300-m deep—are extremely rare. In the 1960s, oil companies using large steam-powered ships entering transatlantic service began using thick ropes to wrangle and redirect much smaller Arctic icebergs before they collided with oil rigs, a practice that is common these days.

Since the 1960s, there have been discussions about iceberg towing to bring freshwater to water-scarce areas. In the 1970s, the U.S. Army and the Rand Corporation, an institution that develops solutions to public-policy challenges, started investigating the use of Antarctic ice as a source of freshwater. At about the same time, Prince Mohammed Al-Faisal of Saudi Arabia became interested in polar research, anticipating that his assembled team of international glaciologists and engineers would find a way to alter the drift of icebergs, potentially bringing them as far as Western Australia. In 1977, Prince Mohammad Al-Faisal sponsored the first large-scale conference on iceberg utilization, which was named as the First International Conference and Workshops on Iceberg Utilization for Fresh Water Production, Weather Modification and Other Applications. Prince Al-Faisal, the President of the Iceberg Transport International Ltd. (a company he founded) in Saudi Arabia, became interested in iceberg transportation as a means of solving the water shortage plaguing the country. His goal was to transport an iceberg within 10 years to the Middle East region to supply water and perhaps use cloud seeding to somewhat alter the arid desert climate. Around the same time when Prince Al-Faisal was embarking on iceberg towing, a French engineer, Georges Mougin, became passionate in the 1970s about the subject and initiated and led the iceberg-harvesting movement. He worked closely with Prince Al-Faisal as Technical Director of Iceberg Transport International Ltd. He was the key participant, main lead, and at the forefront of the conference.

Held at Iowa State University (2–6 October 1977) with 175 participants from around the world, the conference on icebergs addressed various aspects of the feasibility of using icebergs as alternative water and energy resources to address the growing concern on global water and energy shortages. This forward-looking conference discussed the patterns of cooperation in international science and technology and the evaluation of subsidiary effects and concomitant issues and challenges in iceberg utilization. On the technical front, the conference discussed elements of iceberg technology, thicknesses of icebergs, sources and properties of tabular icebergs and towing, handling, processing, and selection of icebergs for towing to water-scarce areas, as well as ecological considerations of iceberg transport from Antarctic waters and energy and freshwater production from icebergs. In addition, the conference participants addressed weather modification, environmental, economic, social, and political implications, and other topics relating to icebergs as a means of freshwater supply. Although the conference ended in skepticism over the eventual possibility of towing an iceberg, the concept remained alive. By 1982, the interest from Prince Al-Faisal and other organizers of the conference had significantly been decreased and later ended, largely due to the lack of support from the government. However, the discussions about iceberg towing continued sporadically over the next two decades, but without a major event or a planned initiative.

Georges Mougin continued working towards his dream of towing an iceberg to water-scarce areas. In 2011, he partnered with a French firm Dassault Systèmes to utilize its advanced 3D modeling system, declassified satellite data, and the relatively new science of oceanic forecasting. The team ran a successful 3D computer simulation of towing an iceberg from Newfoundland, nearly 5,000 km across the Atlantic Ocean, to the Canary Islands, and suggested that iceberg harvesting and towing to a water-scare area is a real possibility. This successful computer simulation and the ever-growing global need for freshwater became the main drivers to the increasing likelihood and common practice of iceberg harvesting.

Olav Orheim, a Norwegian glaciologist, is another key figure in iceberg research. He served as a professor in glaciology at the University of Bergen and as a director of the Norwegian Polar Institute (1993–2005). Prince Al-Faisal also contacted him about iceberg-towing research. Olav Orheim was a central participant in the establishment of the research station Troll in Queen Maud Land in Antarctica. He continues to be interested in the management of polar affairs, climate change, and communicating science to the public.

In recent years, Nicholas Sloane, a salvage master and Director of Resolve Marine who began his career in 1980 working on the tugs of a South African salvage company, has become passionate about towing icebergs from the polar ice caps to water-scarce areas. His work has taken him around the world, wrestling with the wrecks of boats, rigs, and planes from New Zealand to Yemen. Making use of his unusual skill set, he plans to harness and tow an enormous Antarctic iceberg to South Africa and convert it into municipal water. Based on the financial feasibility analysis, he suggests that the iceberg to be towed will have to be big, i.e., about 1,000-m long, 500-m wide, and 250-m deep, and weigh 125 million tons. Such a volume of water would supply about 20% of Cape Town's water needs for a year. Nicholas Sloane has already assembled a team of glaciologists, oceanographers, and engineers. He has also secured a group of financiers to fund the pioneer iceberg towing project called "the Southern Ice Project". The expected cost is more than $200 million, much of it to be put up by two South African banks and Water Vision AG, a Swiss water technology and infrastructure company. Currently, Sloane's team is in the process of discussion to make an agreement with South Africa to buy the Antarctic water if the plan succeeds. Such discussions and possible actions are timely as South Africa has declared a national disaster over the drought that hit its southern and western regions. This latest drought, after 2015 and 2016, resulted in two of the driest years on record. Severe water restrictions are already in place, and Cape Town is at real risk of running dry completely. Authorities have warned that taps could run dry altogether if winter rainfall is not enough to rescue the four million residents of Cape Town. The city is by far the most conveniently located city for a pioneer iceberg towing project, given its relative proximity to Antarctica and the path of the Benguela Current.

Iceberg towing from the polar ice caps to water-scarce Middle East region to end droughts has been in discussion in print and electronic media for several decades. Such discussions have been intensified due to the growing water scarcity. In 2017, the United Arab Emirates (UAE) experienced severe water shortages, and a project was set up to tow an iceberg to the region. Given the pace of development in UAE, groundwater supplies are predicted to run dry in the next 15 years, thereby requiring additional supplies of good-quality water and towing icebergs to the country a likely remedy. Abdulla Alshehi, Managing Director of National Advisor Bureau Ltd. in UAE, is an enthusiastic proponent of iceberg towing and considers iceberg towing a crucial and viable option to increase the UAE's water supply. He considers that iceberg towing may be less costly than the capital investment in desalination plants, which is a key source of freshwater supply for the UAE population. Alshehi believes that iceberg towing projects will alleviate the negative impacts of brine discharges from desalination plants that damage marine life in the Arabian Sea (Euronews 2019). His estimates suggest that the full cost of towing an iceberg from Antarctica to UAE's Fujairah coast will be of the order of US$100–150 million. However, to this date, no proposal, funding, or action on the ground have been initiated (Aquatech 2021).

Actual iceberg towing has not yet been done and remains in its infancy despite several initiatives and extensive discussions about the topic. However, such efforts may materialize soon, with an iceberg towing project realized by moving one from a polar ice cap to a water-scarce country. The timing is pertinent due to the increasing need for freshwater around the world, the continued abundance of icebergs, and the advancing state of the technology and science that would be necessary to make iceberg harvesting possible.

9.4 Research Status

The possibility of utilizing icebergs to provide freshwater to water-scarce areas has been researched for over 50 years. Antarctic icebergs were identified as a realistic source for freshwater, but a specific project was never formally undertaken (Spandonide 2012). Research geared up during 1970s due to the interest of multiple stakeholders— a high-level political figure (Prince Al-Faisal, Saudi Arabia), academia/research (International Conference on Iceberg Utilization for Fresh Water Production), national security and public policy institutions (U.S. Army and the Rand Corporation, USA), and individual initiatives by professionals (Georges Mougin, France; Olav Orheim, Norway). However, skepticism over financial and technological challenges and a lack of funding for projects has resulted in a slowdown of research into harnessing the potential of iceberg towing to address water scarcity (Spandonide 2012).

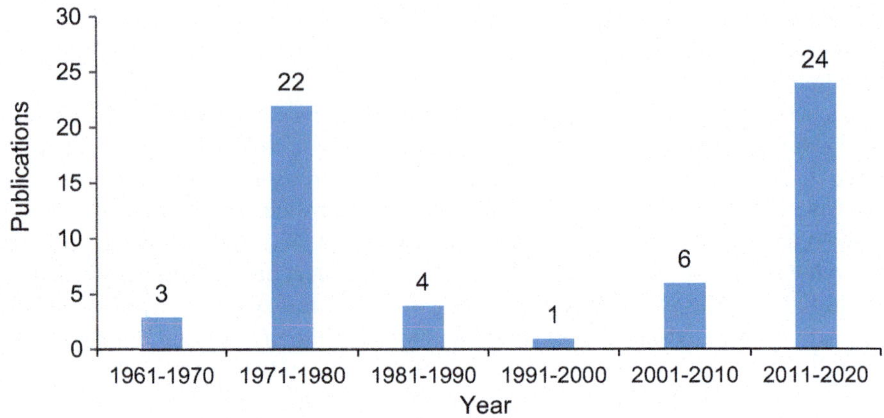

Fig. 9.3 Timeline of publications on iceberg towing, by decade

In the last decade, the frequency of research related to iceberg towing has significantly increased (Fig. 9.3). However, the focus on research is more towards iceberg towing with regards to collision and hazard mitigation. As polar activities and operations increase, there has been a surge in research related to iceberg-hazard modelling assessments. To further amplify the situation, iceberg numbers have increased because of global warming (Bigg et al. 2018). Since 2016, iceberg-towing experiments have been ongoing, such as the towing experiment conducted in the Barents and Kara Seas to assess the oscillation during the towing process over a wide range of towing speeds, and iceberg-water resistance coefficients (Kornishin et al. 2019). Such results can be used to better understand the planning process of iceberg-towing operations. In 2020, during the 39th International Conference on Ocean, Offshore and Arctic Engineering, a 3D simulation model of iceberg towing was presented, prior to implementing iceberg towing, to provide optimal towing configurations to assure the feasibility and sustainability of the towing strategy (O'Rourke et al. 2020). Methods of mapping the underside of icebergs are also being researched due to the potential risks associated with underwater infrastructure damage and marine transportation hazards caused by icebergs (Zhou et al. 2019). Studies to enhance understanding of the balance of heat and freshwater through iceberg-melting processes are being considered through adaptive machine-learning approaches to automatically detect icebergs through high-resolution synthetic aperture radar images, SAR (Barbat et al. 2019). As more data on icebergs is gathered, the opportunity to extrapolate that knowledge to harness the potential of iceberg towing will be valuable in the coming years.

In the print and electronic news media in recent years, the popularity of iceberg towing as a water resource gained traction during the water crisis in Cape Town, South Africa, and in the UAE. The city of South Africa was 90 days away from turning off the taps (Edmond 2019), and the UAE has had ongoing water-management challenges, such as scarcity of groundwater reserves, elevated levels of salts in existing groundwater, and the high cost of producing drinking water. Because South Africa and the UAE are expected to continue experiencing severe water shortages, there is a call for action to find multiple sources of water supply to alleviate such water stresses. Consequently, this has been the spark to trigger actions on the ground, thereby motivating certain individuals and institutions to investigate the operational aspects of iceberg towing. There are discussions on the economics of iceberg towing that consider it competitive with other water supply options in dry areas (Euronews 2019). In recent years, Nicholas Sloane (South Africa) and Abdulla Alshehi (UAE) are the main proponents of implementing large-scale projects on iceberg towing.

In the Northern Hemisphere, people have started utilizing icebergs off the coasts of Canada, Greenland, and Norway at a domestic level. Residents of Newfoundland, Canada, head out to catch ice for their own uses, and in Greenland, a company named Nukissiorfiit uses icebergs to supply water to 700 residents of Qaanaaq (Birkhold 2019). The progress of iceberg harvesting in the north raises concern due to the lack of laws concerning iceberg harvesting as commercialization of icebergs gains traction. Currently, the government of Newfoundland and Labrador is trying to more closely regulate (the profit from) iceberg collection. To lawfully harvest icebergs, one must have a license in Newfoundland. Such licensees must identify or mark the icebergs they intend to collect and stipulate that only one iceberg may be harvested at a time (Birkhold 2019). The regulatory landscape of iceberg towing and harvesting will be subject to change and evolve as corporate entities begin entering the market.

9.5 Trade-offs

There are technical challenges to long-distance iceberg towing. Such challenges must be overcome. For example, the stability of the icebergs being towed and their structural integrity are important factors for an iceberg-towing project. To address such issues, synthetic fiber ropes, which are stronger than steel, can be slung around icebergs at the waterline, but when tugging begins, the rope can slip off or cause the iceberg to roll over. Another reason the towing must be done slowly and carefully is that dragging an iceberg through the ocean can break it apart. The iceberg-towing industry has devised nets for capturing unstable icebergs, but they do not work in every case. The fracture, breakup, and melting of icebergs during transportation may cause a decrease in significant volumes of water initially present in the iceberg at the starting point of its journey from the Northern Hemisphere. In addition, there will be the need for a sufficiently powerful vessel to tow the mass of an iceberg to water-scarce countries in Africa and the Middle East.

Analysis of iceberg stability suggests that many icebergs break into two almost-equal parts after escaping from the relative protection afforded by the pack ice (Spandonide 2009). The secondary smaller icebergs are unstable and ultimately overturn, which accelerates their decay. Suitable icebergs need to be thick, stable, and with little or no cracks. Although not tested, the selected icebergs with a large mass ($\geq$1 and up to 100-million tons) would be wrapped first in a net and then in a mega-bag.

Thus, Spandonide (2009) proposed that iceberg wrapping needs be performed in two steps. The first step is related to wrapping with a net that will protect it from additional calving, splitting or breakage. The net installed below the bag is used to expel the seawater, to manage the air released during ice melting, to enhance the iceberg stability, and to fit the bag to the iceberg shape, by drawing the bag tight around the iceberg. Therefore, the net compression on the iceberg needs to be stronger than the pressure of the seawater. This will improve both the stability of the iceberg and the handling of its melted water. The evolution of the shape of the bag during the melting process could be monitored and the stability of the iceberg controlled. Once the bag is wrapped, the installed net in the bag need to expel the seawater to manage the air released during ice melting and fit to the iceberg shape. The pressure of the net needs to be stronger than the seawater to exert a constant pressure on the sides of the iceberg. The second step of iceberg wrapping is employing a mega bag that will have an appropriate geometry, like a flexible pillow-membrane container, of cubic or cylindrical shapes (Gleick 2001). For iceberg wrapping, a helicopter may need to be dispatched to emplace the bag over the iceberg. The bag can be attached on the top of the iceberg and then by gravity, using its own weight, the bag can slide down the iceberg sides. The bag can also be slipped on horizontally like a giant slipper along the iceberg and then closed. In both cases, the net would serve as a holding structure for the bag, containing a railing system.

Should iceberg harvesting prove to be a successful intervention, the environmental benefits would be extensive in terms of water-resource augmentation, reduced pollution, improvements of water and air qualities, and drought and wildfire mitigation, among others. However, there has been insufficient analysis of the potential adverse environmental effects during any phase of the iceberg harvesting process. Thus, there is a need to undertake comprehensive environmental impact assessments of iceberg towing by considering (1) the impacts to the region of origin; (2) the transit-related impacts to the ocean and the climate; and (3) the delivery site impacts, from offshore processing to onshore processing and domestic distribution (Lewis 2015).

The economics of iceberg towing is an important factor because large investments are needed to implement an iceberg towing project amid skepticism concerning the successful outcome. A single iceberg-towing vessel can cost around $75,000 a day, and to tow a massive iceberg might require several ships simultaneously. The financial feasibility analysis of the iceberg towing to Cape Town suggests that it is an economically attractive option if the iceberg to be towed are large enough and weigh at least 125-million tons. It is important to understand and analyze the economics of action and inaction to overcome the perception of the high costs of iceberg towing projects by undertaking comprehensive analyses of innovative financing mechanisms, the cost of alternate water-supply options, and the economic and social costs. In principle, the projected costs of iceberg harvesting can be broken down into four components: (1) technological innovation; (2) iceberg identification and retrieval; (3) transportation; and (4) arrival site processing and distribution (Lewis 2015).

Beyond technologies and technological innovations, the nontechnical aspects of iceberg towing are as critically important as the water-resource augmentation because an enabling environment is necessary to support such projects that involve harnessing the potential of a water resource via transatlantic transportation. If iceberg harvesting becomes a reality, the practice would be operating in a legal vacuum. There is a lack of a current iceberg-harvesting market, and neither the international community nor any individual nation has promulgated laws or regulations expressly designed to regulate the iceberg-harvesting process. There is a need to create legal instruments and enter into agreements to avoid this potential legal vacuum within the context of towing and harvesting Antarctic icebergs, or at least to minimize the effects on the growth of iceberg harvesting as a new natural-resource industry.

There are two international agreements currently in force that may have legal implications regarding the rights of a country or a private sector entity to obtain, tow, and harvest an Antarctic iceberg (Lewis 2015). First, there is the Antarctic Treaty System, which is an international agreement that serves as the principal governing instrument over Antarctica. Second, there is the Law of the Sea: a centuries-old concept that was codified into international law by the United Nations in 1982 in the United Nations Convention on the Law of the Sea (UN-CLOS). The UN-CLOS has been ratified by over 150 countries and the European community, and only a small group of nations, including the United States, have not ratified it yet (UN-Water 2020).

The Antarctic Treaty System defines Antarctica as all land and ice shelves south of 60°S latitude parallel. The treaty was put in place in 1959 with the signatures of 12 countries. The issue of the potential use of icebergs was raised at several meetings of the Consultative Parties of Antarctic Treaty, but icebergs were eventually not included in the negotiations and not ratified. Currently, icebergs are not included in the 1991 Protocol on Environmental Protection to the Antarctic Treaty, and, consequently, iceberg exploitation is not subject to the current moratorium (Spandonide 2009). The UN-CLOS does not expressly control icebergs or iceberg harvesting. The convention may influence iceberg towing as it recognizes territorial waters and the legitimacy of national sovereignty within them.

Current plans and discussions on iceberg harvesting target icebergs in the open sea north of 60°S. If plans emerge to exploit icebergs within 200 nautical miles of the Antarctic coast, then this would likely create problems between the claimant and nonclaimant states (Spandonide 2009). Another aspect relates to the decline in the area, extent, and volume of sea ice and the melting of the Greenland ice sheet attributed to the increased greenhouse effect caused by the increase in carbon dioxide. Such climate change is having a direct impact on the people that live in the Arctic. Moving icebergs from the Arctic to dry areas may be of concern to the four million Arctic inhabitants.

9.6 Conclusions

Iceberg towing from the polar ice caps to water-scarce areas to end droughts has been in discussion in print and electronic media for several decades, and the topic has been researched intermittently for over 50 years. Antarctic icebergs have been identified as a realistic source of freshwater, but a specific project has never been formally undertaken. Research geared up during 1970s due to the interest of multiple stakeholders, but skepticism over financial and technological challenges and lack of funding for projects led to a slowdown of research into harnessing the potential of iceberg for addressing water scarcity. Frequent droughts and growing water scarcity in recent years have led to renewed interest in towing icebergs from the polar ice caps to dry areas in Africa and the Middle East.

More than 100,000 icebergs melt into the Arctic and Antarctic oceans each year. They decay slowly in oceans despite their potential to provide a substantial, constantly renewable, and potentially environmentally neutral untapped freshwater source. Although towing icebergs from one of the polar ice caps to a water-scarce country in need may not seem like a practical solution to water shortages, recent developments in science and technology have led scientists, scholars, and politicians to consider iceberg harvesting as a potentially viable freshwater source.

The technical feasibility of iceberg towing from Antarctica to the countries in dry areas of Africa and the Middle East can be broken down into four parts: (1) locating a suitable source and supply; (2) calculating the necessary towing power requirements; (3) accurately predicting and accounting for in-transit melt; and (4) estimating the economic feasibility of the entire endeavor. Since transportable icebergs are abundant and exist in various sizes and shapes, the selection of icebergs for long-distance towing is feasible via remote sensing techniques. Primarily found in Antarctica, tabular icebergs are the biggest in size and the most suitable for iceberg towing.

Iceberg towing technology is available as the Canadian oil and gas industry regularly tows icebergs away from offshore platforms when there is a risk of collision. However, such technology has been undertaken and proved effective only on a small scale. There are important drivers to push the use of iceberg-towing technology on a large scale to expedite the efforts to launch iceberg towing practice for water-resource augmentation. Such drivers include, but are not limited to, the increasing need for freshwater around the world, continued abundance of icebergs, and advancements in technology and science for iceberg harvesting.

There are a range of challenges and trade-offs concerning long-distance iceberg towing that need to be addressed in the process of making iceberg towing a successful and sustainable strategy in the long run. Thus, further efforts and refinements of existing work are needed on the following aspects, but not limited to them:

- Cracking, breaking, and overall stability of the icebergs being towed.
- Loss of significant volume of water during the iceberg transportation process.
- Transforming existing small-scale towing technology to a mega-scale iceberg-towing mechanism.
- Environmental impacts, including the carbon footprints of the iceberg-towing process.
- Economics of iceberg towing, while considering innovative financial mechanisms and costs of alternate sources of water.
- Legal instruments and agreements in the context of towing Antarctic icebergs to other continents and countries.

References

Aquatech (2021) Iceberg harvesting solution melts on validation progress. AquaTech. 11 February 2021. https://www.aquatechtrade.com/news/water-treatment/iceberg-harvesting-solution-melts-on-validation-progress/

Barbat MM, Wesche C, Werhli AV, Mauricio MM (2019) An adaptive machine learning approach to improve automatic iceberg detection from SAR images. ISPRS J Photogramm Remote Sens 156:247–259. https://doi.org/10.1016/j.isprsjprs.2019.08.015

Bigg GR, Cropper TE, O'Neill CK, Arnold AK, Fleming AH, Marsh R, Stephens R (2018) A model for assessing iceberg hazard. J Nat Hazard 92:1113–1136. https://doi.org/10.1007/s11069-018-3243-x

Birkhold MH (2019) $166 water could dictate international iceberg law: high-end start-ups and far-flung tycoons are rushing to cash in on the arctic frozen treasures. Regulators are struggling to keep up. The Atlantic, 31 October 2019. https://www.theatlantic.com/science/archive/2019/10/iceberg-water-and-race-exploit-arctic/601147/

Cohen R (1982) Energy from the ocean. Philos Trans Royal Soc London. Ser A, Math Phys Sci (1499):405–437. https://doi.org/10.1098/rsta.1982.0119

Damania R, Desbureaux S, Hyland M, Islam A, Moore S, Rodella A-S, Russ J, Zaveri E (2017) Uncharted waters: the new economics of water scarcity and variability. World Bank, Washington, DC

Edmond C (2019) Cape Town almost ran out of Water. Here's how it averted the crisis. World Economic Forum. 23 August 2019. https://www.weforum.org/agenda/2019/08/cape-town-was-90-days-away-from-running-out-of-water-heres-how-it-averted-the-crisis/

Euronews (2019) Will an Emirati businessman succeed in towing an iceberg to the UAE? Euronews, 5 July 2019. https://www.euronews.com/2019/07/05/will-an-emirati-businessman-succeed-in-towing-an-iceberg-to-the-uae

Gleick PH (2001) Bagged and dragged. Sci American (February 2001):53. http://www.docme.su/doc/1764182/-scientific-american--february-2001---scientific-american

Kornishin KA, Efimov YO, Gudoshnikov YP, Tarasov PA, Chernov AV, Svistunov IA, Nesterov AV (2019) Iceberg towing experiments in barents and kara seas in 2016–2017. Paper presented at the 29th international ocean and polar engineering conference, 16–21 June 2019, Honolulu, Hawaii, USA

Lewis CJ (2015) Iceberg harvesting: suggesting a federal regulatory regime for a new freshwater source. Boston Coll Environ Aff Law Rev 42:439–472. http://lawdigitalcommons.bc.edu/ealr/vol42/iss2/6

O'Rourke JB, Fugelm MK, King T (2020) 3-D simulation of iceberg towing operations: cable modeling and frictional contact formulation using finite element analysis. In: Proceedings of international conference on ocean, offshore and arctic engineering virtual, online. https://doi.org/10.1115/OMAE2020-18788

Spandonide B (2009) Iceberg freshwater sustainable transportation, a new approach. In: Proceedings of the nineteenth international offshore and polar engineering conference, 21–26 June 2009, Osaka, Japan

Spandonide B (2012) Iceberg water transportation from Antarctica to Australia. Ph.D. thesis, School of Geography and Environmental Studies, University of Tasmania, Australia. https://eprints.utas.edu.au/14798/1/Spandonide_whole_thesis.pdf

UN-Water (2020) UN-water analytical brief on unconventional water resources. Geneva, Switzerland. https://www.unwater.org/publications/un-water-analytical-brief-on-unconventional-water-resources/

Winter C (2019) Towing an iceberg: one captain's plan to bring drinking water to 4 million people. Bloomberg, 6 June 2019. https://www.bloomberg.com/news/features/2019-06-06/towing-an-iceberg-one-captain-s-plan-to-bring-drinking-water-to-4-million-people

Zhou M, Bachmayer R, deYoung B (2019) Mapping the underside of an iceberg with a modified underwater glider. J Field Robotic 36:1102–1117. https://doi.org/10.1002/rob.21873

Chapter 10
Ballast Water

Marlos De Souza

Abstract Ships are the transportation engines of the globalized world, responsible for moving around 90% of the global trade. Unfortunately, together with goods, food, and fuel, ships also transport uninvited aquatic organisms that can establish themselves in the receiving port with massive impact on the economy, public health, and the environment. With around 10 billion tons (10 km^3) of ballast water being discharged every year, a United Nations led International Convention on the Control and Management of Ships' Ballast Water and Sediments was adopted in 2004 and entered into force in 2017. The convention created regulatory framework to which the shipping industry and countries must comply. It means that all ships of 400 gross tonnage or more must manage their ballast water in a way that is approved under the convention. A great deal of work has been done by academic and industrial researchers to devise onboard ballast water treatment options based on various approaches. The regulations essentially have created a new unconventional water source based on treated ballast water. Two approaches are used for such treatments: onboard filtration (desalination) and onshore treatment (desalination). As desalinization is applied as a ballast-water treatment, the end-product (desalinated water) is free of invasive aquatic organisms and unhealthy chemical compounds and is usable for other economic activities such as public water supply and irrigation. Recent developments in desalinization processes have made membranes even more efficient, cost-effective, and compact, which is a perfect combination to be used onboard and onshore to produce a reusable, unconventional water from a ship's ballast.

Keywords Ballast-water · Desalination · Filtration · Water-reuse · Ships

10.1 Introduction

The shipping industry accounts for around 90% of the global trade of raw materials, consumer goods, and essential foodstuffs (IMO 2018). As some of these products

M. De Souza (✉)
Land and Water Division, Food and Agriculture Organization (FAO), Viale delle Terme di Caracalla, 00153 Rome, Italy
e-mail: Marlos.DeSouza@fao.org

M. Qadir et al. (eds.), *Unconventional Water Resources*,
https://doi.org/10.1007/978-3-030-90146-2_10

can only be moved by ship, the recent years have experienced a considerable increase in the tonnage capacity in all segments, except general-cargo carriers, reaching in 2019 a carrying capacity of 1.98 billion deadweight tonnage (dwt), 52 million dwt more than the previous year.

Ships face several challenging operational conditions (i.e., rough weather), including safety issues that are crucial to improve the effectiveness of the transportation process (Krata 2013). A ship's stability against capsizing and excessive heeling is one of the most important topics, not only for naval architects but also for ship operators. Throughout history, the stability of ships has been achieved by placing and distributing extra heavy material (also called ballast) in the bottom section of the vessels. Apart from the cargo, the material was used to improve stability and the safe operation of ships by lowering the vessel's center of gravity.

In the early days of navigation, sand was widely used as ballast by ships. However, loading and downloading the sand to and from ships was time consuming and labor intensive. Therefore, other heavy and compact materials, such as roof tiles and rocks, rapidly replaced sand. Due to its easy availability, rocks became the preferred option. Areas around ports were created to extract rocky material, as well as to receive the discharged rocks from vessels during deballasting operations. In Australia, the ballast brought by the colliers and other ships around 1870 were used on the streets of the city of Wollongong (Gardiner-Garden 1975).

Everything changed in the mid-1850s after the coal shippers in England built bulk carriers using water as ballast instead of dry material, becoming the easiest and cheapest option for the shipping industry (Carlton 1985). It has significantly decreased operating time for loading solid materials and dangerous instabilities due to the movement of solid ballast during a voyage (National Research Council 1996). Nowadays, all ships are fitted with ballast tanks, which can be filled with saltwater, freshwater, or brackish water. Ballast water is also used for other purposes rather than stability, such as adjusting the ship's trim, improving maneuverability, increasing propulsion efficiency, reducing hull stress, raising the ship to pass over shallow areas (reducing draft), and lowering it to get under bridges or cranes (reducing air draft) (Cohen 1998).

Ballast-water operations recently discharge around 10 billion tons (10 km^3) of water every year in foreign waters (Yang 2011). The volume of ballast hold in vessels varies according to their size and purpose, ranging from several m^3 in sailing and fishing boats to hundreds of thousands of m^3 in large cargo carriers (i.e., over 200,000 m^3 in large tankers) (National Research Council 1996). Although the use of water as ballast has improved the operability and safety of ships, it has also created serious environmental problems, including the translocation of invasive and harmful aquatic species (marine or freshwater) and chemicals in marine environments. Alien and non-indigenous species have negatively affected countries globally, not only regarding the local ecological equilibrium but also the economy and human health due to passive importation of bacteria, disease agents, and toxic harmful algae through ballast water (Gollasch and David 2019).

Globally, a detailed assessment of the economic impacts of invasive aquatic species has not been systematically done. However, it has been estimated that the direct economic loss due to invasive species may be in the order of $100 billion a year (IMO 2018). Based on almost all the coastal countries of the world with records of invasive species, the economic impact is shared by all.

The International Maritime Organization of the United Nations (IMO) initiated negotiations to develop an internationally binding instrument addressing the translocation of harmful aquatic organisms and pathogens in ships' ballast water after the UN Conference on Environment and Development (UNCED), held in Rio de Janeiro in 1992. The International Convention for the Control and Management of Ships' Ballast Water and Sediments (the Convention) was finally adopted in 2004 and entered into force in 2017. The Convention aims to prevent, minimize, and ultimately eliminate the risks to the environment, human health, property, and resources arising from the transfer of harmful aquatic organisms in ships' ballast water (IMO 2018).

The convention has driven the shipping industry to look for new technologies to treat ballast water efficiently, in accordance with the introduced regulations. Several options have since been developed applying diverse approaches, such as the use of microfiltration, which is largely being used in desalination plants. Therefore, ships fitted with microfiltration technologies will also be able to produce desalinated water, which can be reused as an unconventional water source at the receiving port. For example, a supertanker fitted with desalination technology and carrying 200,000 m^3 of ballast water would be able to supply enough water to a city of 50,000 inhabitants in Brazil (daily average use of around 155 L/capita) for 25 days.

10.2 Technological Interventions

The Convention requires all ships of 400 gross tonnage and above to possess an approved International Ballast Water Management Certificate. There are several options available for ballast water management under the Convention that can be chosen by the ships (Fig. 10.1). The following are the standard regulations of the Convention (Section D) for Ballast Water Management:

Regulation D-1 Ballast Water Exchange Standard—Ships performing Ballast Water exchange shall do so with an efficiency of 95% volumetric exchange of Ballast Water. For ships exchanging ballast water by the pumping-through method, pumping through three times the volume of each ballast water tank shall be considered to meet the standard described. Pumping through less than three times the volume may be accepted provided the ship can demonstrate that at least 95 percent volumetric exchange is met.

All ships using ballast water exchange should conduct ballast water exchange at least 200 nautical miles from the nearest land and in water at least 200 m in depth.

Regulation D-2 Ballast Water Performance Standard—Ships conducting ballast water management shall discharge less than 10 viable organisms per m^3 greater than or equal to 50 μm in minimum dimension and less than 10 viable organisms per milliliter less than 50 μm in minimum dimension and greater than or equal to

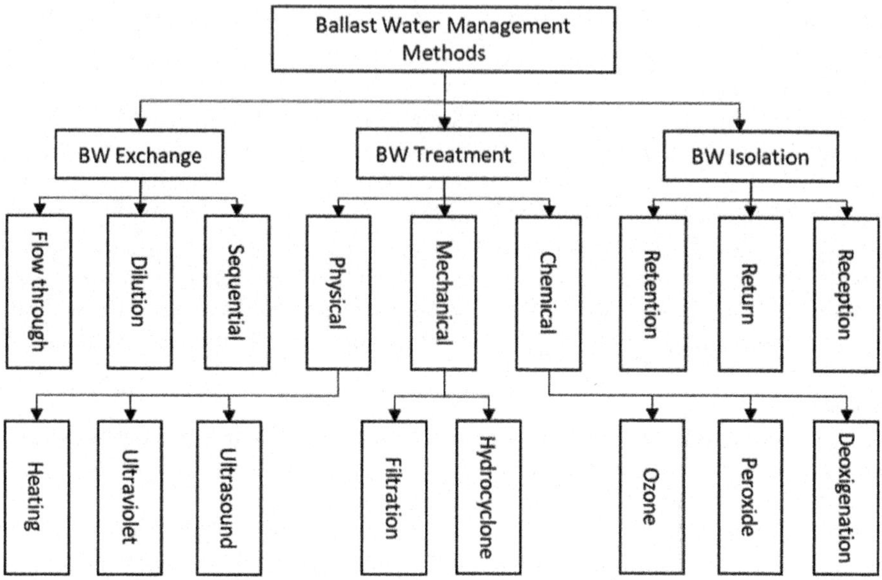

Fig. 10.1 Available options for onboard ballast water (BW) management (Modified from Yongming and Shuhong 2012)

10 μm in minimum dimension; and discharge of the indicator microbes shall not exceed the specified concentrations.

10.2.1 Ballast Water Isolation

Under the Convention, the principle behind the isolation method is that ships can manage their ballast water without deballasting directly into the waters of the destination port. The three accepted options are:

(a) Retention: ships do not need to deballast water as part of their normal operations; therefore, they can retain the water in the ballast tanks. Cruise ships are mainly in this segment because the change in their DWT is usually not very significant during operations so they can keep the same water for months or even years.
(b) Return: the ship is travelling back to its port of origin without deballasting at the destination port. Depending on its operationally, ship may transfer ballast water between its own tanks to allow them to travel back to its port of origin; and
(c) Reception: the receiving port has onshore ballast water-treatment facilities into which the water can be pumped without being discharged directly into the sea. The ballast water will then be treated under the applicable standards described

in the Convention before being either reused for other purposes or discharged back to the environment.

10.2.2 Ballast Water Exchange

Under the Convention (IMO 2018), the exchange method showed in Fig. 10.2 is based on the principle that organisms and pathogens contained in ballast water taken on board from coastal waters will not survive when discharged into deep oceans or open seas because these waters have different temperatures, salinity, and chemical composition. There are three methods stated under the convention for ballast-water exchange:

(a) Sequential Method: A process by which a ballast tank is first emptied and then refilled with replacement ballast water. According to the convention, efficiency is to be at least a 95% volumetric exchange.
(b) Flow-through Method: A process by which replacement ballast water is pumped into a ballast tank, allowing water to flow through overflow or other arrangements. At least three times the tank volume should be pumped through the tank.
(c) Dilution Method: A method by which replacement ballast water is filled through the top of the ballast tank with simultaneous discharge from the bottom at the same flow rate and maintaining a constant level in the tank throughout the ballast-exchange operation. At least three times the tank volume should be pumped through the tank.

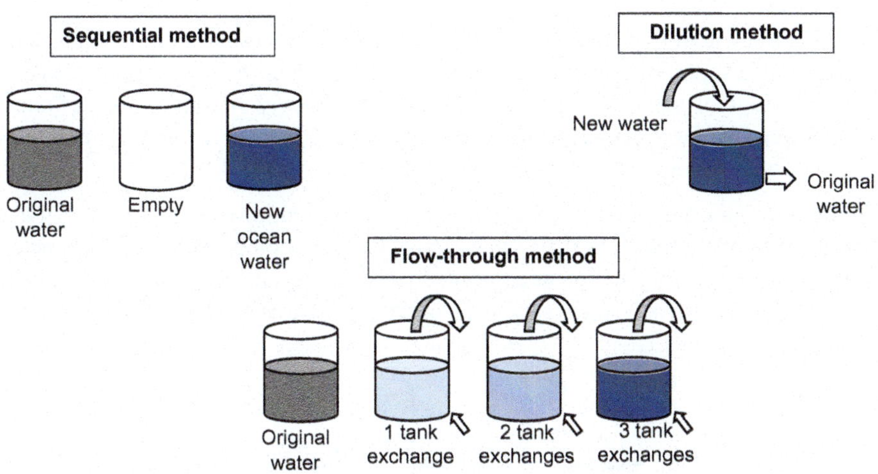

Fig. 10.2 Ballast-water exchange methods; Sequential, Flow-through, and Dilution method

Although these methods are very efficient when conducted properly, ballast-water exchange can be limited by weather conditions, ocean conditions, timing, and the distance to land, making it difficult to always perform.

10.2.3 Ballast Water Treatment Methods

Theoretically, all ships carrying ballast water can manage ballast according to the regulation D1 (ballast-water exchange). Adding to the distance of 200 nautical miles at 200-m deep, exchanging ballast water while enroute is a very complex operation with possibly disastrous consequences if not conducted properly (i.e., structural damages) (Endresen et al. 2004). Therefore, onboard treatment systems are better alternatives for the shipping industry to comply fully with the current regulations in place since they can operate independently of location (i.e., within 200 nautical miles) and some other limiting factors (i.e., time). Currently there are three methods of ballast water treatment: mechanical, chemical, and physical.

Mechanical: During the treatment process, the mechanical separation of aquatic organisms and sediments are divided based on their size (Szczepanek and Behrendt 2018).

(a) Filtration: In the early days of onboard treatment systems, simple filters placed on the ballast-water intake could not prevent small organisms and sediments from entering the ballast tanks (Fig. 10.3). Filters were required to be of a much finer pore size, which made filtration a pretreatment option to improve the performance of secondary treatment systems (Tsolaki and Diamadopoulos 2009).

(b) Hydrocyclone: The operation principle is based on the acceleration of particles and the separation of the light phase from the heavy phase due to different densities of existing materials. Although hydrocyclone has proved efficient to remove large particles, its efficiency was negligible in the elimination of organisms, especially bacteria (Kurtela and Komadina 2010)

Chemical: The principle of chemical treatment is to neutralize microbiological and biological contaminants (Fig. 10.4). Various chemical compounds and approaches are/have been used in isolation or combined with other treatment systems (i.e., filtration). The major disadvantages of the chemical treatments are the generation of

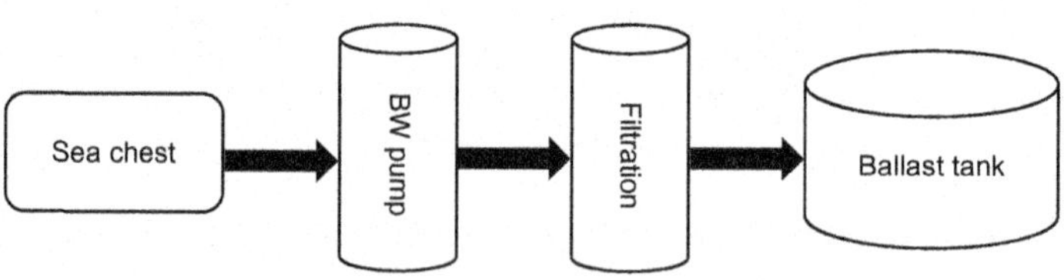

Fig. 10.3 Simple principal scheme for "filtration-only" treatment of ballast water (BW)

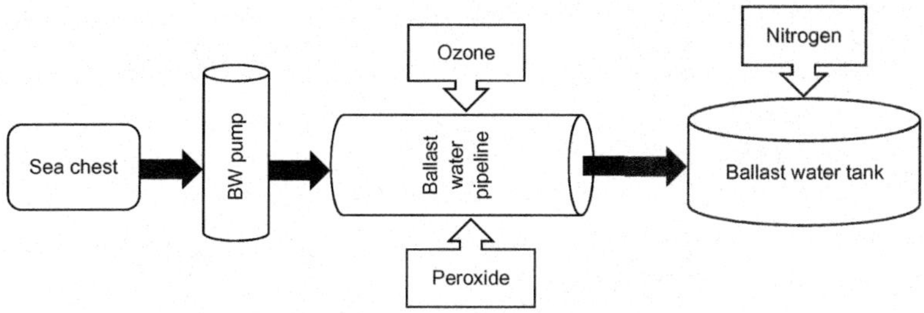

Fig. 10.4 Simple scheme for ballast water (BW) treatment principles for ozone, peroxide, and deoxygenation (by adding nitrogen)

disinfection by-products, the lifetime of the biocides used (i.e., not recommended for short routes), and the need to carry chemical products onboard.

(a) Ozone: It has been used for a long time as a disinfectant in water treatment plants, especially in Europe. Ozone is a very powerful agent to eliminate viruses and bacteria, including spores in freshwater. However, the presence of bromide ions in seawater has added a degree of challenge to achieve initially the same results (Tsolaki and Diamadopoulos 2009). Bromine compounds are the primary biocides generated by ozonation of seawater and are efficient in destroying aquatic organisms, but total residual oxidants can be long-lived in water tanks, making them unsuitable for discharge at ports (Wright et al. 2010).

(b) Peroxide: Hydrogen peroxide (H_2O_2) is an uncharged molecule, which can be used as a disinfectant, by diffusion passes easily through cell membranes. When inside the cells, the reactive and destructive hydroxyl radicals are liberated by H_2O_2 eliminating aquatic organisms (Smit et al. 2009).

(c) Deoxygenation: The principle of this method is based on reducing/removing oxygen from the ballast water tanks, leading to the elimination of aquatic organisms. It can be achieved by creating an anoxic environment by either adding nutrients to the ballast tanks to encourage the growth of bacteria or injecting an inert gas (i.e., nitrogen) to inhibit oxygen from entering (McCollin et al. 2007).

Physical: Physical disinfection is widely applied in freshwater treatment systems. It is based on the application of a variety of physical fields, such as ultraviolet rays and ultrasound for disinfection (Fig. 10.5). Also referred to as 'reagent less' technique, the physical disinfection acts directly on microorganisms without changing the properties and composition of the water or creating unwanted disinfection by-products (Biryukov et al. 2005). Although the physical methods have proved their efficiency in destroying aquatic organisms, usually they are combined with mechanical treatment (i.e., filtration or hydro cyclones) to increase effectiveness.

(a) Ultrasound: Ultrasound generated by converting mechanical or electrical energy into high-frequency vibration causes the formation and collapse of

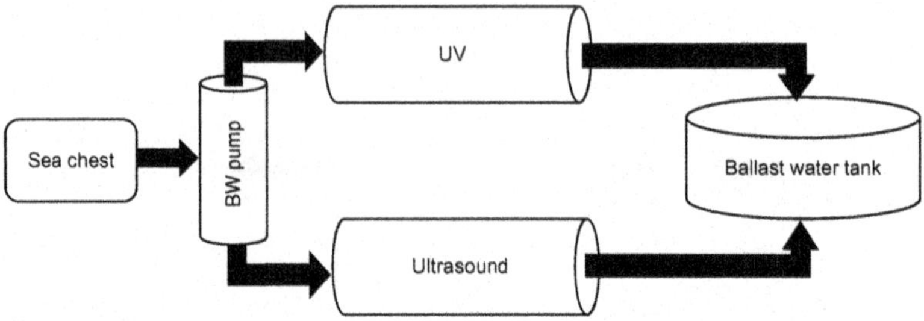

Fig. 10.5 Simple scheme for ballast water (BW) treatment principles using ultraviolet (UV) and ultrasound methods

microscopic gas bubbles in the incoming ballast water, leading to rupture of cell membranes and collisions with other aquatic organisms (Ta et al. 2005).

(b) Ultraviolet: The water-supply sector has been successfully using ultraviolet radiation (UV) for disinfection of drinking water and wastewater. Nucleic acids (DNA and RNA) and cell proteins of aquatic organisms are impacted by UV radiation through photochemical reactions leading to the inactivation of the organisms (Ta et al. 2005). Although UV treatment has been proven to be an effective bactericide and virucide, its effectiveness is related to the size and the morphology of organisms, as well as to the proper dosage application (Tsolaki and Diamadopoulos 2009).

(c) Heating: The heating treatment is based on the increase in the seawater temperature to a level that inactivates the aquatic organisms (Fig. 10.6). The method uses an existing heating system onboard (the engines), which would otherwise be heat that is lost (Mesbahi et al. 2007). Although a promising method, heating has not been a first option for the shipping industry unless used in combination with another method. This is due to the impracticality of heating huge ballast water volumes and the energy costs for heating at the effective temperature (~60–65 °C) and short port stays and voyage periods.

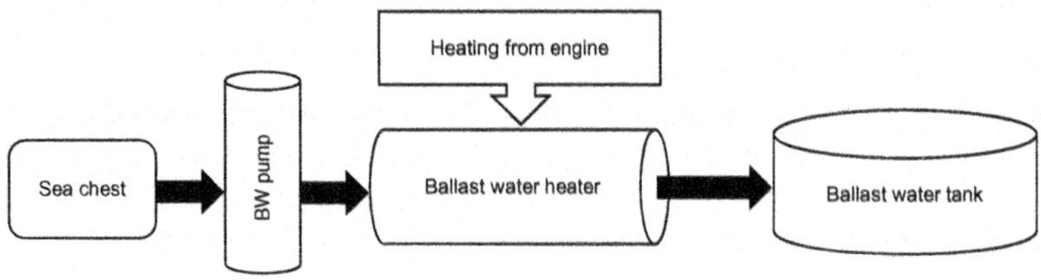

Fig. 10.6 Simple scheme for ballast water (BW) treatment principle using heating systems

10.3 History

Since Charles Darwin's memorable travel around the world on the HMS Beagle (1831–1836), several naturalists have also touched on the issue of invasive species. In 1936, the British ecologist Charles Elton reviewed Nicolaus Peter and Albert Panning monograph on the dispersion of the Chinese Mitten crab (*Eriocheir sinensis*) in Europe. The monograph linked the dispersion to ship's ballast water after two large crabs were found in the tanks of a Hamburg-American steamer in 1932 (Elton 1936). Later, in 1958, Charles Elton published the milestone book entitled *The Ecology of Invasions by Animals and Plants,* which is considered the foundation for all the following work in the field of invasive species (Kitching 2011). In his book, Elton emphasized that ships have been 'the greatest agency of all that spreads marine animals to new quarters of the world' (Fridley 2011).

10.3.1 Development

It was not until 1985, when James Carlton published the '*Transoceanic and interoceanic dispersal of coastal marine organisms: the biology of ballast water*', that addressed ballast-water ecology in detail. Carlton's publication brought light to the modern understanding of patterns and processes of ballast water as a vector of aquatic invasions globally (Davidson and Simkanin 2012). Consequent to this publication, the research field of ballast water as a vector has developed considerably, leading to the development of guidelines for ballast water management and finally the Convention (Fig. 10.7).

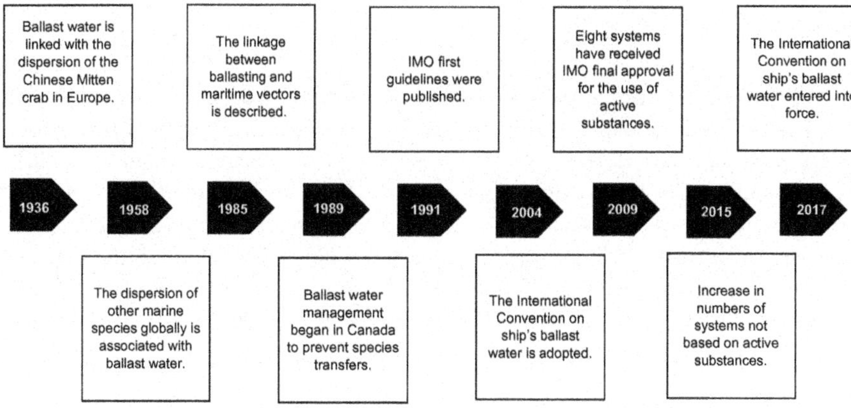

Fig. 10.7 History line of major developments in ballast water management

10.3.2 Progress

As the science behind ballast water as a vector bridged the knowledge gap, researchers began to explore avoiding the translocation of aquatic organisms in ballast tanks. In the beginning, exchanging ballast water at open sea was considered the most practical and feasible way of eliminating invasive species. Unfortunately, ballast water operations at open sea is not an easy task, as discussed previously. Therefore, other practical ways of avoiding the translocation of aquatic organisms were needed.

The first approaches were based on the use of active substances (biocides) that have been used in other sectors to eliminate unwanted organisms (i.e., in water supplies). Although very efficient in killing noxious organisms, active substances also have their own challenges when used onboard a ship, including:

(a) Chemical compounds must be stored onboard and handled by the ship crew.
(b) Possible corrosion of ballast water tanks.
(c) All active substances have a certain lifetime for their capacity to destroy noxious organisms, and the lifetime differs from one substance to another. It is important to note that in short journeys between ports in neighboring countries, the ships using biocides to treat ballast water may be deballasting substances that are still active at the receiving port. This means that the active substance being discharged could also target the local aquatic community.

Such challenges have driven the development of more environmentally friendly treatment systems not based on active substances. The last decade has experienced the development of other options free of active substances that applicable to be used on onboard and onshore.

10.4 State-of-the-Art

Initial developments in ballast water treatment systems were only focused on efficiently deactivating aquatic organisms as requested by the Convention. In this regard, several methods were developed to help the shipping industry to comply with the internationally agreed regulations. In a study made by the IMO in 2015 on the treatment systems approved and commercially accessible for the shipping industry, filtration systems were the most common method and used by 80% of the ships evaluated. This was followed by electrolytic disinfection systems (~40%), ultraviolet irradiation (32%), and the use of chemical biocides with almost 17% (Batista et al. 2017).

10.4.1 Onboard Treatment

Although it is the preferred option for most ships, filtration is usually used as a pretreatment to remove larger-sized classes of organisms and organic particles due to the pore size of filters. This is because the initial filtration technology used onboard to treat ballast water could not successfully deactivate small aquatic organisms (Werschkun et al. 2012) without affecting the ship's operation. The time needed to cope with the large volume of ballast water and the blockage of filters with smaller pore sizes were the principal issues.

The last decades have experienced a rapid development of new filtration technologies, especially microfiltration and reverse osmosis (RO). The RO is the most-used technology for desalination worldwide. The semipermeability of polymers is the basis of the RO process since polymers are highly permeable to water with a relatively lower permeability for dissolved substances producing a high-quality product (Ehteram et al. 2020). However, without pretreatment, RO membranes can be impacted by biofouling that results in membrane deterioration and high-energy costs (Ibrahim et al. 2020). In this regard, microfiltration membranes have proven to be an effective pretreatment to deliver high-quality water for the RO process. Microfiltration membranes separate large molecular weight suspended or colloidal compounds from dissolved solids (Maddah et al. 2018). Guilbaud et al. (2015) assessed the potential of microfiltration membrane treatment for cruise ships and liquid natural gas (LNG) carriers. The study proved the potential of using microfiltration membrane to deactivate aquatic organisms in compliance with current regulations. Although the study also concluded that the microfiltration process is more effective for cruise ships in terms of size and capital cost than for LNG carriers, it highlighted that the situation will change rapidly as manufacturers develop increasingly compact membrane systems.

Being a busy port and located in a water-scarce region, the Emirates are extremely dependent on desalinated water for its economic activities and public supply. Wang and Tsai (2014) investigated the cost and benefits associated with supplying onboard desalinated ballast water brought in by oil tankers and LNG carriers to Abu Dhabi, using waste heat recovered from propulsion. At the receiving port, the desalinated water is transferred to an onshore plant for final processing before it is sold to the end users. Based on three scenarios (high, most likely, and low water demand), the study concluded that the onboard ballast water desalination system generates a saving of $772 million, $718 million, and $602 million when combined with conventional desalination plants. The study showed that integrating desalinated water from ballast operations in Abu Dhabi is economically feasible.

10.4.2 Onshore Treatment Facilities

Ballast water operations bring dissolved and particulate material into the ships. Particulate material is then deposited on the bottom of the ballast tanks during the ship's journey and are not usually discharged during deballasting operations. When sediments accumulate in the tanks to a level that impact the normal operation of a ship, then the sediments must be removed and managed properly in a receiving sediment-management facility during maintenance operations in shipyards (GloBallast 2017). Different from the port facilities specialized in managing sediments, the "reception" facilities are onshore treatment systems designed to treat ballast water from incoming ships before its disposal. Initial developments in onshore facilities were done to address ships unable to be retrofitted to accommodate ballast water treatment systems and/or to attend to ships experiencing failure of their onboard systems.

Although onshore treatment is not new, the concept of treating ballast water at the destination port has not received much attention, especially due to possible high investment in ports' infrastructure. Onshore facilities have several advantages over onboard systems, including (Donner 2010):

(a) Economy of scale: Onshore facilities can operate uninterrupted by serving a multitude of ships, which is more economically rationale rather than running a system onboard only during ballast-water operations;
(b) Ships' crew: Officers and crews of merchant ships may work on several ships from the same company. Although training is provided to them to operate onboard ballast water treatment systems, different ships may operate different systems causing possible mismanagement. The crew members are also not experts in the fields of marine biology or the physical, chemical, or biological processes to treat ballast water that my exist on different ships; and
(c) Monitoring: Onshore facilities can be monitored easily by local regulators making sure that the treatment is achieving the levels of protection required under the current regulations (Pereira and Brinati 2012).

Currently, with the latest technological developments on treatment systems and mobile facilities, onshore systems are getting more attention as an economically viable option and a business opportunity for port operators. Probably the only modification needed for small/medium-sized conventional desalination plants to treat ballast water onshore is the proper management of aquatic organisms as biological waste. As discussed previously in this chapter, the desalination process can deliver not only a water biologically free of aquatic invaders but also a new product (freshwater) that can be sold for other economic activities.

The quantity, timing, and type of ships entering or leaving the port area would define the size of the treatment facility (Tsolaki and Diamadopoulos 2009). Retrofitting a port to install the necessary infrastructure to receive, treat, and finally deliver desalinated ballast water to end users can be an expensive exercise. In this regard, compact filtration treatment systems, which might be seen as too bulky to be placed on ship, can be the solution as mobile onshore treatment facilities. Containerized desalination systems can be placed on barges, making the service mobile and capable of storing the desalinated water to be later transferred to a receiving facility for distribution.

10.5 Major Barriers and Response Options

The ballast water treatment systems were originally projected with the objective of deactivating aquatic organisms that otherwise could become biological invaders at the destination port. Under such an approach, several methods that are not applicable to produce desalinated water have been developed (i.e., ultrasound) as a viable solution. The latest technological developments in seawater treatment (i.e., microfiltration) have been reformulating the prospects for using ballast water as an unconventional water source to supply water for onshore economic activities, especially in regions where water is a scarce commodity. Table 10.1 presents major barriers and respective response options for using ballast water when considered as an unconventional water source.

10.6 Conclusions

The considerable amount of water moved globally by the shipping industry each year as ballast should not be neglected, not only due to its negative impacts but also because of its potential as an unconventional water resource. The impacts on the environment, economy, and public health have been extensively assessed and described by the international literature. As a result, the International Convention for the Control and Management of Ships' Ballast Water and Sediments was developed and is now in force. As defined under the Convention, all ships must manage their ballast water in a way that avoids negative impacts. However, the opportunities for reusing treated ballast water for other means (i.e., irrigation) have been overlooked until recent years.

Recent technological developments in microfiltration have brought a new perspective on the reuse of treated ballast water for other economic activities. The applicability of using seawater-desalination technology as an option for ballast-water treatment onboard and onshore is making ship's ballast water a feasible source of unconventional water. Port cities located in water-scarce countries would benefit greatly by receiving desalinated water from ships and/or onshore treatment facilities to augment their water supply. Ships fitted with desalination systems would be able to offset some of their running costs by selling desalinated water to receiving cities. Ports with onshore ballast water-treatment facilities running desalination systems will also be able to sell the treated ballast for reuse in port cities. Such an approach will give them another revenue opportunity to defray the rates paid by ships to treat their ballast water.

Unfortunately, the onshore treatment of ballast water through desalination processes is still in its infancy, with mainly desktop simulations done in the last decade to demonstrate its economic and technical viability (Donner 2010; Wang and Tsai 2014; Pereira and Brinati 2012 and Pereira et al. 2017). These studies have shown that not only is desalination treatment for ballast water (especially onshore)

Table 10.1 Major barriers and response options for using ballast water (BW) as an unconventional water source

Major barriers	Response options
Onboard treatment	
The primary objective of BW treatment systems was not to produce desalinated water	Current perspectives of desalinated water reuse in port cities under water stress can drive the shipping industry to adapt their ships' treatment systems to also making profit by producing and selling reusable desalinated water
Old filtration technology (mesh size) was not suitable for coping with the volume of ballast water and removing dissolved and particulate salts. For this reason, it was mainly used as a pretreatment option to remove larger particles	New technological developments in microfiltration, which is widely used in modern desalinations plants, are now capable of removing aquatic organisms and dissolved and particulate salts from ballast water
Microfiltration consumes considerable energy to push water against the membranes. It also needs extra room onboard to be able to filter the volume of ballast water entering the BW tanks	Energy recovered from the ship (i.e., waste-heat energy from cooling the engines) can be used to provide the required energy. Nowadays, compact 'containerized' desalination units can be easily fitted onboard
Infrastructure needed to make treatment facilities able to receive and treat BW efficiently and in a timely manner (i.e., connections between the treatment stations and all berths)	Mobile treatment units (i.e., on trucks or barges) with storage capacity could reduce the necessity of major updates in ports' structure. New or renovated ports could include BW treatment facilities as part of the planned infrastructure
Capacity of the treatment system to cope with high volumes of BW in busy ports, which can cause delays in port operations	Busy ports might invest in a more substantial infrastructure to cope with high volumes of BW if selling treated water (desalinated) becomes a business opportunity. In busy ports it would be available only to older ships that cannot be retrofitted with a BW treatment system or to service ships on which the onboard treatment system has failed during the journey. Less busy ports can be a more feasible option as they receive fewer ships
Water authorities	
Lack of knowledge of the potential that BW has as an unconventional water source	Raising awareness of the huge potential of using ballast water as an unconventional water source, especially for port cities located in water-scarce regions
No clear water-management policies that consider unconventional water sources (i.e., ballast water) as an integrated part of the water cycle	Development of new policies integrating unconventional water sources (i.e., ballast water) as a feasible and viable option for water–scarce countries and cities

(continued)

Table 10.1 (continued)

Major barriers	Response options
Urban planning approaches do not consider sourcing water from unconventional sources (i.e., ballast water), especially in port cities located in water-scarce regions	Integration of BW as an unconventional water source at the urban planning level for port cities, especially those in water-scarce regions

a secure and viable option to prevent marine invasions, and they also provided an economic analysis of the investments needed and the financial returns.

Key recommendations/considerations:

- Port cities in water-scarce countries/regions would benefit most if desalinated ballast water from treatment facilities (onboard and onshore) were made available. However, a global–cost benefit analysis overlaying water availability and needs, and the traffic of ships at a port is yet to be done. Such a study would indicate the economically feasible port cities to receive investments in the necessary infrastructure;
- Public policies designed to create/develop a market to desalinate ballast water for reuse in other economic activities (i.e., irrigation) are still missing, including the regulatory frameworks.
- Establishment of financing mechanisms for the private sector to invest in onshore treatment and/or receiving facilities, as well as for the associated infrastructure for treated water distribution, would facilitate the development of the field.
- Mainstreaming the work on unconventional water within the shipping industry (including port operations) will certainly open new business opportunities for ship owners (i.e., recovering costs by selling desalinated water), port operators (i.e., treating and selling desalinated water), and city water managers (i.e., augmenting the water supply portfolio).

References

Batista WR, Fernandes FC, Lopes CC, Lopes RS, Miller W, Ruiz G (2017) Which ballast water management system will you put aboard? Remn Anxieties Mini-Rev Environ 4(3):54. https://doi.org/10.3390/environments4030054

Biryukov AS, Gavrikov VF, Nikiforova LO, Shcheglov VA (2005) New physical methods of disinfection of water. J Russ Laser Res 26(1):13–25. https://doi.org/10.1007/s10946-005-0002-8

Carlton JT (1985) Transoceanic and interoceanic dispersal of coastal marine organisms: the biology of ballast water. J Oceanogr Marine Biol 23:313–371

Cohen AN (1998) Ships' ballast water and the introduction of exotic organisms into the San Francisco Estuary: current status of the problem and options for management. San Francisco Estuary Institute, Richmond, CA, p 81

Davidson IC, Simkanin C (2012) The biology of ballast water 25 years later. Biol Invasions 14(1):9–13. https://doi.org/10.1007/s10530-011-0056-1

Donner P (2010) Ballast water treatment ashore—better for the environment and for seafarers. WMU J Marit Aff 9(2):191–199. https://doi.org/10.1007/BF03195174

Ehteram M, Salih SQ, Yaseen ZM (2020) Efficiency evaluation of reverse osmosis desalination plant using hybridized multilayer perceptron with particle swarm optimization. J Environ Sci Pollut Res 27:15278–15291. https://doi.org/10.1007/s11356-020-08023-9

Elton C (1936) A new invader. J Anim Ecol 5(1):188–192

Endresen Ø, Behrens HL, Brynestad S, Andersen AB, Skjong R (2004) Challenges in global ballast water management. Mar Pollut Bull 48:615–623. https://doi.org/10.1016/j.marpolbul.2004.01.016

Fridley JD (2011) Biodiversity as a bulwark against invasion: conceptual threads since Elton. In: Richardson DM (ed) Fifty years of invasion ecology. Wiley, New York, pp 121–129

Gardiner-Garden CW (1975) Port of wollongong. J Illawarra Hist Soc. Wollongong, p 54. https://ro.uow.edu.au/ihspubs/16

GloBallast (Partnerships Programme) (2017) Guidance on best management practices for sediment reception facilities under the ballast water management convention. GloBallast, Monograph No. 23

Gollasch S, David M (2019) Ballast water: problems and management. In: Sheppard C (ed) World seas: an environmental evaluation. Elsevier Ltd., pp 237–250

Guilbaud J, Massé A, Wolff FC, Jaouen P (2015) Porous membranes for ballast water treatment from microalgae-rich seawater. Mar Pollut Bull 101(2):612–617. https://doi.org/10.1016/j.marpolbul.2015.10.044

Ibrahim Y, Banat F, Yousef AF, Bahamon D, Vega LF, Hasan SW (2020) Surface modification of anti-fouling novel cellulose/graphene oxide (GO) nanosheets (NS) microfiltration membranes for seawater desalination applications. J Chem Technol Biotechnol. https://doi.org/10.1002/jctb.6341

IMO (International Maritime Organization) (2018) Ballast water management convention and BWMS code with guidelines for implementation—2018 Edition. London

Indian Register of Shipping. (2016). A Guide to Ballast Water Management. IRClass Publications from https://www.irclass.org/media/2449/bwm_booklet.pdf

Kitching RL (2011) A world of thought: 'The ecology of invasions by animals and plants' and Charles Elton's Life's work. In: Richardson DM (ed) Fifty years of invasion ecology. Wiley, New York, pp 1–10

Krata P (2013) The impact of sloshing liquids on ship stability for various dimensions of partly filled tanks. TransNav: Int J Marine Navig Saf Sea Transp 7(4):481–489 https://doi.org/10.12716/1001.07.04.02

Kurtela Ž, Komadina P (2010) Application of hydrocyclone and UV radiation as a ballast water treatment method. J Promet-Traffic Transp 22(3):183–191. https://doi.org/10.7307/ptt.v22i3.274

Maddah HA, Alzhrani AS, Bassyouni M, Abdel-Aziz MH, Zoromba M, Almalki AM (2018) Evaluation of various membrane filtration modules for the treatment of seawater. J Appl Water Sci 8(6):150. https://doi.org/10.1007/s13201-018-0793-8

McCollin T, Quilez-Badia G, Josefsen KD, Gill ME, Mesbahi E, Frid CL (2007) Shipboard testing of a deoxygenation ballast water treatment. Mar Pollut Bull 54(8):1170–1178. https://doi.org/10.1016/j.marpolbul.2007.04.013

Mesbahi E, Norman RA, Vourdachas A, Quilez-Badia G (2007) Design of a high-temperature thermal ballast water treatment system. Proc Inst Mech Eng Part M J Eng Marit Environ 221(1):31–42. https://journals.sagepub.com/doi/10.1243/14750902JEME39

National Research Council (1996) Stemming the tide: controlling introductions of nonindigenous species by ships' ballast water. National Academies Press. https://doi.org/10.17226/5294

Pereira NN, Brinati HL (2012) Onshore ballast water treatment: a viable option for major ports. J Dep Naval Archit Ocean Eng Escola Politécnica, Marine Pollut Bull 64(11):2296–2304. https://doi.org/10.1016/j.marpolbul.2012.07.026

Pereira NN, Brinati HL, Antunes RA (2017) Onshore reception facilities for ballast water. J Ship Sci Technol. https://doi.org/10.25043/19098642.148

Smit MG, Ebbens E, Jak RG, Huijbregts MA (2009) Time and concentration dependency in the potentially affected fraction of species: the case of hydrogen peroxide treatment of ballast water. J Environ Toxicol Chem 27(3):746–753. https://doi.org/10.1897/07-343.1

Szczepanek M, Behrendt C (2018) Assessment of selected on-board ballast water treatment systems in terms of technical and operational parameters. New Trends Prod Eng 1(1):461–467

Ta N, Viitasalo S, Sassi J, Rytkönen J, Leppäkoski E (2005) Ozone, ultraviolet light, ultrasound and hydrogen peroxide as ballast water treatments-experiments with mesozooplankton in Low-Saline Brackish water. J Marine Environ Eng 8(1)

Tsolaki E, Diamadopoulos E (2009) Technologies for ballast water treatment: a review. J Chem Technol Biotechnol 85(1):19–32. https://doi.org/10.1002/jctb.2276

UNCTAD. (2019). UNCTAD Handbook of Statistics 2019. Geneva

Wang Y, Tsai IT (2014) Cost benefit analysis of on-board desalinated ballast water from oil and natural gas trade as a source of water supply for Abu Dhabi. Int J Trade Econ Financ 5(4):355

Werschkun B, Höfer T, Greiner M (eds) (2012) Emerging risks from ballast water treatment. Federal Institute for risk assessment. https://www.bfr.bund.de/cm/350/emerging-risks-from-ballast-water-treatment.pdf

Wright DA, Gensemer RW, Mitchelmore CL, Stubblefield WA, van Genderen E, Cooper WJ (2010) Shipboard trials of an ozone-based ballast water treatment system. Mar Pollut Bull 60(9):1571–1583. https://doi.org/10.1016/j.marpolbul.2010.04.010

Yang M (2011) Shipping and maritime transport. In: Nriagu JO (ed) Encyclopedia of environmental health, 1st edn. Elsevier Science, pp 33–40

Yongming S, Shuhong S (2012) The study of ships ballast water replacement monitoring at sea based on MCU. Procedia Environ Sci 12:199–205. https://doi.org/10.1016/j.proenv.2012.01.266

Part VI
Developing New Water

Chapter 11
Desalinated Water

Nikolay Voutchkov

Abstract As desalination technology advances and water scarcity becomes prevalent in most arid and semi-arid coastal regions of the Middle East, North Africa, North America, Australia and Europe, policy makers around the world are adopting desalination as a long-term solution for closing the gap between water supply and demand in future years. Large-scale desalination plants are widely accepted as an economically viable alternative source of water supply for coastal urban centers worldwide. At present, 107 million m^3 per day of desalinated water is used to supply to approximately 5% of the world's population. It is projected that the worldwide production capacity of desalination plants will double by the year 2030 and cost of desalinated water will be reduced by half. Ocean-brine mining has been gaining momentum over the last five years and is expected to yield commercially viable products that are likely to completely offset the cost of desalinated water production in the next decade. This chapter provides an overview of the status of desalination and discusses key barriers and solutions associated with its wider adoption as an unconventional water supply alternative, including technological advances, freshwater production costs, energy use, environmental impacts, and institutional challenges.

Keywords Desalination · Concentrate · Brine · Reverse osmosis · Energy · Environmental impacts

11.1 Introduction

The water industry today faces multiple challenges—from accelerated population growth to exhaustion of traditional freshwater resources and long-term water scarcity driven by climate change. In response to these challenges, the water supply-planning paradigm over the last 10–15 years has been evolving towards building an environmentally sustainable diversified water portfolio where low-cost, conventional water

N. Voutchkov (✉)
Water Globe Consultants, LLC, 824 Contravest Lane, Winter Springs, Florida, USA
e-mail: nvoutchkov@water-g.com

M. Qadir et al. (eds.), *Unconventional Water Resources*,
https://doi.org/10.1007/978-3-030-90146-2_11

sources (e.g., rivers, lakes, and dams) are balanced with more costly but also more reliable and sustainable unconventional water-supply alternatives such as water reuse and desalination (World Bank 2019).

Approximately 97.5% of the world's water resources exist in the oceans and seas. Freshwater contributes the remainder, of which approximately 70% is in the form of polar ice and snow and the balance of 30% lies in groundwater, rivers and lakes, and air moisture. At present, less than 1% of the available water resources worldwide are naturally fresh and can be directly utilized to produce drinking water by applying conventional water treatment technologies.

Most water resources of the world are saline and can be converted into freshwater using more costly, elaborate, and energy-intensive methods for salt separation commonly referred to as desalination technologies. These technologies facilitate the removal of salt and other undesirable compounds (e.g., radioactive materials, heavy metals, organic toxins, and pathogens) from natural saline water sources or wastewater to produce freshwater. Brackish water, with salt content between 800 and 10,000 mg per liter (mg/L) and ocean and seawater, with a salinity between 30,000 and 50,000 mg/L, are the main saline water resources used for freshwater production at present.

Although desalination of brackish water is less costly and energy intensive than that of seawater, its ability to provide a sustainable long-term solution to world's water-supply challenges is very limited because the available brackish water resources on planet Earth are only 0.2% of the world's total water resources. In addition, most of the known brackish water aquifers near large, urbanized centers worldwide are already utilized, and these aquifers have limited capacity and very slow recharge rate (i.e., measured in months or years), which in most cases mainly depend on rain events. This makes brackish water desalination unsustainable in the long term as the main unconventional water supply source for rapidly growing municipalities and urban regions of the world.

In contrast, seawater desalination produces drought-resilient, sustainable, and reliable long-term water resource. It is practically limitless in terms of availability, and, as compared to water reuse and conservation, it creates a new source of freshwater supply, rather than just being a tool for more efficient use of existing water resources. Desalination is thus drought resilient, and it is a rational solution to address climate change and rapid population growth, along with industrial growth-related water-supply risks.

Furthermore, desalination can be a strategically valuable tool to address exogenous risks such as water dependency on other countries. Singapore, for example, opted for large-scale desalination to reduce its dependence on increasingly costly imported water from Malaysia. The stable, efficient supplies of urban and industrial water that desalination provides can help governments manage a range of economic, social, and political risks (World Bank 2019).

This chapter provides an overview of the status of desalination and discusses key barriers and solutions associated with its wider adoption as an unconventional water supply alternative, including technological advances, freshwater production costs, energy use, environmental impacts, and institutional challenges.

11.2 Technological Interventions

Seawater and brackish waters are typically desalinated using one of two types of water treatment technologies—thermal evaporation (distillation) and membrane separation. In thermal distillation processes, freshwater is separated from the saline source by evaporation. In Reverse Osmosis (RO) desalination, freshwater is produced from saline source water by its pressure-driven transport through semi permeable membranes. The main driving force in RO desalination is the pressure that is needed to overcome the naturally occurring osmotic pressure, which in turn is proportional to the source-water salinity.

Besides thermal distillation and RO, two other desalination technologies currently applied in the industrial and municipal sector are electrodialysis (ED) and ion exchange (IX). Electrodialysis is an electrically driven desalination process where salt ions are removed from the source water by exposure to direct electric current. The main driving force for ED separation is electric current, which is proportional to the salinity of the source water. Ion exchange is the selective removal of salt ions from water by adsorption on an ion-selective resin media. The driving force in this desalination process is the ion charge of the IX resin, which can selectively attract and retain ions of opposite charge contained in the saline source water.

Table 11.1 provides a general indication of the range of source-water salinity for which distillation, RO separation, ED, and IX can be applied cost effectively for desalination. For processes with overlapping salinity ranges, a lifecycle cost analysis for the site-specific conditions of a given desalination project is typically applied to determine the most suitable desalination technology for the project (Voutchkov 2012).

11.2.1 Thermal Desalination

All thermal desalination technologies apply distillation (heating of the saline source water) to produce water vapor, which is then condensed into low-salinity water. The principle of evaporation is based on water molecules requiring less heat to be turned from liquid to vapor than the dissolved solids contained in the water. Since the

Table 11.1 Desalination process applicability

Separation process	Range of Source-Water Total Dissolved Solids Concentration for Cost-Effective Application (mg/L)
Distillation	20,000–100,000
Reverse osmosis	50–46,000
Electrodialysis	200–3,000
Ion exchange	1–800

energy for water evaporation is practically not dependent on the source water-salinity concentration, thermal evaporation is suitable for desalination of highly saline waters and brine.

This is one of the reasons why thermal desalination has been widely adopted by all Middle Eastern countries, including Saudi Arabia, Oman, Qatar, the United Arab Emirates, Bahrain, and Kuwait. The Red Sea, Arabian (Persian) Gulf, Gulf of Oman and the Indian Ocean are among the most saline water bodies. At present, around 85% of the world's thermal desalination plants are in the Arabian Peninsula, half in Saudi Arabia (World Bank 2019).

The three most-used types of thermal desalination technologies are multistage flash distillation (MSF), multi-effect distillation (MED), and vapor compression (VC). Each class of these technologies has evolved over the past 40–60 years towards improvements in efficiency and productivity. Over 90% of the thermal desalination plants worldwide apply MSF or MED evaporation processes. Thermal desalination technologies generate water with significantly lower salinity (10–25 mg/L) than RO membrane-separation processes (100–300 mg/L). The desalinated water also has a very low content of pathogens and other contaminants of concern, such as boron, bromides, and organics.

11.2.2 Membrane Desalination

Membrane desalination is a process for the separation of minerals from the source water using semi-permeable membranes. Two types of commonly used technologies for membrane desalination currently are RO and ED. Reverse osmosis is a process where the product water (permeate) is separated from the salts contained in the source water by pressure-driven transport through a semi permeable membrane. In ED systems, salts are separated from the source water by applying direct current.

Reverse osmosis is a process where water containing inorganic salts (minerals), suspended solids, soluble and insoluble organics, aquatic microorganisms, and dissolved gases (collectively called source water constituents or contaminants) is forced under pressure through semi permeable membranes that are designed to selectively allow for water to pass through at a much higher rate than the transfer rate of any other water constituent.

Depending on their size and electric charge, most water constituents are retained (rejected) on the feed side of the RO membrane, and the purified water (permeate) passes through the membrane. Reverse osmosis membranes can reject particulate and dissolved solids of practically any size. However, they do not reject gases well due to their small molecular size. As a result, the RO membrane systems produce two streams—fresh water with a salinity of less than 500 mg/L and a highly salinity waste stream referred to as concentrate, reject, or brine.

While RO membranes can retain both particulate and dissolved solids, they are designed to primarily reject soluble compounds (mineral ions). The RO membrane structure and configuration are such that these membranes cannot store or remove

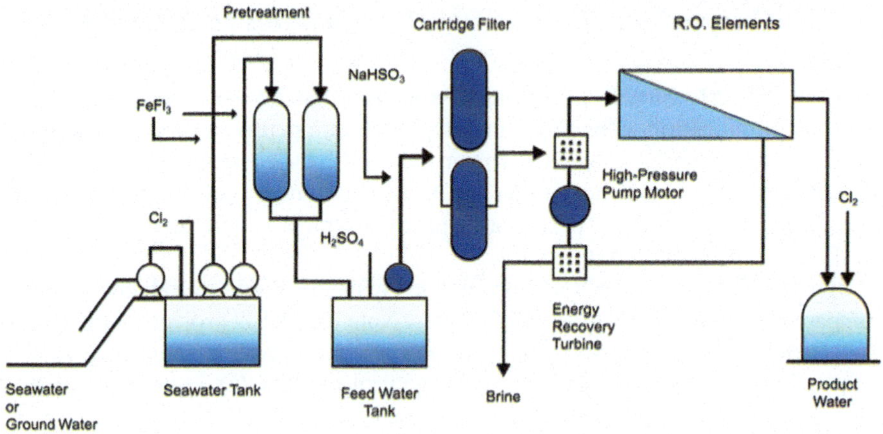

Fig. 11.1 Schematic of a typical seawater reverse-osmosis desalination plant

large amounts of suspended solids from their surface. If left in the source water, the solid particulates would accumulate and quickly plug (foul) the surface of the RO membranes, not allowing the membranes to maintain a continuous steady-state desalination process. Therefore, the suspended solids (particulates) contained in source water used for desalination must be removed before they reach the RO membranes.

At present, practically all reverse osmosis desalination plants, such as that shown in Fig. 11.1, incorporate two main treatment steps designed to sequentially remove suspended and dissolved solids from the source water. The purpose of the first step—source seawater pretreatment—is to remove the suspended solids and to prevent some of the naturally occurring soluble solids from turning into solid form and precipitating on the RO membranes during the salt separation process. Typically, pretreatment of saline surface-source water is accomplished by clarification, using lamella settlers and dissolved air flotation clarifiers (DAFs), and/or granular media or membrane filtration (Voutchkov 2017).

The second step of the RO system separates dissolved solids from the pretreated source water, producing low-salinity freshwater suitable for human consumption, agricultural uses, and industrial and other applications. Once the desalination process is complete, the freshwater produced by the RO system is further treated for corrosion and health protection and disinfected prior to distribution for final use. This third step of the desalination-plant treatment process is referred to as posttreatment. The permeate generated by RO is stabilized by the addition of lime and carbon dioxide to provide an adequate level of alkalinity and hardness for protection of the product water-delivery and distribution system against corrosion. The conditioned water is stored and disinfected prior to delivery to the final users.

11.2.3 Comparison of Alternative Desalination Technologies

Over the past 20 years, RO membrane separation has evolved more rapidly than any other desalination technology, mainly due to competitive energy consumption and lower water-production costs. The analysis of the specific energy demand data presented in Table 11.2 indicates that the all-inclusive energy consumption for fresh-water production of thermal desalination plants is typically more than double that of brackish and seawater desalination.

Brackish Water RO (BWRO) desalination yields the lowest overall energy use as compared to other desalination technologies. Note that MED projects built recently have been completed at costs comparable to similarly sized Sea Water RO (SWRO) plants. However, for most medium and large projects, SWRO desalination is more cost competitive than thermal desalination technologies.

Table 11.3 presents costs for medium- and large-sized plants, the most common types of desalination technologies for various seawater sources for relatively new desalination plants. The cost of water production by thermal desalination (MSF, MED) is not sensitive to source water quality, which makes these technologies competitive in the Arabian (Persian) Gulf and the Red Sea. The costs provided in Table 11.3 are derived based on the analysis of actual data from more than 50 SWRO desalination plants worldwide built between the years 2000 and 2020.

Figures 11.2 and 11.3 show the magnitude of the main capital cost components of thermal and SWRO desalination plants, while Figs. 11.4 and 11.5 reflect on O & M costs. As seen in Fig. 11.4, thermal energy in the form of steam is a significant portion of the O & M cost of thermal desalination plants

Table 11.2 Energy use of alternative desalination technologies

Process type/energy use	MED	MSF	VC	BWRO	SWRO
Steam pressure (atm)	0.2–0.4	2.5–3.5	Not needed	Not needed	Not needed
Electric energy equivalent					
kWh/m^3	4.5–6.0	9.5–10.0	NA	NA	NA
kWh/1,000 gal	17.0–22.7	35.9–46.0	NA	NA	NA
Electricity consumption					
kWh/m^3	2.0–8.0	3.2–4.0	8.0–2.0	0.3–2.8	2.5–4.0
kWh/1,000 gal	4.5–6.8	12.1–15.1	30.3–45.4	1.1–10.6	9.5–15.1
Total energy use					
kWh/m^3	5.7–7.8	12.7–15.0	8.0–12.0	0.3–2.8	2.5–4.0
kWh/1,000 gal	25–29.5	48–56.7	30.3–45.4	1.1–10.6	9.5–15.1

Note NA—Not applicable

Table 11.3 Seawater desalination costs for various technologies and sources

Desalination plant type	Capital costs (Million \$/MLD)		O & M costs (\$/m^3)		Cost of water production (\$/m^3)	
	Range	Average	Range	Average	Range	Average
MSF	1.7–3.1	2.4	0.22–0.30	0.26	1.02–1.74	1.44
MED-TVC	1.2–2.3	1.6	0.11–0.25	0.14	1.12–1.50	1.39
SWRO Pacific and Atlantic Oceans	1.1–3.3	2.2	0.20–0.85	0.38	0.50–2.50	1.50
SWRO Mediterranean Sea	0.8–1.8	1.3	0.25–0.74	0.35	0.41–1.31	0.86
SWRO Arabian (Persian) Gulf	0.6–1.6	1.1	0.35–0.85	0.60	0.31–1.82	1.07
SWRO Red Sea	0.7–2.3	1.5	0.38–0.96	0.57	0.47–1.88	1.18

Note costs for medium-, large- and mega-size desalination plants

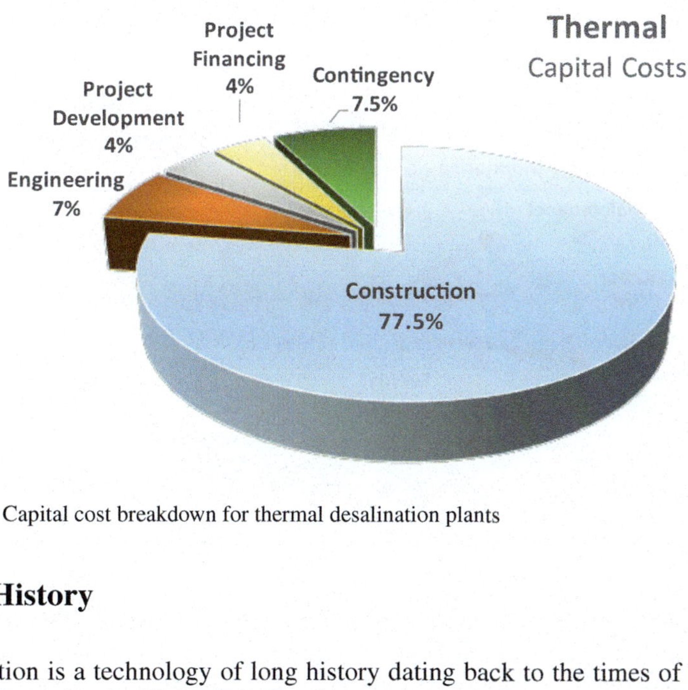

Fig. 11.2 Capital cost breakdown for thermal desalination plants

11.3 History

Desalination is a technology of long history dating back to the times of the Greek philosopher Aristotle (384–322 BC), who mentioned thermal evaporation as a means for Greek sailors to produce freshwater during long voyages at sea. Filtering and distillation technologies evolved over the years to yield the first dry land-based steam distillation—desalination plant in England in 1869.

By 1907, Saudi Arabia has introduced the first two thermal desalination plants in Jeddah. In the first half of the 20th century, several small thermal evaporation systems, mainly applying MSF distillation technologies, were built in the Middle East, the Caribbean, the US, and North Africa. In the early 1960s, a new thermal

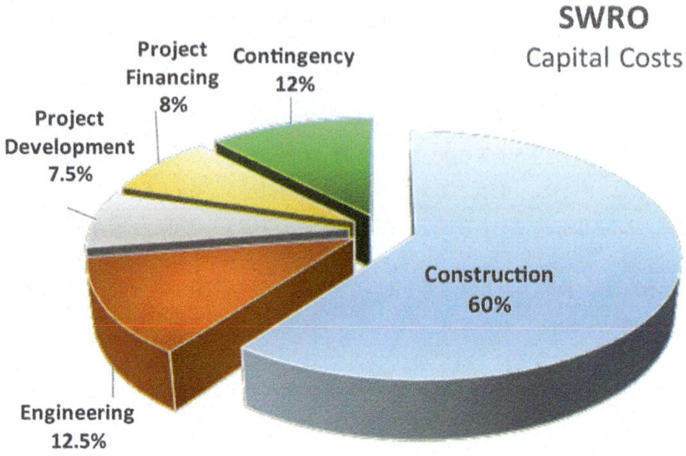

Fig. 11.3 Capital cost breakdown for seawater reverse-osmosis (SWRO) desalination plants

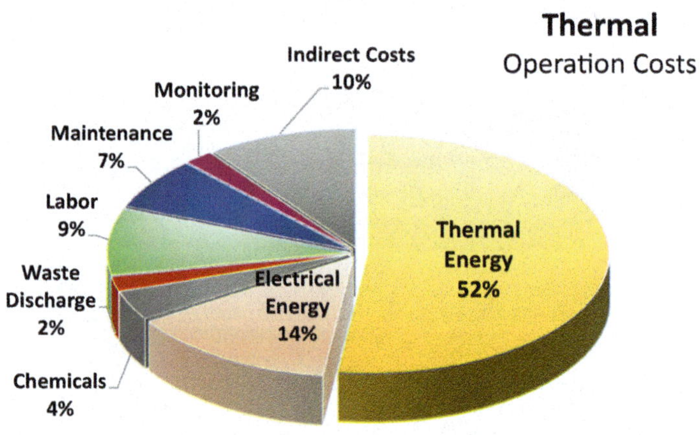

Fig. 11.4 Operation and maintenance cost breakdown for thermal desalination plants

desalination technology was introduced to the market—the MED. This technology, however, did not receive wide commercial application until the 1990s.

In the mid-1960s, desalination using thermal evaporation had become a water-supply option with improved feasibility and with wider application, especially in the Middle East. The creation of the Saline Water Conversion Corporation (SWCC) in Saudi Arabia in 1974 initiated a new era in the use of desalination for municipal water supply on a large scale. At present, SWCC is the largest producer of desalinated water in the world, with total desalination plant capacity of 5.6 million m^3/day—5% of the world's production of desalinated water in 2019 (IDA 2019a).

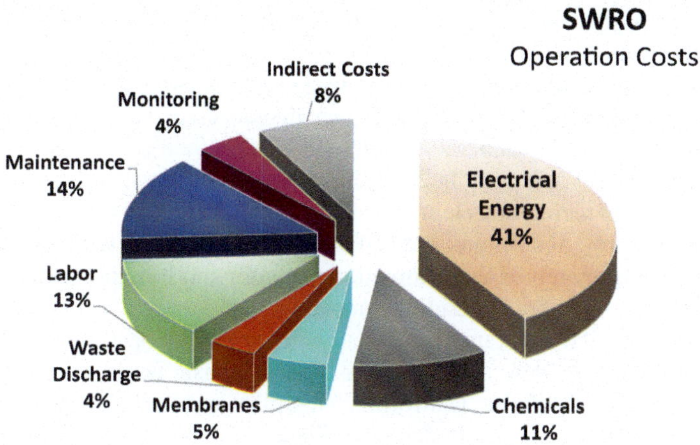

Fig. 11.5 Operation and maintenance cost breakdown for seawater reverse-osmosis (SWRO) desalination plants

The first prototype of reverse-osmosis membranes was developed in 1959 and advanced in the early 1960s in the University of California, Los Angeles, by Loeb and Sourirajan (Buros 1980). These membranes were made of cellulose acetate, and the pressure needed for the separation of freshwater from seawater was over 80 bars. At present, SWRO membrane desalination systems typically use between 50 and 70 bars of pressure to produce freshwater, depending on the source seawater's salinity and temperature. Brackish water desalination plants operate at membrane feed pressures of 5–30 bars and process saline water of TDS concentrations between 800 and 10,000 mg/L.

In the late 1970s, the first nanofiltration (NF) membrane elements were developed as an alternative to RO membranes tailored for saline waters with a low sodium and chloride content and a high content of bivalent ions such as calcium, magnesium, and sulfates. Such elements require two-to-four times lower feed pressure/energy for membrane desalination than SWRO membranes and are well suited for softening (removal of calcium and magnesium from the source water) and for removal of organics that are precursors to disinfection by-products and/or color from the source water (AWWA 2008).

RO and NF elements are made of two alternative materials: polyamide and cellulose acetate and its derivatives. The first commercially available RO membranes were made of cellulose acetate in a hollow fiber configuration and launched by DuPont in 1967. In 1976, DuPont introduced the 8-inch B-10 hollow-fiber seawater membrane elements which, along with later models, dominated the desalination market until the late 1980s. In 2000, DuPont exited the membrane market ending the dominance of the use of hollow-fiber cellulose-acetate elements for desalination. Since then, the main supplier of such elements is Toyobo and its main user in SWCC. In 2019

SWCC made the decision to replace the hollow-fiber SWRO elements of all its plants with spiral wound.

The desalination industry currently uses spiral-wound thin-film composite-polyamide RO elements. These elements were first developed in 1963 and introduced on the market in 1975 by General Atomics (ROGA) in their present standard size of 8-in diameter and 40-in length. In 2002, Koch Membrane systems launched the first 18-in diameter RO membrane element (Mega-Magnum). In the following five years, Hydranautics, Torey, and Filmtec created 16-inch elements. All these large-diameter RO elements have received very limited application and to date 8-in RO elements still dominate the membrane desalination market.

In Spain, the first seawater desalination plant began operation in Lanzarote (Canary Islands) in 1964. In 1965, the first brackish water-desalination plant was inaugurated in Coalinga, California. This plant used tubular cellulose-acetate membranes. A decade earlier, Ionics (now Suez) built the first commercial electrodialysis reversal desalination plant at the same location.

Energy recovery devices introduced to the desalination market in the mid-1980s allowed the use of residual energy left in the brine after the salt separation process to be recovered and applied to pump new seawater into the RO system, thereby reducing the total energy consumption for desalination by 30–40%. The technology for recovery of energy from the brine of SWRO desalination plants evolved on a parallel track with the development of new, more productive, higher salt-rejection and lower-energy membranes. In 1988, the US company Pump Engineering Inc. (now part of Energy Recovery International–ERI) installed the first turbocharger energy recovery system at a desalination plant in the Virgin Islands.

In 1990, Calder (now part of Flowserve) installed the first isobaric chamber-based pressure exchanger. Prior to that, Calder's Pelton wheel energy-recovery systems dominated the mid- and large-sized SWRO plant energy-recovery equipment market. In 1997, ERI installed their first commercial pressure exchanger (PX) unit at a SWRO desalination plant in Lanzarote, the Canary Island. Within 10 years, the ERI pressure exchangers have become the most widely used devices in the desalination industry. At present, over 80% of all SWRO plants worldwide have adopted this technology for energy recovery.

The use of desalination for production of fresh drinking and industrial water has gained significant momentum over the past two decades. The number and size of desalination projects worldwide have been growing at a rate of 5.0–7.5% per year since 2010, which corresponds to an addition of between 3.0 and 4.5 million m^3 per day of new freshwater production capacity installed annually.

11.4 Status

At present, desalination plants provide drought-proof water supply to nearly 5% of the world's population located in the most arid urban coastal municipalities of the Middle East, Europe, Africa, Australia, and the Americas (Jones et al. 2019).

Approximately 44% of this capacity is in the Middle East, which has experienced the highest growth of new desalination capacity over the last 30 years and is attributed to the relative lower cost of energy in the region. Recently, other regions of the world have also been experiencing accelerated growth—in particular, Singapore, China, India, the United States, and Latin America.

Worldwide production of freshwater by desalination through mid-July 2019 was 107 million m^3/day (IDA 2019)—more than four times the production in 2000. Approximately 74% of the existing 20,000 desalination plants in operation at present use membrane RO technology for salt separation; 21% apply thermal evaporation; and 5% employ other salt-separation technologies, such as electrodialysis and ion exchange to produce freshwater. By the year 2024, the total worldwide desalination plant capacity is projected to reach 130 million m^3/day and by year 2030 to exceed 200 million m^3/day (GWI 2020).

After 2015, most Middle Eastern countries have drastically reduced the construction of new thermal desalination plants and have refocused on the use of SWRO membrane desalination due to its lower energy demand and operational flexibility. These countries are also taking the leadership role in the use of renewable power sources for desalination and in ocean-brine mining and beneficial reuse of brine.

Over the last two years, this region of the world has yielded some of the lowest cost of water desalination projects on record (Table 11.4). Unit energy costs in the region are also the lowest in the world ($0.025–0.050/kWh), which contributes significantly to the lower overall costs of freshwater production.

Other factors contributing to the record low desalination costs are the significant economy of scale associated with the construction of mega (> 400,000 m^3/day) desalination plants and the very low costs of labor and construction materials in the Middle Eastern countries. Unit labor costs for construction workers with limited qualifications are approximately 8-to-10 times lower than in the USA, Europe, and Australia, while the labor rates of highly qualified construction professionals such as crane operators and super duplex steel pipe welders, are 15-to-20 times lower.

The construction of mega-size desalination facilities in the Middle East is a distinct recent trend also observed in other parts of the world, such as the Mediterranean, Australia, Europe, and the US. Taking advantage of the economy of scale cost offered

Table 11.4 Desalination projects with the lowest cost of water production at present

Project	Location	Capacity (m^3/day)	First-year cost of water ($/$m^3$)
Hassyan	United Arab Emirates	545,000	0.306
Sorek 2	Israel	548,000	0.413
Jubail 3a	Saudi Arabia	600,000	0.440
Yanbu 4	Saudi Arabia	450,000	0.470
Tawellah	United Arab Emirates	909,200	0.495
Shuqaiq 3	Saudi Arabia	450,000	0.521
Rabigh 3	Saudi Arabia	600,000	0.531

using larger-sized plant equipment, piping and structures is a trend likely to continue in the future.

The accelerated growth of desalination over the past decade is driven by advances in membrane technology and material science. Recent technological advancements, such as pressure-exchanger based energy-recovery systems, higher-efficiency RO membrane elements, nanostructured RO membranes, innovative membrane vessel configurations, and high-recovery RO systems, are projected to further decrease the energy and costs for seawater desalination.

The steady trend in the reduction of desalinated water-production energy and costs coupled with the increasing costs of conventional water treatment and water reuse, driven by more stringent regulatory requirements, are expected to accelerate the current trend of reliance on the ocean as an attractive and competitive water source.

This trend is likely to continue in the future and to further establish ocean-water desalination as a reliable drought-proof alternative for most coastal communities worldwide in the next 10 years. At present, desalination provides approximately 10% of the municipal water supply of the urban coastal centers in the US, Europe, Israel, and Australia, and over 60% of the drinking water of the Gulf Cooperation Countries; by 2030 the contribution of desalination to the local water supply is expected to exceed 25 and 95%, respectively (Daigger et al. 2019).

11.5 Major Barriers and Response Options

The key challenges associated with the wider use of desalination as compared to conventional water supply are: (1) higher costs of water production; (2) greater energy demand and associated carbon footprint; (3) potential environmental impacts associated with desalination plant concentrate management; and (4) institutional and regulatory challenges.

11.5.1 Cost of Water Production

At present, the production of desalinated water is more costly than that of treatment of conventional freshwater resources such as rivers and lakes, as well as water reclamation and indirect potable reuse (Table 11.5).

However, low-cost freshwater resources are limited to less than 2.5% of the water available on the planet, and, in large, urbanized centers of most developed countries worldwide, such traditional freshwater resources are near depletion, while new sources are not readily available to sustain long-term population growth, industrial development, and the quality of life.

Table 11.5 Cost of water for production of alternative freshwater production methods (modified from Daigger et al. 2019)

Water supply alternative	Cost of water production ($/m^3)
Conventional treatment of surface water	0.2–0.4 (0.30)[a]
Water reclamation	0.3–0.6 (0.45)
Indirect potable reuse	0.5–0.8 (0.65)
Direct potable reuse	0.7–1.2 (0.95)
Brackish water desalination	0.4–1.5 (0.95)
Seawater desalination by reverse osmosis	0.5–2.5 (1.50)

[a] Figures in parenthesis indicate average values of the cost of water production

It is also interesting to note that, at present, the cost of water for direct potable reuse is comparable to that of BWRO desalination and slightly lower than the cost for SWRO desalination. However, it is expected that the cost of drinking-water production by direct potable reuse will increase in the future, considering that the waste discharges from such plants contain high concentrations of non-biodegradable, environmentally damaging substances such as pharmaceuticals and endocrine disruptors, which, if treated by the same means that are currently used for production of portable water from wastewater, will be more than double the expenditures needed for direct potable reuse.

While conventional processes, such as sedimentation and filtration, have seen modest advancement since their initial use for potable water treatment several centuries ago, now more efficient seawater-desalination membranes and system configurations, as well as equipment improvements, are introduced frequently. Like computers, the RO membranes of today are many times smaller, more productive, and cheaper than the first working prototypes. The future improvements of the RO membrane technology are forecast to encompass:

- Development of membranes of higher productivity, salt, and pathogen rejection; and reduced trans-membrane pressure, and fouling potential.
- Improvement in membrane resistance to oxidants, elevated temperature, and compaction.
- Extension of the useful life of membranes beyond 10 years.
- Integration of membrane pretreatment, advanced energy recovery, and SWRO systems.
- Integration of brackish- and seawater-desalination systems.
- Development of a new generation high-efficiency pumps and energy-recovery systems.
- Replacement of costly stainless steel with plastic components to increase plant longevity and decrease the overall cost of water production.
- Reduction in membrane element costs by complete automation of the entire production and testing process.

Table 11.6 Forecast of energy use and costs for seawater desalination plants (modified from Daigger et al. 2019)

Parameter for well-operating desalination plants	2020	2025	2030
Total electrical energy use (kWh/m^3)	3.5–4.5 (4.00)	2.8–3.2 (3.00)	2.1–2.4 (2.25)
Cost of water ($/m^3)	0.4–2.5 (1.45)	0.3–1.0 (0.65)	0.2–0.5 (0.35)
Construction cost ($/MLD)	0.8–2.2 (1.50)	1.0–1.8 (1.40)	0.5–0.9 (0.70)
Membrane productivity (m^3/membrane)	28–48 (38)	55–75 (65)	95–120 (108)

[a] Figures in parenthesis indicate average values

- Development of methods for low-cost continuous membrane cleaning making possible reductions in downtime and chemical cleaning costs.
- Creation of technologies for the extraction of valuable minerals from brine (brine mining).

These technological advances are expected to ascertain the position of SWRO treatment as a viable and cost-competitive process for potable water production and to reduce the cost of freshwater production from seawater by 25% by 2022 and by up to 60% by 2030 as shown on Table 11.6 (Daigger et al. 2019).

The rate of construction of new desalination plants in coastal urban centers will depend on the magnitude of water stress and availability of lower-cost conventional water resources. In the future, desalination is likely to be adopted as the main water supply in most arid and semi-arid regions of the world, such as in the Middle East, North Africa, the western United States, Australia, and in locations of concentrated industrial demand for high-quality water such as Singapore, China, India, and northern Chile.

11.5.2 Power Use

Salt separation from seawater requires a significant amount of energy to overcome the naturally occurring osmotic pressure exerted on the reverse-osmosis membranes. Seawater reverse osmosis (SWRO) desalination is several times more energy intensive than the conventional treatment of freshwater resources. Table 11.7 presents the energy use associated with various water-supply alternatives (Voutchkov 2019).

The energy needed for seawater desalination is approximately 8-to-10 times higher than that for production of freshwater from conventional sources such as rivers, lakes, and freshwater aquifers. Energy use for water reclamation is significantly lower than that for seawater desalination.

Even though the carbon footprint for production of desalinated water is higher than that of production of drinking water from traditional freshwater resources, it is smaller than many other human activities that improve the quality of life, such as food refrigeration, heating of water for domestic use, driving a personal vehicle, or flying. The average carbon footprint of producing desalinated water for one person

Table 11.7 Energy use for alternative freshwater production methods

Water supply alternative	Energy use (kWh/m^3)
Conventional treatment of surface water	0.2–0.4
Water reclamation	0.5–1.0
Indirect potable reuse	1.5–2.0
Direct potable reuse	1.7–2.4
Brackish water desalination	1.0–1.5
Seawater desalination by reverse osmosis	2.5–4.0

is 0.11 tons CO_2/year, which is only 3.7% of the sustainable carbon footprint of one person of 3 tons/ CO_2/year (IDA Forum 2019b).

Currently, most desalination plants worldwide are supplied by power generated from fossil fuel. However, several recently constructed SWRO desalination plants in Australia have implemented wind-driven power generation projects, which produce as much power as used by the desalination plants. Over the last five years, several Middle Eastern countries have taken the initiative to develop a robust portfolio of renewable power-generation plants to provide electricity for seawater desalination (World Bank 2019).

Solar and wind power are the most abundant renewable-energy sources worldwide at present. Key advantages of solar, as compared to wind, powered desalination plants in the Middle East are the high intensity and reliability of the power source (e.g., solar irradiation), and their relatively lower construction and O & M costs. However, as with wind farms, a key challenge of solar power-supply facilities is the need for a large amount of land to accommodate the renewable-energy equipment to supply power to SWRO desalination plants.

Outside of the Middle East, wind-generated power has found wider use than solar power for desalination project supply because of its availability as an energy-generation source. Even in countries such as Saudi Arabia and UAE, solar power intensity adequate for steady power generation occurs only for six seven hours per day and for less than 80% of the time.

As a rule of thumb, the land area needed for a photovoltaic (PV) field to power 1,000 m^3/day SWRO plant is 10 ha, while a wind farm for the same size SWRO plant requires 20 ha. This is approximately 50 times and 100 times higher, respectively, than the land needed to construct the SWRO desalination plant itself. The total capital cost for construction of a solar power plant to supply the entire amount of electricity needed for seawater desalination plant at present is typically 60–80% of the capital cost of the desalination plant itself. More detailed discussion of the feasibility of linking renewable power and desalination projects is provided elsewhere (World Bank 2012).

Solar power-driven desalination projects under development at present encompass indirect or direct coupling of conventional SWRO, MSF, or MED desalination plants with either concentrated solar power-generation technologies (CSPs) or PV cells (Blanco et al. 2011). The most promising combinations of solar power and desalination technologies are PVs with RO and ED systems and CSPs with MSF or MED systems (Moser et al. 2013; Shatat et al., 2013; Pinto and Marques, 2017).

Currently, PV-based SWRO solar desalination is the main focal point of research and full-scale desalination project implementation because of the significant decrease in solar-panel costs over the last five years. At present, conventional SWRO desalination plants powered through the electric grid remain economically more competitive than PV-powered RO or CSP-powered MED configurations, as well as to other combinations of desalination technologies and alternative power sources (Fiorenza et al. 2003; Moser et al. 2013).

Selecting the most suitable renewable energy-driven desalination technology depends on the size of the plant, the source water, and product water quality, the availability of access to the electric power grid, and the type of renewable power technology (Ghaffour et al. 2015).

Desalination based on the use of renewable energy sources can provide sustainable long-term production of freshwater and is expected to become economically attractive soon. This is because the costs of renewable energy-production technologies continue to decline and the costs of fossil fuel continue to rise over time. In addition, environmental externalities associated with fossil-fuel based electricity generation (e.g., the need to offset the desalination plant's carbon footprint) may offset the difference in energy and water production costs (Karagiannis and Soldatos 2008; Gude 2016).

In parallel with the exploration of renewable power alternatives, the world's leading research centers in the US, Saudi Arabia, and Europe are working on the development of a new generation of energy recovery devices, high-pressure pumps, and membranes that aim to bring the total energy consumption of desalination plants to less than 2.45 kWh/m^3 and the energy demand of the reverse-osmosis desalination system below 1.8 kWh/m^3.

These advancements will result in the reduction of the total energy consumption and carbon footprint of desalination plants by over 30%. The new technologies are tailored to fit equally well in both existing desalination plants and future reverse-osmosis facilities.

The new generation energy-recovery devices under development at present are designed to reuse over 98.5% of the energy remaining in the brine after membrane separation. Such energy-recovery efficiency will significantly exceed the performance of the existing commercially available best-in-class energy-recovery technologies and will address some of the flaws of these technologies, such as brine mixing and equipment complexity.

Research and development in the next generation of high-pressure pumps for SWRO systems are projected to have a disruptive impact on reducing energy use. State-of-the-art high-pressure pumps used for desalination currently have an efficiency between 75 and 83%. The new generation of pump technologies under development is targeting an efficiency of 95% or more. As high-pressure pumps consume between 70 and 75% of the total energy in desalination plants, this dramatic improvement of pump efficiency will yield an unprecedented reduction in the desalination plants' carbon footprints.

11.5.3 Concentrate Management

Like conventional water treatment plants and water reclamation facilities, desalination plants generate source water treatment by-products. The main desalination plant by-product is concentrated source water typically referred to as concentrate or brine.

Desalination concentrate contains dissolved compounds found in the original saline source water (minerals, organics, metals, etc.), which are rejected by the reverse osmosis membranes. Typically, seawater concentrate has a salinity of 50,000–70,000 mg/L, while concentrate from BWRO plants has a salt content of 4,000–20,000 mg/L. Usually, concentrate constitutes 90–95% of the total desalination plant-discharge volume (Voutchkov and Kaiser 2020).

Backwash water is the second largest discharge stream from desalination plants and is generated during the periodic cleaning of the pretreatment filters. This stream contains solid particulates and other compounds removed from the source water prior to desalination and usually contributes 3-5% of the plant discharge. Membrane cleaning water, which contains a low concentration of spent detergent, is produced intermittently (usually one-to-two times per month) in very small quantities (0.1% or less) compared to concentrate flows—it is produced when the membranes are cleaned.

Concentrate from seawater desalination plants typically has the same color, odor, oxygen content, and transparency as the source seawater from which the concentrate was produced. Therefore, concentrate discharge to surface water bodies (ocean, river, etc.) does not typically change its physical characteristics or have aesthetic impact on the aquatic environment, except for its density.

Desalination treatment processes do not cause depletion of the natural oxygen content of the source seawater used to produce freshwater. In fact, the backwashing with a mix of air and water of the filters used for pretreatment of the seawater enriches the oxygen content of the plant discharge and prevents the occurrence of hypoxia (low content of oxygen) in the discharge area.

There is no relationship between the level of salinity and biological or chemical oxygen demand of the desalination plant concentrate. Over 86% of the minerals that comprise the concentrate's salinity are sodium and chloride, and they are not food sources or nutrients for aquatic organisms. The dissolved solids in the concentrate discharged from seawater desalination plants are not of anthropogenic origin as compared to pollutants contained in discharges from industrial or municipal wastewater-treatment plants, water reclamation facilities, or plants for indirect or direct potable reuse.

Membrane desalination processes do not change the temperature of the desalination plant discharge because the process of desalination does not involve heating of the source seawater to produce freshwater.

Chemical-Free Desalination: The state-of-the-art desalination processes employed in contemporary desalination plants use a very limited amount of chemicals. All chemicals added in various treatment processes of desalination plants are of food-grade quality, biodegradable, and specifically selected not to cause aquatic-life toxicity. Therefore, the discharges from seawater desalination plants are neither toxic nor harmful for marine life and are engineered to dissipate rapidly and without permanent alterations to the surrounding marine ecosystem (Mickley and Voutchkov 2016; IDA 2019b).

Over the past five years, many countries with large desalination plants such as Saudi Arabia, Australia, Israel, and Spain have initiated the implementation of comprehensive programs for green desalination, which aims to reduce both the amount and the types of chemicals used in the production of desalinated water. These programs will ultimately convert all existing facilities into chemical-free seawater desalination plants by implementing the latest advancements of desalination technology and science.

Desalination plants used to continuously chlorinate their intake seawater using sodium hypochlorite to suppress the growth of marine life in the intake piping and on the RO membranes. Such practice was abandoned by most desalination plant operators close to a decade ago, and currently chlorination is used only one-to-two times per month for a period of 6-8 hours. In addition, some desalination plant operators do not apply any disinfectants to the intake seawater because they prefer to use the pretreatment system of the plant for control of biofouling instead of chemicals.

Ferric chloride and ferric sulfate are the most-used coagulants for the pretreatment of seawater at present. These chemicals used to be added at a constant rate and at a relatively high dosage. At present, the desalination industry has adopted a real-time monitoring of the content of solids in the seawater and automated adjustment of the coagulant dosage proportionally to the actual content of suspended solids in the water. This operational strategy, introduced over the last 10 years at most plants worldwide, has reduced the use of coagulant to less than one half of what it once was.

Acids and flocculants were used to optimize the chemistry of water treatment in many desalination plants until a decade ago. Most advanced desalination plants and skilled plant operators no longer use acids and flocculants for pretreatment—instead, they rely on optimized pretreatment system design and operation to manage water chemistry.

Until 2010, antiscalants and sodium hydroxide were commonly applied in many desalination plants worldwide, mainly to prevent the scaling associated with the removal of boron from desalinated water. Since 2014, when the WHO increased the drinking water guideline limit for boron from 0.5 mg/L to 2.4 mg/L, most desalination plants discontinued the addition of sodium hydroxide and antiscalants.

The desalination industry is constantly developing and adopting new chemical-free, renewable energy-based technologies. The next step in this development process is to use chemicals extracted from the brine for post-treatment of the desalinated water instead of using commercially supplied calcium compounds, such as the lime from limestone.

Beneficial Use of Concentrate: Brine generated from desalination plants can be used as a source of valuable minerals, such as calcium, magnesium, and sodium chloride. Rare-earth elements, including lithium, strontium, thorium, and rubidium, can also be extracted from brine

Recent stresses on the availability and supply in the global market of rare-earth elements have exposed sustainability challenges in conventional extraction and production. These metals are used to fabricate critical components of numerous products, including airplanes, automobiles, smart phones, and medical devices. There is a growing realization that the development and deployment of clean-energy technologies and sustainable products, processes, and manufacturing industries of the 21st century will also require large amounts of rare metals and valuable elements, including platinum group metals, such as lithium, copper, cobalt, silver, and gold.

11.5.4 Institutional Challenges

At present, the water sector of most countries implementing desalination projects do not have a sound and comprehensive institutional, legal, and policy framework with respect to desalination. Although in recent years some countries enacted few new water-regulation laws specifically dealing with desalination, and they usually address desalination projects only indirectly.

The main types of institutional issues that hinder the wider use of desalination are:

- Lack of country-wide desalination project implementation plans and policies.
- Complex environmental regulations and policies.

In most countries, except for Saudi Arabia, Israel, Cyprus, and Malta, there are no official long-term strategic government plans specifically focusing on the staged development of new desalination plants. Desalination project-related planning policies are usually driven by overall country water demand and are established at the level of the statewide government utilities or agencies responsible for the country's water supply.

The creation of national or regional programs and policies for planning and implementation of desalination projects will be very beneficial for the development of this new water resource and its coordinated implementation along with other alternative water resources, such as reclaimed water and measures for controlling water demand such as water conservation.

Similarly, most countries worldwide lack regulatory requirements specifically pertaining to the operation of desalination plants and minimization of their environmental impacts (Mickley and Voutchkov 2016).

In May 2015, the State of California introduced the first-in-the-world regulations specifically pertaining to the environmental impacts of seawater desalination plant intakes and outfalls (SWRCB 2015). These regulations incorporate stringent environmental requirements and mitigation measures intended to promote the use of subsurface intakes (wells) instead of open seawater intakes and to significantly increase the environmental mitigation requirements for desalination plant operations. Such stringent regulations are projected to increase the cost of production of desalinated water in California by 20–25%. Since their introduction, such regulations have practically resulted in the suspension of implementation of new large-scale seawater desalination projects in California.

11.6 Conclusions

While desalination currently provides only around 5% of water supply worldwide, it is expected that in the next decade the construction of new desalination plants will be more than double (IDA 2019). This can be attributed to the impact of climate change, increased demand due to population growth, limited availability of new, inexpensive terrestrial water sources, and advances in membrane technology that are projected to further reduce the cost and energy needed for desalination.

Near and long-term desalination technology advances are projected to yield significant decreases in the costs of production of desalinated water by 2030. Innovative technologies, such as nanoparticle enhanced membranes, biomimetic membranes, and forward osmosis, as well as beneficial extraction of valuable minerals from the brine generated by desalination plants, are aimed at reducing energy consumption by 20–35%, reducing capital costs by 20–30%, improving process reliability and flexibility, and greatly reducing the volume of brine discharge. Soon, the brine-derived commercial products, such as valuable metals and salts, are expected to create an adequate revenue to partially, and over time, fully subsidize the production of desalinated water, making desalination the lowest-cost unconventional water-supply resource worldwide.

Key recommendations for future research include:

- Develop low-cost plastic materials for the production of high-pressure piping and other plant components that can replace the use of costly duplex stainless steel and improve the reliability, performance, and water-production costs.
- Create new generation membranes with a uniform molecular pore structure that can enable an increase in membrane productivity by up to a factor of 20.
- Advance brine concentration and mining technologies for the extraction of metals and salts of high commercial value.

- Generate non reverse-osmosis technologies of salt separation that make possible further reductions in energy demand and costs for the production of desalinated water.
- Accelerate the use of renewable power sources for reliable and cost-effective desalination.

References

Al-Karaghouli A, Kazmerski LL (2013). Energy consumption and Water production cost of conventional and renewable-energy-powered desalination processes. J Renewable and Sustainable Energy Review 24, 343–356https://doi.org/10.1016/j.rser.2012.12.064

AWWA (American Water Works Association) (2007) Reverse osmosis and nanofiltration. manual of water supply practices: M46, 2nd edn. AWWA, Denver, Colorado

Blanco J, Alarcón-Padilla DC, Zaragoza G, Palenzuela P, Guillén E, Ibarra M (2011, April) Solar energy and desalination processes: status and development perspective. In: Proceedings of the water desalination conference in the Arab Countries (ARWADEX) Riyadh, Saudi Arabia

Buros O (1980) The USAID desalination manual. IDEA Publications, Englewood, New Jersey

Daigger GT, Voutchkov N, Lall U, Sarni W, Machado F, Mimmi L (eds) (2019) The Future of Water: a collection of essays on "disruptive" technologies that may transform the water sector in the next 10 years—discussion Paper No. IDB-DP-657. Inter-American Development Bank-Water and Sanitation Division, Washington D.C

Fiorenza G, Sharma VK, Braccio G (2003) Techno-economic evaluation of a solar powered water desalination plant. J Energy Conserv Manag 44(14):2217–2240

Ghaffour N, Bundschuh J, Mahmoudi H, Goosen MF (2015) Renewable Energy-driven desalination technologies: a comprehensive review on challenges and potential applications of integrated systems. J Desalin 356:94–114. https://doi.org/10.1016/j.desal.2014.10.024

Gude VG (2016) Desalination and sustainability–an appraisal and current perspective. Desalination 89:87–106. https://doi.org/10.1016/j.watres.2015.11.012

GWI (Global Water Intelligence) (2020) Desalination water report. GWI, vol 36, no 28. Media Analytics, Oxford, U.K

IDA (International Desalination Association) (2019a) IDA water security handbook 2019–2020. International Desalination Association. Media Analytics, Oxford, U.K

IDA (International Desalination Association) (2019b, October) Desalination impacts on the environment–myths and misconceptions. In: Presentation of IDA energy and environmental committee at IDA World Congress, Dubai. https://doi.org/10.13140/RG.2.2.22335.05286

Jones E, Qadir M, van Vliet MT, Smakhtin V, Kang S (2019) The state of desalination and brine production: a global outlook. J Sci Total Environ 657:1343–1356. https://doi.org/10.1016/j.scitotenv.2018.12.076

Karagiannis I, Soldatos PG (2008) Water desalination cost literature: review and assessment. J Desalin 223(1–3):448–456. https://doi.org/10.1016/j.desal.2007.02.071

Mickley M, Voutchkov N (2016) Database of permitting practices for seawater concentrate disposal. Water Environment Reuse Research Foundation, Alexandria Virginia

Moser M, Trieb F, Fichter T, Kern J (2013) Renewable desalination: a methodology for cost comparison J Desalin Water Treat 51(4–6):1171–1189. https://doi.org/10.1080/19443994.2012.715446

Pinto FS, Marques RC (2017) Desalination project economic feasibility: a standardization of cost determinants. J Renew Sustain Energy Rev 78:904–915. https://doi.org/10.1016/j.rser.2017.05.024

Shatat M, Worall M, Riffat S (2013) Opportunities for solar water desalination worldwide: review. J Sustain Cities Soc 9:67–80. https://doi.org/10.1016/j.scs.2013.03.004
SWRCB (State Water Resources Control Board) (2015) California Ocean Plan 2015. Sacramento, California: SWRCB; State of California; California Environmental Protection Agency
Voutchkov N (2012) Desalination engineering: planning and design. McGraw Hill, United States, New York
Voutchkov N (2017) Pretreatment for reverse osmosis desalination. Elsevier, Netherlands, Amsterdam
Voutchkov N (2019) Desalination project cost estimating and management. CRC Press—Taylor & Francis Group, United States Florida
Voutchkov N, Kaiser G (2020) Management of concentrate from desalination plants. Elsevier, Amsterdam
World Bank (2012) Renewable energy desalination: an emerging solution to close the water gap in the middle east and North Africa-MENA Development Report. World Bank, Washington, DC. https://openknowledge.worldbank.org/handle/10986/11963
World Bank. (2019) The role of desalination in an increasingly water-scarce world-technical paper of water global practice. World Bank, Washington, DC. https://openknowledge.worldbank.org/handle/10986/31416

Part VII
Promoting the Enabling Environment

Chapter 12
Governance of Unconventional Water Resources

Renée Martin-Nagle and Christina Leb

Abstract Governance principles concerning the development and utilization of freshwater resources derive from a variety of factors: historical and current patterns of usage, cultural and administrative norms and practices, and legal principles and policies. While governance of domestic surface water has benefitted from millennia of evolving practice, the legal principles regarding the development and utilization of transboundary freshwater resources emerged only during the mid-20th century and were codified towards the end of that century. The emergence of general rules and principles governing the use of groundwater resources is more recent still, as their systematic development began much later than the development of surface water. Similarly, the governance frameworks for unconventional water resources (UWR) are not yet well-developed, and legal gaps in regulating their exploitation and use have not yet been filled. This chapter presents an overview of legal theories and principles that are relevant to the design of the governance framework for both domestic and transboundary UWRs. The chapter focuses on legal aspects and attempts to predict how rights and obligations for various forms of UWR will emerge under current principles and practices.

Keywords Law · Stakeholders · Sovereignty · Rights · Governance

---

Christina Leb: The views presented in this article are those of the author alone and in no way reflect the position of the World Bank, its Board of Executive Directors, or any of its member countries.

R. Martin-Nagle (✉)
Environmental Law Institute, A Ripple Effect pllc, 600 West Highland Avenue, Ebensburg, PA 15931, USA
e-mail: Renee@arippleeffect.net

C. Leb
World Bank, 1818 H Street NW, Washington, D.C., WA 20433, USA
e-mail: cleb@worldbank.org

M. Qadir et al. (eds.), *Unconventional Water Resources*,
https://doi.org/10.1007/978-3-030-90146-2_12

12.1 Introduction

Attempting to predict a governance regime for unconventional water resources (UWR) presents a daunting task on several levels. To begin, the various types of UWR span a wide spectrum of incarnations, ranging from water vapor floating in tropospheric clouds to water droplets saturating porous rocks in deep underground aquifers to freshwater icebergs floating in the oceans. Unless all freshwater resources can be regulated as a single resource, crafting one set of rules to address all the forms of UWR will be impossible. To further complicate the analysis, definitions of what constitutes a natural resource governance regime also span a wide spectrum. Under the most structured definition, governance can mean a precise set of government-imposed laws and regulations, but governance can also refer to an informal understanding of acceptable behaviors (Jiménez et al. 2020). Finally, traditional notions of sovereign rights play an important role. Nations have long claimed exclusive ownership over natural resources that lie within their borders (Roth 2004), but a fluid resource such as freshwater famously does not respect man-made artificial borders, and the development of principles regarding shared transboundary resources proceeds slowly and unevenly. Even if a clear set of global rules regarding ownership and the usage of transboundary resources existed, each nation may have its own governance principles and laws for exploitation of UWR, and perhaps a different set of laws for each type of UWR, which is currently the case in nations such as the US. Multiplying the number of nations and their formal and informal legal regimes by the types of UWR produces several governance permutations, far too many to explore in an efficient manner. Therefore, this chapter will discuss only globally relevant legal principles and governance structures and will not address national or subnational principles or structures except for illustration.

The multitude of complexities surrounding UWR poses obvious hurdles for presenting a proposed governance regime for UWR in a single chapter, and the analysis must necessarily be limited in scope. This chapter will begin by explaining the definition of governance that will be used and summarizing the extent and limitations of sovereign rights. The chapter will then proceed to summarize accepted key principles of international water law that feature in any discussion of global water governance. Finally, the chapter will provide predictions for how sovereignty will attach to each UWR, before concluding with recommendations for designing domestic legal regimes for each UWR.

12.1.1 What is Governance?

Water's hydrological cycle makes it unique. Water can evaporate from the territory of one state, be transported to another state via clouds, fall as precipitation on a third state, and then travel to more states via surface water, groundwater, ships, and icebergs. No other natural resource has such an alternating, meandering character, and policymakers attempting to craft a governance regime for UWR must determine when and to what extent sovereign rights and duties attach to each form of UWR.

Generally, legal governance regimes attempt to allocate rights and duties among the stakeholders (Reed et al. 2009), but determining rights and duties for UWR may prove difficult due to the fact that the stakeholders may also be unconventional. For example, stakeholders of nonrenewable forms of UWR, such as deep groundwater, include both current and future generations. Water from cloud-seeding and fog harvesting is theoretically renewable, but causing atmospheric water to precipitate prematurely may cause unintended scarcity for stakeholders in the downwind territories.

Without an agreed structure for development and utilization, access to and exploitation of a vital natural resource such as freshwater can result in anarchy, as fears of scarcity may encourage users to engage in a race to obtain, and establish rights to use, as much of the resource as possible before it is depleted. In his well-known essay, Garrett Hardin coined the phrase "tragedy of the commons" to describe the gradual exhaustion of a shared resource through lack of controls over its usage (Hardin 1968), but single users can deplete a resource as well. On the other hand, Elinor Ostrom, drawing on her research on the utilization of groundwater as a shared resource, won a Nobel Prize for advancing the theory that users of a common resource will develop formal and informal structures to ensure that the resource is preserved (Ostrom 1990). When predicting a legal regime for governing UWR, both Hardin's tragedy of the commons and Ostrom's theories of cooperation may provide guidance.

For the purposes of this chapter, the term governance will refer to an agreed set of principles that will guide assignment of rights and obligations regarding utilization of UWR. Ideally, a well-crafted governance regime serves to encourage collaboration while reducing and perhaps eliminating conflicts. While mature governance regimes feature not only rights and obligations but also precise rules and delegation of authority (Abbott et al. 2000), principles regarding ownership and utilization of UWR are only now beginning to be discussed. Thus, precise rules and assumption and delegation of authority cannot be expected to emerge until later stages in the evolution of UWR governance.

12.1.2 The Role of sovereignty

Sovereignty over a natural resource connotes ownership and control, and the debate about the rights and obligations of sovereign entities with respect to natural resources has endured for millennia. Initially, states were deemed to have absolute power within their own borders, and even in modern times they have retained the right to enact laws and regulations affecting their own domestic activities. However, the post-WWII plethora of nation-states has resulted in a myriad of shared borders that has consequently multiplied the number of shared natural resources, resulting in increased opportunities for transboundary impacts, collaborations, and conflicts. Through legal opinions such as the 1941 *Trail Smelter* arbitration between the US and Canada and the 2018 dispute between Costa Rica and Nicaragua that went before the International Court of Justice, as well as the multitude of declarations and treaties that were produced by the 1992 Earth Summit, the global community has accepted that territorial sovereignty does not permit states to engage in acts that would result in significant transboundary harm to their neighbors. Conversely but symmetrically, no state enjoys the security of absolute territorial integrity and must accept some degree of transboundary impacts simply by being part of a community of nations (Mickelson 1993). These primary limitations on sovereignty have been joined by others, including responsibilities to protect human rights and ecosystems, even in purely domestic territory (Jackson 2003).

Freshwater has been utilized long enough for agreed principles to have developed regarding sovereign rights and obligations pertaining to the resource. Surface water in rivers and lakes that are entirely situated within the borders of a single nation and do not flow from, or subsequently flow into, the territory of another state is subject only to the sovereignty of that nation. Surface water shared with one or more other nations limits the sovereign rights of each of the transboundary nations. To reduce conflicts, principles such as equitable and reasonable use (ERU), prevention of significant harm to a neighboring state, and the general duty to cooperate arose to serve as guides for cooperative management of the resource (Leb 2013). Sovereignty over groundwater follows roughly the same principles that apply to surface water, although not necessarily in an equivalent manner, and the rule of capture, which grants ownership to the party that first possesses the resource (Onorato 1968), is still widely practiced (Martin-Nagle 2020a). Although nations have shared transboundary surface water and groundwater for millennia, the principles governing conjunctive use have only been clarified in the past century (Martin-Nagle 2020b). Two global treaties embody the agreed principles of shared freshwater resources: the 1997 UN Convention on Non-Navigational Uses of International Watercourses and the 1992 UNECE Convention on the Protection and Use of Transboundary Watercourses and International Lakes. While these two conventions have not yet received universal support, together they represent the most current expression of the principles of international water law and limitations on sovereign rights and will be discussed in Sect. 12.2.

12.2 Overall Principles to Guide the Development of UWRs

As mentioned at the outset, given the diversity of characteristics and methods of extraction regarding what are considered UWR, as a group UWRs hardly fit under one uniform governance regime. However, as water is a vital resource and due to its very nature a shared resource, several legal principles become relevant with respect to governance of UWRs. As with all shared resources, it is important to consider the sustainable management and development of UWRs to ensure that future generations can continue to benefit from this life-giving resource. This section highlights a set of accepted key principles relevant to sustainable water resources management, and also to UWRs more particularly. It discusses their origin and scope as well as their application to UWR management and development. Modalities of how these legal principles intervene with respect to individual UWRs are summarized at the end of this section in Table 12.1.

12.2.1 Good neighborliness and no Significant Harm

The principle of good neighborliness and the related due-diligence obligation not to cause significant harm to others apply in the national as well as the international context. These principles originated in the generally accepted maxim of *sic utere tuo ut alienum non laedas,* meaning that everyone has a right to use her/his own property as long as that use does not adversely affect the rights of others. In the international context, the principle of good neighborliness applies not only to resources under unrestricted territorial sovereignty of a state but also, and even more importantly, to any resources shared with others, since shared resources are subject to limited sovereignty rights (Lammers 1984). The Charter of the United Nations harkens to this principle in its role of facilitating peaceful relations between states (Preamble and Article 74).

With respect to UWR, the principle of good neighborliness and the related obligation to prevent significant harm guide the use and development of these resources. Any capture of UWR in the national or transboundary context needs to be done in consideration of any adverse effect on the potential rights of others to the use of these resources. For example, sludge or brine that results from water recycling or desalination needs to be disposed of in such a manner so as not to pollute other areas, ecosystems, or waters which others have a lawful right to use and an expectation of good quality. The exploitation of groundwater, be it entirely domestic or through transboundary aquifers, must consider the impact on the rights of other users of the aquifer. Similarly, the harvesting of atmospheric water needs to consider the impact on those who may otherwise depend on it, even if their rights may not be firmly established. Experimentation with cloud seeding and weather-modification techniques to harvest or modify atmospheric water flows gave rise to several lawsuits in the US in the 1950s and 60s, with claimants at times successfully asserting resulting loss of precipitation on their lands (Corbridge and Moses 1968).

Procedures have been put into place in national and international legal frameworks to facilitate compliance with the principle of good neighborliness and related obligations. Thus, the legal requirement to conduct environmental impact assessments is present in most national legal frameworks, and a customary international law obligation to assess potential transboundary impact has been confirmed by the International Court of Justice (ICJ) (ICJ 2010).

12.2.2 Protection of the environment and the precautionary approach

The principle of good neighborliness has also been linked to the principle of environmental protection that has found expression in numerous international declarations, such as Principle 2 of the 1992 Rio Declaration on Environment and Development (Leb 2013). The ICJ has further confirmed that the principle of environmental protection forms part of customary international environmental law (ICJ 1996, 1997, 2010). Under the ICJ's interpretation of that principle, states have the right to exploit their own resources, and, at the same time, they have "the responsibility to ensure that activities within their jurisdiction and control respect the environment of other states or of areas beyond national control" (ICJ 1996). Through its decisions, the ICJ expanded this principle beyond its initial conceptualization in the Rio Declaration, which limited state responsibility to avoiding damage to other states' territories and other areas and now mandates respect for the environment more generally. The development of UWR therefore must not only comply with national laws and obligations related to the protection of the environment but is also governed by the equivalent customary law principle in the international context.

Given that the development and use of certain UWRs are still in their early stages, the impact of their development on the environment is not yet fully known. This is particularly the case for the employment of weather-modification techniques for the harvesting of atmospheric waters and, to a certain extent, also with respect to fog harvesting and iceberg towing. The adoption of a precautionary approach to the exploitation of these UWRs therefore becomes particularly important. The precautionary approach emphasizes a focus on and bias towards caution. According to this principle, where there is a potential risk of serious or irreversible damage, actors cannot use the lack of scientific certainty as an excuse to postpone measures to prevent environmental damage (UNCED 1992). First pioneered by Germany in domestic legislation in 1974, the precautionary approach was initially recognized as regional customary law in Europe (Sirinskiene 2009) but has since become widely recognized also as a customary principle of international environmental law that must be observed (McIntyre 2019).

The legal principles on the protection of the environment and the precautionary approach are building blocks to realize sustainable development, which is an objective enshrined in Principles 3 and 4 of the 1992 Rio Declaration (UNCED 1992). The objective of sustainable development is to develop and utilize natural resources, including freshwater, while balancing current environmental and economic needs against the needs of future generations. While opinions on the weight of sustainable development as a legal principle diverge, it is widely agreed that the development of water resources needs to be guided by this concept (Birnie et al. 2009).

One approach that supports achievement of sustainable development of freshwater and that is gaining increasing traction is water stewardship. This approach, which is currently employed primarily by global private companies concerned about impacts of water scarcity on supply chains, is defined "as the use of water that is socially and culturally equitable, environmentally sustainable and economically beneficial, achieved through a stakeholder-inclusive process that includes both site- and catchment-based actions" (Alliance for Water Stewardship 2020). Water stewardship, based on the idea of responsibly managing something that is not owned, is a particularly compelling approach to managing water resources, including UWR, because it is attune to the transboundary nature of the hydrologic cycle and the fact that in several domestic legal systems water is not owned until captured.

12.2.3 The equitable and reasonable use principle and intergenerational equity

The fact that current generations hold the planet and its natural resources in trust for their offspring and future generations is one of the reasons that the focus on sustainable development is so important. The objective is to protect the life-support systems that Earth should provide for future generations; each generation is allowed equitable use of the planet's resources in its own time and has a duty to ensure equitable use across generations (Brown Weiss 1990). The increased use of UWR is evidence that we are reaching the limits of this generation's patterns of exploiting freshwater.

The concept of intergenerational equity is not new. It appeared in Principle 2 of the 1972 Stockholm Declaration on the Human Environment and has since been included in multiple multilateral environmental agreements (UNCHE 1992; UNFCCC 1992; UNECE 1992; UNWCC 1997, among others). The principle of environmental protection and the precautionary approach are key components to achieving this objective. Intergenerational equity also finds expression in the principle of equitable and reasonable use. This principle, which partially originated in US domestic law (McCaffrey 2007), is a key customary principle of the law of transboundary water resources. It obliges states that share cross-border water systems, including deep aquifer systems, to utilize these waters in an equitable and reasonable manner, considering the interests of other concerned states. The principle of equitable and reasonable utilization applies to all forms of utilization, including extraction, allocation, distribution, and consumptive and non-consumptive uses, as well as the protection of water resources. Nation-states have to consider various factors: the multiple characteristics of a shared water system and criteria relating to uses geographic, hydrological, climatic, and ecological factors; social and economic needs of riparian states; the population dependent on the shared water resources; the effects of uses on the multiple riparian states; and economic efficiency of use of the shared resources among others. The criteria also include the requirement to account for existing, as well as potential future, uses of shared waters (UNWCC 1997).

The application of the principle of equitable and reasonable use to all shared surface and groundwater resources, including deep groundwater, is well established in international law and is based on the consideration that the use of water by one state may impact the use of water by another. Given there is currently only one treaty that deals with use of atmospheric water, namely the 1978 Convention on the Prohibition of Military and Other Hostile Use of Environmental Modification Techniques, courts, tribunals and scholars have so far had little opportunity to debate and decide whether the equitable and reasonable use principle should also apply to this form of water. It can be argued, however, that the harvesting of atmospheric water should follow this principle where harvesting may cause an adverse impact, as may be the case with cloud seeding and its impact on downwind communities.

Table 12.1 Application of legal principles to UWR

Type of unconventional water resource	Legal principles and relevance				
	Good neighborliness (Potentially impacted neighbors)	No significant harm (Risk of Harm)	Protection of the environment (Potential negative impact)	Precautionary approach (Reasons for applicability)	Equitable and reasonable use (Interested stakeholders)
Cloud seeding	Downwind users[a]	Risk of less rain and/or floods for downwind stakeholders	Interference with natural rainfall patterns	Impact not fully known	Downwind users
Fog water harvesting	Downwind users	Risk of less moisture and rain for downwind stakeholders	Withdrawal of water from nature	Impact not fully known	Downwind users
Micro-catchment rainwater harvesting	Users of surface water or groundwater not being replenished by rainwater	Reduction in replenishment of surface water or groundwater	Withdrawal of water from ecosystem	Impact of large-scale use not fully known	Users of surface water or groundwater not being replenished by rainwater
Offshore fresh–brackish groundwater	Transboundary interest holders	Risk of abstracting neighbor's share; damage to marine ecosystems	Risk of salt intrusion; damage to seabed ecosystems	Not available for future generations	Future generations and other existing users

(continued)

Table 12.1 (continued)

Type of unconventional water resource	Legal principles and relevance				
	Good neighborliness (Potentially impacted neighbors)	No significant harm (Risk of Harm)	Protection of the environment (Potential negative impact)	Precautionary approach (Reasons for applicability)	Equitable and reasonable use (Interested stakeholders)
Onshore deep groundwater	Transboundary interest holders	Risk of abstracting neighbor's share	Little risk	Not available for future generations	Future generations and other existing users
Municipal wastewater recycling	Downstream users	Risk of accidental release and pollution	Sludge disposal; risk of accidental release and pollution	Potential health and environmental impacts from recycled wastewater	Possible issues regarding distribution
Agricultural drainage water	Downstream users of drainage water or a mix of drainage water and freshwater	Potential impact on downstream users of drainage water or a mix of drainage water and freshwater	Change to flow patterns	Potential health impacts not fully known	Existing downstream users
Iceberg towing	[Ownership and use rights not yet determined]	Risk of accidents and changes to ecosystem	Change in natural salinity levels at "parking"; impacts on ecosystem at source and destination; risk of accidents	Impact not fully known	Possible issues regarding distribution
Ballast water	Stakeholders at source and destination	Risk of invasive species at destination	Risk of invasive species at destination	Impact not fully known	Users of source water

(continued)

Table 12.1 (continued)

Type of unconventional water resource	Legal principles and relevance				
	Good neighborliness (Potentially impacted neighbors)	No significant harm (Risk of Harm)	Protection of the environment (Potential negative impact)	Precautionary approach (Reasons for applicability)	Equitable and reasonable use (Interested stakeholders)
Desalinated water	Coastal populations	Impact of brine disposal and on ecosystem	Impact of brine disposal and on ecosystem; emissions	Long-term impact not fully known	Possible issues regarding distribution of desalinated water

[a] "Users" are defined as those who currently use the resource, hold use rights, or have otherwise current and future interests in the usability of the resource.

12.3 Sovereign Rights and UWR

Any proposal for governance principles for UWR must begin with determining ownership of the resource because ownership carries the power to allocate rights, duties, and usage. Certain types of UWR, such as atmospheric water that has not yet touched land, could be viewed as a commonly held resource subject to international law or domestic law, depending on the location of the water vapor. The absence of examples of successful governance of a global commons (Stern 2011) favors a governance structure for UWR based on principles of international and domestic law rather than theories regarding common pool resources. This section will address how theories of national sovereign ownership will apply to each of the types of UWR, and Sect. 12.4 will explore the principles that could guide a domestic governance regime.

12.3.1 Harvesting water from the atmosphere

A threshold question regarding sovereign ownership over atmospheric water revolves around whether sovereignty can extend to forms of freshwater such as fog, clouds, and rain that have not yet touched sovereign land.

Only one treaty is directly applicable to atmospheric water: the 78 nations party to the 1978 Convention on the Prohibition of Military and Other Hostile Use of Environmental Modification Techniques agree to refrain from using weather modification as a means of warfare. Since the treaty is silent regarding weather modification for other purposes, its applicability to civilian utilization of atmospheric water is questionable. The 1944 Convention on International Civil Aviation grants each state "complete and exclusive sovereignty over the air space above its territory", which includes its coastal waters, but that convention has thus far been interpreted to apply only to aircraft over-flights. To further complicate the analysis, policymakers must determine whether, similar to wind currents and the high seas, atmospheric water should be considered as a global commons beyond national jurisdiction. Since evaporation from the oceans produces much of the fog, rain, and clouds that pass over land, an argument can be made that water vapor should be viewed as a common resource until it condenses and touches the earth (Quilleré-Majzoub 2004).

If transpired and evaporated water vapor is indeed a shared resource (Simms 2010), then perhaps the law of capture applies to atmospheric water, and ownership rights should be awarded to whoever captures and possesses the resource. The hydrocarbon industry discarded the rule of capture for shared reserves in favor of collaboration to develop a reserve as a unit. Unitization of a discrete reservoir or field helps to ensure fair and equitable allocation between the parties, but unitizing fog and clouds will prove to be difficult since the quantities of water produced are variable and unpredictable. Regardless, where water resources have the potential to be a common resource, open dialogue in pursuit of equitable arrangements reduces the potential for conflict.

With these concepts in mind, each form of unconventional water resource that is harvested from the atmosphere will be examined separately.

12.3.1.1 Rain enhancement through cloud seeding

Although the effectiveness of cloud seeding has been questioned (Levin et al. 2010), successful cloud seeding has an impact on the hydrological cycle by causing water to precipitate prematurely, artificially enhancing rainfall in one area to the potential detriment of "downstream" neighbors. When the effects of rain enhancement are felt only within the border of one state, then that state would clearly have the sovereign right to regulate cloud seeding over its own territory. For example, in the US each of the 50 states has the right to regulate cloud seeding, but all cloud-seeding operations must be reported to the federal government (Vélez-León 2017). In the international context, when cloud seeding has a deleterious transboundary impact, by artificially causing either less or more rain to fall in the "downstream" state, then, depending on the severity of the consequences, the impacted nation-state may have a cause of action under the international law principle that imposes a due-diligence obligation to prevent significant harm to a neighboring state.

12.3.1.2 Fog water harvesting

The legal analysis for governance of fog water harvesting will probably follow a very similar pattern as that for rain enhancement. However, unlike clouds, fog's contact with land supports a claim of sovereign ownership. States could therefore claim sovereign rights over fog that crosses and touches their territory, with the same limitation that disruption of a natural pattern causing significant transboundary harm to a neighboring state would give rise to a claim for damage by the harmed state.

12.3.1.3 Micro-catchment rainwater harvesting

As indicated in Chap. 6, micro-catchment rain harvesting generally involves either capturing rain from rooftops or diverting the rainwater that runs off a catchment area into a reservoir or into the root zone of a cultivated area. Generally speaking, capturing rain that would naturally fall onto a state's sovereign territory should imbue that state with sovereign rights to that natural resource. However, once again, the analysis is complicated by the fluid nature of water. Rainwater serves to recharge aquifers and river basins, and occasionally those aquifers and rivers are shared resources. In the event that rain harvesting in one state impedes or prevents rain from recharging a transboundary aquifer or river and the diminished water supply causes significant harm in a neighboring state, then once again the damaged state would have a claim against the rain-harvesting state. Generally, however, rainwater has not yet been viewed as a shared water resource.

12.3.2 Deep groundwater

This book defines deep groundwater as subsurface groundwater that is not renewable within 50 years and that is found at a depth up to one km. The sovereignty analysis for such deep, nonrenewable groundwater is much simpler than that for atmospheric water, since, through the use of technical processes that are often costly, the location and volume of fresh groundwater can be measured with some degree of accuracy, and the resource is largely stationary. However, due to the evolutionary trajectories of international water law, the governance regime for freshwater changes at the coastal shoreline, leading to the possibility for different governance analyses for deep groundwater depending on whether it is located onshore or offshore.

12.3.2.1 Onshore deep groundwater

Onshore deep groundwater has been generally acknowledged as a natural resource subject to the sovereignty of the state under whose land it is located (Martin-Nagle 2011). However, unlike transboundary hydrocarbon reserves, which are generally exploited cooperatively as a unit by the states sharing ownership rights, transboundary groundwater reserves are often exploited according to the law of capture. The four fully-ratified treaties for transboundary groundwater provide for information sharing and joint committees rather than allocation and utilization (Martin-Nagle 2020a). The 2015 agreement between Saudi Arabia and Jordan regarding a portion of the shared

Al Sag/Al Disi aquifer limits withdrawals from a portion of the aquifer in order to preserve the groundwater but does not allocate distribution.

12.3.2.2 Offshore fresh brackish groundwater

Governance of offshore deep groundwater will follow a different path to a potentially different conclusion. The vast majority of sovereign nations have ratified the UN Convention on the Law of the Sea, whose provisions grant to coastal states sovereign rights to natural resources located within the seabed of their continental shelves. Beginning at the low-tide line, the length of a state's continental shelf where it has exclusive rights extends for 200 nautical mi, an expanse known as the Exclusive Economic Zone (EEZ). If a coastal state can establish, through a complex set of calculations, that its continental shelf extends further than the EEZ, it can claim exclusive rights to seabed natural resources for up to an additional 150 nautical mi, although these resources are subject to benefit-sharing provisions. Due to saltwater intrusion into the seabed over millennia, fresh groundwater will not be found outside of the EEZ. Sovereign rights to groundwater in a state's area of exclusivity have therefore been well-established, but once again a governance vacuum exists regarding ownership of transboundary reserves. Since a well-established practice of collaborative joint development exists for transboundary offshore hydrocarbons, states developing shared offshore deep groundwater will probably eschew the rule of capture and employ cooperative governance structures for another fluid resource residing in their continental shelves (Martin-Nagle 2020a).

12.3.3 Reusing water

12.3.3.1 Municipal wastewater

Logically, sovereign rights to ownership and control of municipal wastewater and agricultural drainage water will generally not be subject to the same complex analyses as sovereign rights to atmospheric water and deep nonrenewable groundwater. Municipal wastewater results from freshwater that has been diverted and utilized within a sovereign state and is generally collected and treated by infrastructure located in that same sovereign territory. At all points in the process, the water has probably been held within the territory of a single sovereign, making the determination of ownership and control simple and straightforward and subjecting the municipal wastewater to domestic laws and regulations. Occasionally, wastewater treatment may be shared by cities in different sovereign states, and in that case shared utilization of the facility and/or the treated wastewater would be negotiated and agreed to by the states.

12.3.3.2 Agricultural drainage water

Governance of agricultural drainage water could follow a similar analytical path. Water that is abstracted, either from surface water reserves or an aquifer that lies within a single sovereign state, can be viewed as belonging to that entity for reuse or other purposes. Irrigation water that originates from a shared watercourse or aquifer may follow the rule of capture but would be subject to the international law principles of equitable and reasonable utilization, no significant harm and cooperation. If capture of agricultural drainage water prevents a downstream sovereign state from receiving water that it formerly utilized and that loss causes significant harm to the downstream state's territory, then under international law principles the downstream state would have a claim against the upstream state.

12.3.4 Moving water physically

12.3.4.1 Water transportation through iceberg towing

Analysis of sovereignty over icebergs floating in the ocean begins with identification of the original location of the glacier from which the icebergs calved. If an iceberg forms from a glacier that has calved into a coastal state's territorial waters or EEZ and remains in that state's territorial waters or EEZ, then that iceberg belongs to the coastal state from which it originated for as long as it remains in the state's territorial waters or EEZ.

Under LOSC and generally accepted practice, no state can claim sovereign rights over any resource in an area beyond national jurisdiction (ABNJ), such as the high seas (Beckman 2019). Although isotope testing may be able to prove the origin of an iceberg, once an iceberg has passed out of any state's EEZ and into an ABNJ, the rule of capture will apply. However, an ABNJ is not completely lawless, and both LOSC and customary law provide that, when a ship is in an ABNJ, the state under whose laws the ship is registered (the "flag state") has exclusive jurisdiction over that ship, its passengers, and cargo, and any activities taking place on board (Honniball 2016). Thus, even though an iceberg in an ABNJ is subject to the rule of capture, the laws of the ship's flag state will determine ownership and control of the captured iceberg. Pursuant to LOSC, once a ship passes within 24 nautical mi of a coastal state's low-tide line, in an area known as the contiguous zone, the coastal state is permitted to enforce its customs, fiscal, immigration, and sanitary laws. Whether an iceberg being towed into a nation's contiguous zone will violate any customs or sanitary laws remains to be seen.

A complication arises with any iceberg that may be harvested from the Antarctic area, which is subject to a multilateral agreement. The 1959 Antarctic Treaty includes the 1991 Protocol on Environmental Protection to the Antarctic Treaty, known as the Madrid Protocol, which designates the Antarctic area as 'a natural reserve devoted to peace and science', with strict environmental protection measures included in the annexes to the Protocol. In addition, the Protocol forbids '[a]ny activity relating to mineral resources, other than scientific research'. Water in frozen form meets the literal definition of a mineral (Spellman and Stoudt 2013), so towing an iceberg from the Antarctic for commercial purposes may be prohibited under the Protocol, which has 38 parties.

12.3.4.2 Ballast water

Ownership of ballast water depends on where the water was loaded. Freshwater from a land-based location will originally be subject to the sovereign jurisdiction of that state, but a commercial transaction can easily transfer ownership to the ship owner or a third party as the water is loaded into the ship to serve as ballast. Because it is loaded in one location and discharged in another, ballast water has acted as a vector to transport harmful invasive species such as the Zebra mussel (GEF-UNDP-IMO 2017). In recent years both international and domestic regulations have been adopted in an attempt to manage the environmental impacts of ballast water. On the international level, the 2004 International Convention for the Control and Management of Ships' Ballast Water and Sediments, which entered into force in 2017, requires any ship engaged in international transport to manage its ballast water to a predetermined standard, to follow a ship-specific ballast water-management plan, to keep a record book, and to carry an international ballast water management certificate. The requirements apply to ships that are registered in a flag state that has ratified the convention, and, as of March 2020, the convention has 84 parties representing 91% of global tonnage. To prevent harmful organisms from being discharged in sensitive coastal areas, the convention requires exchanges of ballast water to take place at sea. Originally, the convention applied only to new ships but its application to all ships above 400 gross tonnage is being phased in, and future plans include mandating on-board ballast water treatment for every ship. In addition to internationally mandated regulations, many coastal states have their own domestic laws, so any transport of ballast water as an UWR will require planning and compliance with a myriad of regulations and requirements.

12.3.5 Desalinated water

Analysis of sovereignty over desalinated water would combine analyses for municipal wastewater and agricultural drainage water, as well as the analysis for deep offshore groundwater. To begin the desalination process, water is withdrawn from the sea within the 12-nautical mi territorial sea that, under the UN Convention on the Law of the Sea, is considered to be the exclusive sovereign domain of the coastal state. After being withdrawn, the seawater is then piped to a desalination plant that is on the sovereign land of the coastal state and distributed from there. Thus, at every point in the process of desalination, the water has been captured and utilized by the coastal state within its territory, giving that state exclusive claim to ownership, control, and distribution. Although the desalinated water may have direct benefits for only one state, access to this additional resource may prevent conflicts through reducing reliance by that coastal state on transboundary freshwater resources (Aviram et al. 2014).

12.4 Developing a Domestic Governance Regime

Once sovereign rights to ownership, possession, and utilization of the various forms of UWR have been established, the relevant sovereign entity is entitled to form a system of laws and regulations that will apply within its borders in accordance with its own domestic system. Ideally, this domestic governance regime will conform to treaty obligations and agreed international law principles.

In the domestic realm, creation of a governance regime for water is complicated by several factors. Water is critical to multiple sectors, including agriculture, industry, energy, and other utilities, as well as to ecosystems and development efforts. These multiple uses across a wide spectrum result in a broad array of stakeholders and lay fertile ground for overlapping and conflicting laws and regulations. Paradoxically, the multiplicity of users and regimes also creates scenarios where no legal regime applies or where there are overlapping legal regimes, especially where international, regional and domestic laws and regulations are all in effect.

Where then to start designing or even predicting a domestic governance regime for water resources whose utilization has not yet become mainstream? In 2015 the 37 members of the Organization for Economic Co-operation and Development (OECD) produced a set of 12 water governance principles that were adopted by 66 water-related organizations and have been endorsed by more than 150 water-focused actors (Akhmouch et al. 2018). Although the OECD members come mainly from Europe and developed countries, their Principles of Water Governance (OECD 2015) could be viewed as creating a global platform given the large number of supporters.

Due to the localized nature of water resources and the multiple sectors and levels of stakeholders, the OECD principles indicate that one size does not fit all circumstances, and they advocate for designing water governance regimes by taking a bottom—up approach that involves as many stakeholders as possible. According to the OECD, the key water governance dimensions that underpin a sound regulatory framework and encourage innovative governance practices are effectiveness, efficiency, and trust and engagement, and the basic principles of good governance are legitimacy, transparency, accountability, human rights, the rule of law and inclusiveness. Broad stakeholder involvement and technical capacity are critical components of a robust and resilient governance structure. Ultimately, the goals of a sound governance regime are to enable development and distribution of water resources in a cost-effective manner with a clarity and fairness that provide sound guidance and reduce opportunities for conflict.

Given these guidelines that recommend involvement of and direction from local stakeholders in crafting water governance regimes, the approach for domestic governance of UWR will necessarily depend largely on the type of resource, its uses and users, and its geographic scope. In all cases knowledge regarding the most effective means of utilizing the resource will be necessary, and, for some UWRs, capacity-building and transfer of technology will be critical. Since the various forms of UWR have varying geographical impacts, stakeholders, and technical requirements, each of them may require its own governance regime. Most importantly, the range of stakeholders and level of technical requirements must be understood.

Since the OECD principles were adopted so recently, their impact on domestic legal regimes for freshwater cannot yet be gauged, but a recent article analyzed the extent to which the principles have already been incorporated into six domestic legal regimes for freshwater: the European Water Framework Directive, the Australian National Water Initiative, the New Zealand National Water Policy, the Brazilian National Water Policy, the South African National Water Policy and the Lisbon Charter. The authors found that of the regimes studied, all of them except the South African National Water Policy, already largely incorporated the OECD principles (Neto et al. 2018). Since the principles already appear in domestic governance structures, nations wishing to craft a legal regime for UWR may well look to the OECD for guidance.

Inclusion of stakeholder input is critical for a successful water governance regime (Megdal et al. 2017). Utilization of atmospheric water can involve both local and regional stakeholders, and the different forms of atmospheric water utilization require different levels of capacity-building. Fog harvesting draws water from a small, localized area, and thus the community of stakeholders would be local residents who utilize the resource mainly for drinking water. Allowing those communities to determine the circumstances under which fog may be harvested may support broader adoption of the practice. Capacity-building may be required, but the technology would be easily transferable at minimal cost. Transboundary impacts would be limited to the immediate downwind areas and could affect ecosystems reliant on moist air. Micro-catchment rainwater harvesting would impact a similarly local group of stakeholders that would include residents utilizing the water for domestic purposes and farmers using it for irrigation. The techniques being implemented do not require large capital investments, but some capacity-building and transfer of knowledge would enhance the effectiveness and efficiency of the technique. For both fog harvesting and micro-catchment rainwater harvesting, locally crafted governance regimes, assisted by guidance on governance and capacity-building from national or international entities, would be most effective.

Cloud seeding, on the other hand, can have much wider impacts in spite of the relatively modest volumes (Friedrich et al. 2020), and stakeholders could include not only the landholders and users who are the intended beneficiaries, but also people, entities, and ecosystems in the region who are downwind of the prevailing weather patterns and could receive either more or less rain than normal (Quinton 2018). In addition, while the technology is straightforward, meteorological and chemical expertise should be engaged for any cloud seeding exercise to reduce the odds of unintended and damaging consequences. Thus, a governance regime for cloud seeding should include input from local as well as regional stakeholders and should mandate the inclusion of meteorologists and chemical experts. Consideration should be given to management on a national level, with analysis on a case-by-case basis of potential transboundary impacts.

Determining a governance regime for both onshore and offshore deep non-renewable groundwater will involve similar stakeholders for each and a much greater level of technical expertise and funding than for atmospheric water. The stakeholders for both types of deep groundwater would be the owner(s) of the land where the groundwater lies and where the extraction equipment will be sited, as well as any consumers of the water, such as residents and agricultural and industrial users. Since the deep groundwater is non renewable, ecosystems will not be affected by an absence of water on which they had depended, but nearby ecosystems will be affected by the drilling and extraction processes. Future generations that could have relied on the water will be impacted, although identifying a person or entity to represent that stakeholder group would be difficult. Offshore deep groundwater may have greater effects on the surrounding ecosystems due to the sonar used to locate the reserves, the deposition of drill cuttings, and the chemicals used in the drilling process (Martin-Nagle 2020a). If marine denizens are disturbed or dislocated, the local fishing industry may be an additional stakeholder. The level of necessary capacity-building would be high, and in most cases the drilling operation would have to be contracted out, perhaps with a local workforce being trained in operation and maintenance. Since deep groundwater can be considered to be stationary, the administrative entity responsible for the area under which it lies should be the primary body to develop a legal regime. In some cases, the deep groundwater may be a transboundary resource, in which case the relevant administrative entities should ideally collaborate on forming a governance regime.

The stakeholders for recycled wastewater could include a wide spectrum of interested parties, complicating the design of a representative governance regime. Since the wastewater would be collected, recycled, and distributed, several administrative units and their respective citizens could be involved. The recycling plant may be a public–private partnership or may be public or privately owned, and the interests of the owners must be considered. Wastewater recycling is necessarily a highly capital-intensive, technology-driven process, so the financing parties and technical experts should also be consulted when an inclusive governance regime is being designed. Finally, the list of ultimate users could extend to multiple sectors, such as agriculture, industry, energy, and private citizens. Bringing all interests together, or even having them represented, will be a daunting but worthwhile exercise. Issues regarding transfer of technology and know-how and capacity-building should be addressed in the early stages in order to ensure self-reliance for the plant owners and operators.

Desalination likewise involves numerous and varied stakeholders, such as the administrative entities in charge of the marine area from which the seawater is drawn, the owner(s) of the equipment and the land where the plant is located, and the areas through and to which the treated water is transported. In addition, desalination has a high environmental impact due to the fossil-fuel emissions, brine disposal (Jones et al. 2019), and ingestion of marine organisms during seawater intake, so environmental interests should be represented. The volumes of freshwater produced by desalination are sufficient to meet domestic and municipal needs (Jones et al. 2019), so those users and utilities must be considered as stakeholders as well. Since desalination, like wastewater recycling, is highly capital-intensive and requires specialized technology and infrastructure, financing parties and technical experts should be among the stakeholders contributing to the design of a governance regime. In addition, technology transfer and capacity-building will be critical to ensure the proper functioning of the plant by local entities.

Recapture and reutilization of agricultural drainage affects several administrative entities who should be involved in designing a governance regime—the administrative area(s) from which the drainage originates, the area(s) on which it is recaptured, the area(s) where it is utilized, the area(s) where the saline water is eventually stored, and any downstream area(s) that will no longer receive the drainage as runoff. The group of stakeholders also includes the beneficiaries of the recaptured drainage, such as those who utilize it for agriculture or energy. While the level of capital investment and technology is not as high as for other forms of UWR, knowledge transfer and capacity-building are still important to avoid depositing an excess of salt on the land.

The governance regime for icebergs involves very few stakeholders, but most of them are nation-states: the nation from where the iceberg calved, the treaty parties for any icebergs originating in the Antarctic, the flag state of the towing vessel, any nation into whose territorial waters the iceberg enters, and the party or parties purchasing the iceberg and utilizing the resulting freshwater. Whether the nation-states develop an overarching governance regime or interact with the ship owner on a case-by-case basis probably depends on how frequently this form of UWR is utilized. Aside from purchasing the towing vessel and equipment, significant capital investments will not be required, but knowledge of seamanship is of course essential.

Utilization of ballast water as an UWR would seem to be more of a private transaction, where the water is purchased from a utility or other seller in one port and carried to a buyer in another. Therefore, aside from the treaty restrictions to prevent transport of invasive species that were noted earlier, the domestic laws of the originating nation and the destination nation would be solely applicable. Since ship owners and operators are already familiar with ballast, no additional technology would have to be generated or transferred for the transportation of ballast water, although freshwater as ballast water will require more careful handling and storage to retain its character and avoid contamination.

A threshold issue may be to consider which UWRs will be utilized first, and that will be a function of the potential volumes, the required technology, the development and environmental costs, the potential transboundary impacts, and the balance between current and future needs. For example, deep groundwater, recycled wastewater, and desalination produce significant quantities of freshwater, but each of them also requires significant capital investment in highly technical processes and incurs ongoing costs for operation and maintenance. On the other end of the spectrum, fog harvesting does not produce large volumes of freshwater, but implementation of the technique is fairly easy and would require little capital investment, making it attractive in certain undeveloped rural areas. The environmental impacts of each type of UWR will differ. Desalination produces substantial environmental impacts through emissions and brine disposal. Extraction of deep groundwater from land has very little immediate environmental impact, but extraction of offshore deep groundwater would cause a myriad of environment impacts on the seabed and surrounding area. Ultimately, a combination of ease of access, projected volumes and capital requirements will probably determine which UWRs will be adopted most readily. Once a type of UWR has been utilized enough to attract sufficient attention and/or cause conflict, a legal regime to manage its production and usage will doubtless follow.

12.5 Conclusion

Given the fact that UWR have only begun to be recognized and accepted as significant contributors to alleviating freshwater crises, the limited guidance on developing governance regimes for these resources should come as no surprise. The task of crafting a set of rules for rights and obligations for UWRs is further complicated by their great variety, which suggests that no universally applicable governance regime would be appropriate. Nevertheless, certain legal principles that have evolved for more commonly used freshwater resources can provide guidance in formulating governance regimes for UWRs.

When designing a governance regime for each UWR, the analysis must begin with identifying its location in order to determine the sovereign(s) whose claims to rights of ownership and/or use over the freshwater will be recognized. Next, the type and extent of the impacts resulting from the UWR's development and utilization must be recognized to identify both stakeholders and transboundary interests. Finally, the generally accepted legal principles for freshwater resources should be consulted—good neighborliness, reasonable and equitable use, cooperation protection of the environment, and the precautionary principle—as well as concepts and approaches such as intergenerational equity, sustainable development, and water stewardship. While the governance regimes for UWRs may vary widely, all of them should aspire to a goal of providing freshwater for current uses, while minimizing the impact of development on citizens, neighbors, ecosystems, and future generations.

Key recommendations for designing governance regimes for UWRs:

- Treat each UWR individually;
- Identify the entity/entities with primary rights to use the UWR;
- Identify and consult interested stakeholders;
- Use customary legal principles as guidance to determine stakeholder rights and obligations;
- Protect the interests of citizens, neighbors, ecosystems and future generations, and

- Implement regimes locally while retaining national authority and considering any transboundary context.

References

Abbott KW, Keohane RO, Moravcsik A, Slaughter A-M, Snidal D (2000) The concept of legalization. Int Organ 54(3):401–419. https://doi.org/10.1162/002081800551271

Akhmouch A, Clavreul D, Glas P (2018) Introducing the OECD principles on water governance. J Water Int 43(1):5–12. https://doi.org/10.1080/02508060.2017.1407561

Alliance for Water Stewardship (2020) About the Alliance for Water Stewardship. https://a4ws.org/about/. Accessed 31 July 2020

Aviram R, Katz D, Shmueli D (2014) Desalination as a game-changer in transboundary hydro-politics. J Water Policy 16(4):609–624. https://doi.org/10.2166/wp.2014.106

Beckman RC, McCreath M, Roach JA, Sun Z (eds) (2019) High seas governance. Brill Nijhoff, Leiden, Netherland. https://doi.org/10.1163/9789004373303

Birnie P, Boyle A, Redgewell C (2009) International law and the environment, 3rd edn. Oxford University Press, New York

Brown Weiss E (1990) Our rights and obligations to future generations for the environment. AJIL 84:198–207. https://scholarship.law.georgetown.edu/facpub/1627/

Corbridge JN, Moses RJ (1968) Weather modification: law and administration. Nat Resour J 8(2):207–235. https://digitalrepository.unm.edu/nrj/vol8/iss2/2/

Friedrich K, Ikeda K, Tessendorf SA, French JR, Rauber RM, Geerts B, Parkinson S (2020) Quantifying snowfall from orographic cloud seeding. Proc Natl Acad Sci USA 117(10):5190–5195. https://doi.org/10.1073/pnas.1917204117

GEF-UNDP-IMO GloBallast Partnerships Programme (2017) The globallast story: reflections from a global family: globallast monograph no. 25. International Maritime Organization; United Nations Development Programme; Global Environment Facility, London UK.

Hardin G (1968) The tragedy of the commons. Science 162(3859):1243–1248. https://doi.org/10.1126/science.162.3859.1243

Honniball AN (2016) The exclusive jurisdiction of flag states: a limitation on pro-active port states? J Int J Mar Coast Law 31:499–530. https://doi.org/10.1163/15718085-12341410

ICJ (International Court of Justice) (1996) Legality of the threat or use of nuclear weapons, advisory opinion, ICJ reports, 226ff. https://www.icj-cij.org/en/case/95

ICJ (International Court of Justice) (1997) Gabčíkovo-Nagymaros project (Hungary/Slovakia), Judgment, ICJ reports, 7ff. https://www.icj-cij.org/en/case/92/judgments

ICJ (International Court of Justice) (2010) Case concerning Pulp Mills on the River Uruguay (Argentina *v.* Uruguay), Judgment, ICJ reports, 14ff. https://www.icj-cij.org/public/files/case-related/135/135-20100420-JUD-01-00-EN.pdf

Jackson JH (2003) Sovereignty-modern: a new approach to an outdated concept. Am J Int Law 97(4):782–802. https://doi.org/10.2307/3133680

Jiménez A, Saikia P, Giné R, Avello P, Leten J, Lymer BL, Ward R (2020) Unpacking water governance: a framework for practitioners. Water 12(3):827–848. https://doi.org/10.3390/w12030827

Jones E, Qadir M, van Vliet MTH, Smakhtin V, Kang S (2019) The state of desalination and brine production: a global outlook. J Sci Total Environ 657:1343–1356. https://doi.org/10.1016/j.scitotenv.2018.12.076

Lammers JG (1984) Pollution of international rivers: a search for substantive rules and principles of law. Martinus Nijhoff Publishers, Dordrecht, Lancaster

Leb C (2013) Cooperation in the law of transboundary water resources. Cambridge University Press, Cambridge UK

Levin Z, Noam Halfon N, Alpert P (2010) Reassessment of rain enhancement experiments and operations in Israel including synoptic considerations. J Atmos Res 97:513–525. https://doi.org/10.1016/j.atmosres.2010.06.011
Martin-Nagle R (2011) Fossil aquifers: a common heritage of mankind. J Energy Environ Law 2:39–60
Martin-Nagle R (2020a). Governance of offshore freshwater resources. Brill Nijhoff, Leiden, Netherlands. https://doi.org/10.1163/9789004421042
Martin-Nagle R (2020b) Offshore freshwater aquifers: toward equitable distribution. Wires Water 7(5):e1463. https://doi.org/10.1002/wat2.1463
McCaffrey S (2007) The law of international watercourses: non-navigational uses, 2nd edn. Oxford University Press, New York
McIntyre O (2019) Environmental protection and the ecosystem approach. In: McCaffrey S, Leb C, Denoon R (eds) Research handbook on international water law. Edward Elgar Publishing, pp 126–145
Megdal SB, Eden S, Shamir E (2017) Water governance, stakeholder engagement, and sustainable water resources management. J Water 9(3):190–196. https://doi.org/10.3390/w9030190
Mickelson K (1993) Rereading trail smelter. Canad Yearb Int Law 31:219–233
Neto S, Camkin J, Fenemor A, Tan P, Baptista JM, Ribeiro, M, Elfithri R (2018) OECD Principles on Water Governance in practice: an assessment of existing frameworks in Europe, Asia-Pacific, Africa and South America. J Water Int 43:60–89. https://doi.org/10.1080/02508060.2018.1402650
OECD (Organisation for Economic Co-operation and Development) (2015) OECD principles on water governance. https://www.oecd.org/governance/oecd-principles-on-water-governance.htm
Onorato WT (1968) Apportionment of an international common petroleum deposit. Int Comp Law Q 17:85–101
Ostrom E (1990) Governing the commons: the evolution of institutions for collective action. Cambridge University Press, Cambridge, UK. https://doi.org/10.1017/CBO9780511807763
Protocol on Environmental Protection to the Antarctic Treaty (adopted 4 October 1991, entered into force 14 January 1998) 402 UNTS 71, 30 ILM 1455.
Quilleré-Majzoub F (2004) À qui appartiennent les nuages ? Essai de définition d'un statut des nuages en droit international public. J Annuaire Français De Droit International 50:653–667. https://doi.org/10.3406/afdi.2004.3813
Quinton S (2018 February) Why 'Cloud Seeding' is increasingly attractive to the thirsty west. Pew Trusts. https://www.pewtrusts.org/en/research-and-analysis/blogs/stateline/2018/02/20/why-cloud-seeding-is-increasingly-attractive-to-the-thirsty-west
Reed MS, Graves A, Dandy N, Posthumus H, Hubacek K, Morris J, Stringer LC (2009) Who's in and why? A typology of stakeholder analysis methods for natural resource management. J Environ Manag 90(5):1933–1949. https://doi.org/10.1016/j.jenvman.2009.01.001
Roth B (2004) The enduring significance of state sovereignty. Fla Law Rev 56:1017–1050. https://digitalcommons.wayne.edu/lawfrp/188/
Simms V (2010) Making the rain: cloud seeding, the imminent freshwater crisis, and international law. Int Lawyer 44(2):915–937
Sirinskiene A (2009) The status of precautionary principle: moving towards a rule of customary international law. Jurisprudence 4(118):349–364
Spellman FR, Stoudt ML (2013) The handbook of geoscience. The Scarecrow Press Inc., Lanham, Maryland
Stern PC (2011) Design principles for global commons: natural resources and emerging technologies. Int J Commons 5(2):213–232. https://doi.org/10.18352/ijc.305.
UNCED (United Nations Conference on Environment and Development) (1992, August) Rio Declaration on Environment and Development, UN Doc. A/CONF.151/26 (vol. I)
UNCHE (United Nations Conference on the Human Environment) (1992) Stockholm declaration on the human environment. In: Report of the united nations conference on the human environment, Stockholm, June 1972. United Nation, New York, pp 5–16

UNFCCC (United Nations Framework Convention on Climate Change) (1992, May) United Nations Treaty Collection 1771 UNTS 165, New York

United National General Assembly (UNGA) (1962) Resolution 1803 (XVII) on the "Permanent Sovereignty over Natural Resources". Accessed 14 Dec 1962

UNWCC (United Nations War Crimes Commission) (1997, May) Convention on the law of non-navigational uses of international watercourses.United Nations War Crimes Commission UN Doc. A/RES/51/299, New York

UNECE (United Nations Economic Commission for Europe) (1992, March) Convention on the protection and use of transboundary watercourses and international lakes, Helsinki.UNECE Water Convention 31 ILM 1312, Geneva, Switzerland

Vélez-León A (2017) Rain on demand: regulating weather modification throughout the United States. George Wash J Energy Environ Law 8(2):148–162

Chapter 13
Economics and Innovative Financing Mechanisms in a Circular Economy

Francesc Hernández-Sancho and Águeda Bellver-Domingo

Abstract Increasing water shortage forces arid and semi-arid regions worldwide to reuse reclaimed water for several purposes. Four aspects have been identified as key points to successfully implement water-reuse projects: (i) increase in the quantity of treated wastewater motivated by new regulations; (ii) technical improvements in water regeneration systems that lead to producing high-water quality at affordable costs; (iii) institutional and societal context focused on water-reuse regulations; and (iv) use of economic incentives to promote and ensure water reuse for various purposes. Although the objectives of water reuse are highly desirable, there are some challenges to be addressed. Private as well as public water companies are currently searching for opportunities in water reuse and hence to expand the variety of uses. This shift needs to be accompanied by analyzing the demand and market potential, as well as identifying feasible business models. The main aim of this chapter is to explore innovative financing instruments to promote water reuse and make this an attractive and sustainable option in many areas of the world.

Keywords Water reuse projects · Circular economy · Non-action cost · Water tariffs · Economic incentives · Financing mechanisms

13.1 Introduction

Reclaimed water is becoming an essential part of integrated water-resource management. According to the European Commission (2020), in the European Union (EU), about one billion m^3 of treated urban wastewater is reused annually, which represents 2.4% of the treated effluents and less than 0.5% of the EU's annual freshwater withdrawals. However, the water reuse potential is higher, on the order of 6 billion

F. Hernández-Sancho (✉) · Á. Bellver-Domingo
Institute of Local Development (ILD-WATER), Water Economics Group, University of Valencia, Avda. Tarongers S/N, 46022 Valencia, Spain
e-mail: francesc.hernandez@uv.es

Á. Bellver-Domingo
e-mail: agueda.bellver@uv.es

M. Qadir et al. (eds.), *Unconventional Water Resources*,
https://doi.org/10.1007/978-3-030-90146-2_13

m^3 (European Commission 2020). Through water reuse not only can agricultural water demand be met, but other uses can also be implemented, such as environmental (indirect recharge of groundwater bodies or environmental flow of rivers and wetlands), and industrial. In many countries water reuse is a key part of their water management, while, at the same time, it is true that there are some technical and institutional (regulatory) issues that still have to be addressed to avoid undesirable impacts on both the environment and public health.

It is well-known that the main benefit of water reuse is the provision of an additional water source that reduces the dependency on conventional water sources (European Commission 2020). Furthermore, reclaimed water is regulated by law to meet a specified minimum quality. Through this requirement, reclaimed water has a suitable quality to be used for various purposes (such as irrigation or environmental flow), ensuring low environmental impacts in receiving water bodies—see Table 13.1 (Hernández-Sancho et al. 2006; Raso 2013). Considering the framework of an integrated water management strategy—under a catchment scale—water reuse benefits (such as the increase in water availability) need to be addressed to contributing to both, enhancing a region's water resources, and minimizing wastewater outflow. However, there are some barriers that need to be addressed to achieve suitable implementation of water reuse. Firstly, reclaimed water production is more expensive than conventional wastewater treatment due to the quality requirements regulated by law. To achieve these requirements, some investments that increase the cost of reclaimed water production need to be addressed. Secondly, the social perception of reclaimed water is not always good. In some cases, reclaimed water is considered a "risky water" due to its origin (e.g., the wastewater treatment plant). Hence, authorities have the responsibility to increase public awareness about the safety of using reclaimed water.

Considering these barriers, the use of reclaimed water as unconventional water source can be encouraged by meeting three fundamental objectives of integrated water resources management (Hernández-Sancho et al. 2015b): (i) environmental sustainability by reducing the discharge of pollutants into water bodies, thereby improving the quantitative and qualitative status of the water bodies and reducing the need for chemical fertilizers; (ii) economic efficiency alleviating water scarcity by encouraging water efficiency, improving conservation, reducing wastage, and balancing long-term water demand and water supply; and (iii) contributing to food security by supporting the production of more food.

Despite the significant benefits associated with reclaimed water, economic variables have been identified as a major barrier for their implementation (Molinos-Senante et al. 2014). Specifically, there are some challenges to be addressed. Private as well as public water companies currently search for opportunities to valorize the reclaimed water and hence expand its use. From a socioeconomic point of view, there are four key issues relevant to the successful implementation of water reuse projects: (i) the increase in the quantity of wastewater treated due to new regulations; (ii) technical improvements in water regeneration systems that lead to producing high-quality water at affordable costs; (iii) the institutional and societal context focused on water reuse regulations; and (iv) the use of economic incentives to promote and ensure water reuse in various sectors.

Table 13.1 Identification of externalities from water reuse projects (Modified based on Hernández-Sancho et al. 2006)

Typology of externalities	Units
Avoids constructing facilities to store freshwater	€[a]
Avoids drinking-water treatment costs	
Avoids water distribution costs	
Reuse of nitrogen in agriculture	kg of N
Reuse of phosphorus in agriculture	kg of P
Reuse of sludge in agriculture and gardening	kg
Reuse of thermal energy	Watt
Increase the water volume available	m^3
Guarantees supply during water scarcity times	% Confidence
Water quality suitable for various uses	kg waste
Biological risk associated with wastewater reuse	People exposed
Chemical risk associated with wastewater reuse	People exposed
Increases in the level of river flows	m^3
Avoids overexploitation of water resources	Aquifer level, m
Avoids water pollution	Waste eliminated, kg
Enables recovery of wetland and river ecosystems	Users
Increases offensive odors and noises due to pollution	Number of people
Decreases the value of land nearby	€
Raises social awareness about water reuse	Number of people

[a] *1 € = 1.17 USD*

Consistent with the key issues related to water reuse, the aim of this chapter is to address the market opportunities of water reuse projects through a circular economy approach. Specifically, this chapter highlights that the reclaimed water promotion needs to consider the socioeconomic situation of each area and its environmental conditions and impacts, focusing on the internalization of environmental externalities to achieve full cost recovery through a suitable reclaimed water-tariff structure. To achieve this aim, information about the environmental and economic approaches of water reuse, as well as the different tariffs and economic arrangements, are addressed. Through understanding this chapter, decision-makers will be able to identify the market opportunities for water reuse projects, as well as the importance of inclusion of the environmental externalities related to reclaimed water production.

13.2 Market Opportunities for Water Reuse in the Circular Economy

The circular economy has become an important concept in environmental management during recent years. Specifically, the circular economy is a framework that changes the classical economic model to achieve a zero-waste economy through the revaluation of multiple streams of productive process (Smol et al. 2020). Wastewater management is also included in the circular economy framework through adding value to the effluents from wastewater treatment plants, i.e., producing reclaimed water. Promoting the use of reclaimed water adds value to the effluent and transforms wastewater treatment plants into unconventional water sources. From an environmental point of view, a circular economy for the water sector means reducing water stress through increasing water supply, as well as reducing the environmental impacts of effluent disposal through the improvement of effluent quality specified by law. On the other hand, from an economic point of view, both the circular economy and reclaimed water are opportunities to implement new technologies and projects through which managers can achieve efficient and innovative management of wastewater treatment facilities. This section analyses the various market opportunities for water reuse, considering the importance of effective pricing policies and tariff systems. Achieving this aim implies including all the variables that affect water reuse, such as competition with other water sources. The best option for promoting water reuse (and reducing conventional water resources consumption) is designing a suitable tariff in which value becomes an economic opportunity to the stakeholders. Through these principles, a model for water reuse under the circular economy approach can be established.

Reclaimed water should be recognized as an important part of an integrated water cycle management strategy to recharge unconventional water resources for indirect potable applications, to directly substitute potable applications for industry and irrigation, and to reduce the environmental impact of discharges (European Union 2016). Reuse is already a key part of water management in some areas, but several barriers remain to be addressed to make sure it has no negative social or environmental impacts. In addition, safe reuse practices require water quality guidelines and appropriate training.

Identifying market opportunities for reclaimed water comes from knowing the water reuse framework. Figure 13.1 shows both the water reuse framework and the issues that need to be addressed to promote the use of reclaimed water. It is necessary to implement new technologies and facilities that ensure suitable quality of reclaimed water. This issue has a direct impact on the economic dimension of unconventional water sources (Hernández-Sancho and Molinos-Senante 2015). Hence, the other challenges of promoting water reuse are related to economic tools to integrate unconventional water sources from an institutional context. These challenges mainly are: (i) developing the appropriate tariff structures and the financial instruments and mapping feasible business models; (ii) creating institutional and societal strategies and procedures to promote water reuse, enabling a level playing field for

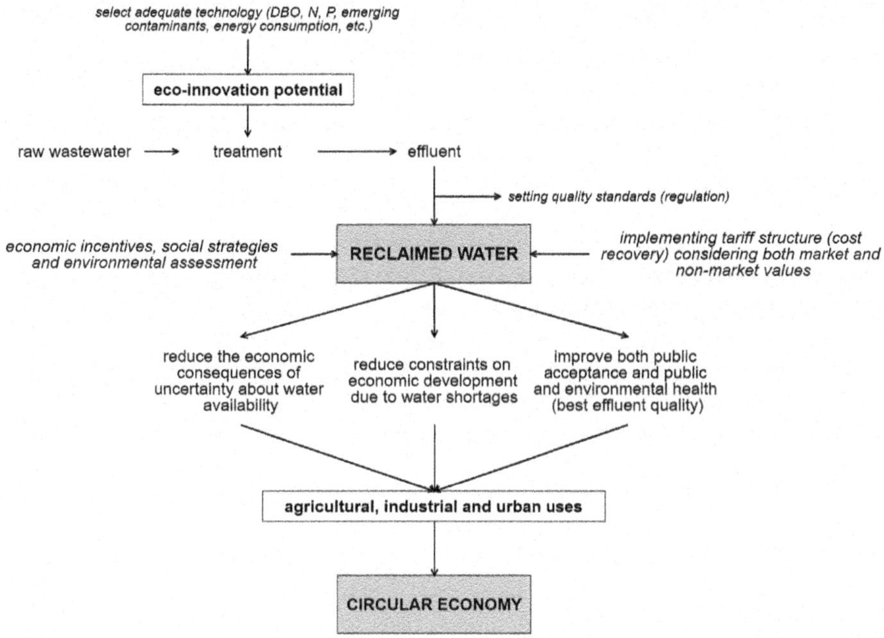

Fig. 13.1 Water reuse framework

all water use sources; and (iii) quantifying both the market and non-market values of unconventional water sources. The significant benefits related to water reuse are clear and accepted by governments and users from both the economic and environmental points of view.

13.2.1 Who Pays for Water Reuse?

The question of who should pay for water reuse is a complex issue that needs to be addressed. Regulation allows various water uses, such as golf-course irrigation and industrial use, although agricultural irrigation is the most widespread use. The economics of these uses are different because water reuse systems are generally designed ad hoc in line with the characteristics of the final users (Hernández-Sancho and Molinos-Senante 2015). Water prices should reflect the financial, operational, and maintenance costs, as well as the capital costs, of providing water services. This idea has been formalized in Europe through the Water Framework Directive (WFD, Directive 2000/60/EC) that demands the establishment of a full-cost recovery principle for water services. However, in Europe, only England and Wales, Germany, and the Nordic countries have achieved cost recovery of their water services. In the case of US wastewater treatment utilities, most either recover less than 25% of their operating costs, or they are unaware of how much they are recovering. This situation is not desirable and requires corrective action.

Considering the water-regeneration costs and the tariffs paid by water users, in most cases, some degree of subsidy is needed to recover the full costs of reclaimed water. For example, the Italian Legislative Decree 152/2006 orders that, to promote water reuse, tariffs for industrial users must be discounted. In Israel, water reuse projects for agricultural purposes are highly subsidized. The Israeli state pays for transporting, pumping, and ponding regenerated water and for upgrading it to a 'high quality level'. In any case, the subsidy is less costly than treating wastewater to a level of quality suitable for discharge into surface water. In the US, in addition to the price paid by water users, water utilities receive revenues to meet the operating costs for reclaiming water (Hanjra et al. 2015). While water reuse projects are fully justified in terms of objectives, it is not always possible to defray costs by charging tariffs. Moreover, it would be relevant to know if only water users should pay or should all beneficiaries contribute to the costs.

It has been shown that until now the principle of cost recovery is not being met in almost all water reuse projects. The first step to improve the application of this economic principle is to identify the barriers that prevent policymakers from establishing higher water-reuse tariffs. Three assumptions are needed to develop a reasonable price strategy: (i) the political climate accepts the 'polluter pays' principle; (ii) the water users are likely to be responsive to price changes; and (iii) there are no critical thresholds being approached, such as a level of extraction where irreparable damage is likely to occur. The polluter-pays principle is accepted by urban and industrial users accustomed to charges for sanitation services. However, this principle is not accepted by many farmers due to the lack of experience (Hernández-Sancho and Molinos-Senante 2015). In this field, the response of farmers to changes in water tariffs is conditioned by aspects such as the existence of water rights, the productivity of the crops, and the existence of water markets. Therefore, before applying a pricing policy, it is necessary to study them case by case (Dinar 2000). It is essential to understand the possible interferences, positive and negative, with other farming policies. In general, to take measures in relation to water prices, which include regenerated water, it is vital to analyze demand elasticity for each use (Hernández-Sancho et al. 2015b). There can also be situations where the increase of the drinking-water price has a strong effect on water reuse. However, some water users have an inelastic water-demand response if drinking water price increases, such as agricultural users. This may indicate a greater potential for substitution with reclaimed water in response to price increases (Mudgal et al. 2015).

Another key aspect that hampers the full-cost recovery is the low price of drinking water, which is subsidized in most cases. To encourage the use of regenerated water, tariffs should be significantly lower than those for drinking water. Tsagarakis and Georgantzis (2003) have shown that the willingness to use regenerated water by farmers was strongly motivated by the price differential between conventional and reclaimed water. Therefore, in almost all water reuse proposals, the tariff for regenerated water ranges from 0 to 25% of drinking water rates. This issue is related to the social perception about reclaimed water.

Menegaki et al. (2009) analyzed the willingness to pay for reclaimed water for irrigation purposes through a survey for a sample of 1,004 people. The results highlighted that 40% of respondents would not use reclaimed water for irrigation and would not consume the agricultural products irrigated with reclaimed water. Despite that, there are several treatment technologies that allows to go beyond the quality requirements, because reclaimed water is perceived as more risky than conventional water for irrigation uses. It is necessary to improve social acceptance of reclaimed water through the demonstration of its potential benefits such as mitigation of water scarcity; energy savings; positive environmental impacts from reduced fertilizer use (where possible); and local economic development (Kirhensteine et al. 2016). All these issues, together with the suitable tariff scheme, contribute to full–cost recovery of water reuse and reinforce the circular economy framework.

Many water-reuse proposals would not be carried out if the costs were only to be paid by private users. Currently, entities or companies only participate in water-reuse projects if regenerated water is used for high-productivity uses, such as golf-course irrigation, despite that water-reuse projects are financially feasible. For this reason, most water-reuse proposals have been developed based on subsidies and grants. Low drinking-water rates, which in most cases are subsidized, make regenerated water uncompetitive. If the principle of cost recovery were implemented in the cases of drinking water and water reuse, significant changes in tariffs would be produced, improving the competitiveness of the regenerated water. In this context, governments should participate in such projects since they generate positive externalities that improve the welfare of everyone (Hernández-Sancho et al. 2015a, 2017).

The participation of the public administration in water reuse projects can be justified for several reasons. For example, water reuse projects may often be considered as an application of the precautionary principle since they may avoid damages to water ecosystems. In other cases, water reuse may prevent the construction of large and expensive infrastructure. When new facilities are planned, it is necessary to predict the operation and maintenance costs of water reclamation plants. Not only total costs are important but also the relationship between costs and the quality of the water for each water regeneration process. This information would help the administration and water management companies in the decision-making process (Raso 2013). Although the objectives of water reuse are very desirable, there are some issues to be addressed that can be used as a guideline for financial arrangement for reclaimed water, such as:

- Tariff structures. It should ideally reflect both the long-term fixed nature of the investment and a volumetric element that provides some incentive for consumers to conserve water.
- Costs to be recovered from consumers. It should include depreciation, renewal, and maintenance costs, as well as the cost of financing long-term investment, so that the benefits are shared between current and future generations in a sustainable manner.

Currently, no financial instruments exist in the EU to stimulate water reuse (Kirhensteine et al. 2016). This source still competes with (subsidized) potable water.

In the US, many agencies sell regenerated water at rates 60–85% that of their potable supply to encourage industry and local communities to participate. In Australia, the commercial price of regenerated municipal water is shown to be rarely viable mainly because of the capital costs of water distribution. The basic idea, although not always considered by policy makers, is that pricing for water demand management, pricing to encourage the use of regenerated water, and pricing for cost recovery are not simultaneously achievable.

The pricing policy to encourage the reuse of reclaimed water cannot be adopted in isolation. From the point of view of the circular economy, it is essential to act globally on water prices from all sources. It makes little sense to strictly apply the principle of cost recovery to water reuse projects while drinking water is still subsidized. The same economic principles must be applied to all water sources so that they 'compete' on equal terms. The principle of full-cost-recovery pricing that accounts for environmental externalities represents an ambitious goal. Nevertheless, it is necessary to start introducing policies and mechanisms aimed at facilitating this objective. Moreover, awareness campaigns, education, and dissemination of results from previous experiences are needed to help change attitudes and encourage water reuse.

Water reuse pricing aimed at controlling water demand, and simultaneously encouraging water reuse and cost recovery, is almost impossible. If this type of tariff is applied both for the supply of drinking water and for reused water, including externalities, then water reuse projects would strongly benefit. In the case of reclaimed water, many of the extraction and distribution related externalities would be avoided and thus not included in the cost. Another issue to consider is the level of treatment required for water reuse. It is well known that, depending on the destination and use, reclaimed water should meet different quality criteria. Hence, the type of treatment required and the cost associated varies. Reclaimed water in Spain's Segura River basin, for example, is sold to irrigators at around $0.14/m^3. This price represents a fraction of the estimated cost including capital, operational, and environmental expenditures ($0.47/m^3) (GWI 2012; BIO by Deloitte 2015).

Identifying the market opportunities of water reuse begins with a better understanding of costs, prices, tariff systems, and project benefits along with a higher awareness of water resource management. From there, decision makers can develop policies to promote cost-effective investments in water reuse. The cost of the regenerated water and the tariffs paid by water users illustrates that, in most water reuse proposals, the principle of cost recovery is not met. However, such projects may also generate positive externalities contributing to improving the welfare of the whole society, e.g., concerning public health, the environment, and water availability. From this respect, governments may contribute to fund and maintain these types of projects. A framework for the costs and financial, institutional, and societal arrangements for water reuse projects would help countries, water associations, and commercial water companies to focus on new (commercially viable) water reuse projects and opportunities.

Reclaimed water represents a substantial market value since it contributes to balance long-term water demand and water supply. Moreover, it is more environmentally sustainable. However, the economics is still the weak point of water reuse projects. According to the Water Framework Directive (Directive 2000/60/EC), the price of the regenerated water should reflect the financial costs, operation and maintenance costs, and capital costs of providing water services (Hernández-Sancho and Molinos-Senante 2015). Regardless of the level of treatment required for reusing the water, the structure of the proposed tariff is the same: what are the items included in the fixed and volumetric charges? If reclaimed water is generated from secondary treatment, according to the polluter-pays principle, this treatment (investment, operation, and maintenance costs) should be paid by polluters and not by the users of reclaimed water.

It would be highly recommended to elaborate guidelines for the development of a socially acceptable and economically viable water-tariff system. Both issues are difficult to achieve due to the large volume of information needed. However, the more information researchers and decision-makers have, the more accurate the development of the tariff will be. From an economic point of view, reclaimed water tariffs need to be based on information about environmental, social, and economic issues of the study area to introduce policies and mechanisms that achieve competitiveness of the reclaimed water. This information would be especially useful for water authorities, stakeholders, and policy makers for supporting decisions contributing to promote water reuse projects. It would be very useful to mobilize the practitioners at the national, regional, and local level for building adaptation schemes according to the goals of sustainable management of water resources and climate change adaptation in the water sector and to develop their managerial and technical capacity to improve service provision. Hence, new approaches for exchanging ideas and new practices and learning from new experiences are needed. The latter implies connecting organizations and people through new structures, technological solutions, financial tools, and more opportunities for interactions and support.

From the perspective of the circular economy, the analysis of water reuse should consider the cost and benefits of reclaimed water, as well as the costs of alternative water supply options such as drinking, desalination, or stormwater. Hence, it is possible to determine a ranking of cost-effective solutions for guaranteeing water demand. One of these solutions is the use of tariffs to internalize the environmental externalities of water reuse and the costs of technologies used to treat the effluents. A more transparent full-cost pricing of all water sources is required. In this sense, a higher cost for drinking water (full-cost tariff) could be a factor driving some utilities to develop or expand their reclaimed water programs. It is fundamental that pricing for drinking water also takes the cost recovery principle into account. Otherwise, reclaimed water will not be competitive.

Achieving the full-cost tariff in both drinking and reclaimed water needs to consider not only the economic issues of the water production process, but also the related environmental externalities. Until now, these externalities have not been considered due to a lack of awareness. Currently, governments and decision-makers are aware of environmental problems (such as water scarcity, pollution), and the inclusion of such externalities in the water planning and development process has been promoted. However, the inclusion of externalities in both drinking and reclaimed water pricing process is difficult because these externalities are often unperceived by society. Therefore, it will be essential to develop public awareness campaigns about the true cost and benefits of both sources of water.

13.3 Environmental Benefits and Non-action Cost in Water Reuse

Water reuse may have many important benefits while enabling the achievement of the Sustainable Development Goals. The most obvious benefit is the provision of an additional dependable water resource. The second is the reduction of environmental impacts by reducing or eliminating wastewater disposal, which results in the preservation of water quality downstream. Therefore, in the framework of an integrated water management strategy on a catchment scale, the benefits of water reuse should always be assessed considering that it contributes to both enhancing a region's water resources and minimizing the wastewater outflow. In addition, using recycled water for irrigation can reduce the need for fertilizer, thanks to the nutrients it contains (BIO by Deloitte 2015; European Union 2016).

Considering the relevance of water reuse in the agricultural sector, the European Commission has developed a new regulation for irrigation to stimulate and facilitate its implementation starting June 2023 (European Commission 2020). Although irrigation with reclaimed wastewater is in itself an effective purification (a sort of slow-rate land treatment), appropriate treatment must be performed for the protection of public health, the prevention of nuisances during storage, and the prevention of damage to the crops and soils. So far, in only a few countries worldwide (the US, Australia, Israel, and Japan), wastewater recycling and reuse is well enough established to have led to the drafting of specific regulations or guidelines.

The suitable management of both raw and treated wastewater provides significant benefits to the environment and society. This benefit can be considered as the avoided costs of wastewater treatment. For that reason, all actions focusing on ensuring and/or improving the effluent quality involve a benefit to the environment. On the contrary, an action or measure that is not being implemented—in terms of wastewater management—results in further costs. It means that there is a benefit that has not been achieved due to the discharge of wastewater effluent lacking suitable quality. Hence, reclaimed water has an implicit benefit related to environmental and socioeconomic issues. Specifically, the environmental benefit is represented by the lower quantity of pollutants that is being discharged into receiving water bodies. On the other hand, the socioeconomic benefit is represented by the recreational uses of healthy water bodies and the potential of reuse of this effluent to meet the water demand and reduce water stress (Ancev et al. 2017; Bellver-Domingo et al. 2019).

13.3.1 Monetary Valuation of Non-action Cost

When a new technological action is proposed, it is always necessary to overcome the barriers that usually discourage the implementation of this type of action. The non-implementation generates inefficiencies, which generate high economic, social, and environmental costs. Hence, these costs of non-action should be considered when the viability of the proposed action are assessed (Hernández-Sancho et al. 2017). Specifically, the costs of not using reclaimed water are the sum of the: (i) cost of not guaranteeing the water supply; (ii) irrigation constraints in arid and semi-arid regions; (iii) overexploitation of groundwater; and (iv) lack of water flow in rivers, as well as the lack of quality of their waters.

Only through the monetary valuation of these costs—using a reliable methodology, such as shadow prices (Bellver-Domingo and Hernández-Sancho 2018; Bellver-Domingo et al. 2017, 2018)—decision-makers will be able to obtain complete information about the high cost of non-action in terms of water reuse. Shadow pricing is a methodology that allows researchers to quantify the monetary value of environmental externalities that lack a reference market value, such as wastewater pollutants (Bellver-Domingo et al. 2017). Through the monetary valuation of environmental externalities, decision-makers have the information about the non-action costs of water pollution because the monetary values obtained act as a proxy for the non-action costs. Specifically, the advantage of the monetary value obtained is its inclusion into cost–benefit analysis (Bellver-Domingo et al. 2017). This approach highlights the existence of positive effects derived from the improvement in the quality of treated wastewater expressed in terms of the reduction of environmental damage. In the literature there are examples that use shadow prices to quantify the environmental benefits of removing pollutants from effluents, such as nitrogen, phosphorus, BOD, salts (electrical conductivity), and emerging contaminants (Bellver-Domingo et al. 2018).

Monetary valuation is necessary to implement an integral vision of water management while considering the water management challenges with a short-term approach. Specifically, it is not recommended to consider water reuse management as a specific isolated action with a temporary nature; a structural approach is required. In other words, a wastewater treatment plant should be considered as an unconventional water source. And, as such, it should be included within the climate change strategy, helping to reduce the water scarcity effects. From a territory-wide point of view, water reuse proposals need to guarantee their own viability, through the inclusion of several facilities in the same basin with multiple users and beneficiaries. Hence, wastewater treatment plants should be included into water planning, as well as in the tariff system to finance the required investments to achieve a suitable water quality.

The water reuse challenge is to quantify not only the market benefits of its implementation, but also the non-market benefits related to the environmental externalities of us reclaimed water. For that purpose, the use of monetary valuation methods has been implemented. Valuation of these benefits is nevertheless a barrier to overcome to justify suitable investment policies and financing mechanisms for promoting water reuse. The benefits of water reuse should be estimated, and the option of water reuse should be compared to other alternatives. From a methodological point of view, monetary valuation of the externalities of the wastewater treatment process is used to justify technical improvements and investments in wastewater treatment plants. This approach provides a practical method to quantify the environmental benefit in monetary units (also known as action costs) derived from both the reduction of water pollution and the reclaimed water use. Considering the literature published on monetary valuation of action costs, the shadow-price methodology is a useful tool that optimizes the decision-making process in the water cycle (Hernández et al. 2015a; Bellver-Domingo and Hernández-Sancho 2018).

Under the circular economy approach, the assessment of water reuse economics should consider the costs and benefits of reclaimed water, as well as the costs of alternative water supply options such as desalination water or stormwater. Hence, it is possible to determine a ranking of cost-effective solutions for guaranteeing water demand management and reducing water stress in arid and semi-arid areas. Designing a tariff structure is required to finance the reclaimed water needed and encourage its use because if the benefit is global, the payment should also be. In the design stage, tariff increases should be avoided, especially for small consumers, thus contributing to the efficiency principle promoted by the Framework Directive. Since the reclaimed water proposals may generate positive externalities to improve the society welfare, e.g., concerning health, environment, and water availability, governments may contribute to fund and maintain water reuse proposals.

Considering the issues presented in this chapter, if the cost-recovery principle were implemented both in the drinking water and water reuse sectors and reflected in their respective tariff systems, the competitiveness of reclaimed water may be significantly improved, strengthening reclaimed water and the circular economy in any country. From a social point of view, the establishment of a new tariff system for reclaimed water should be accompanied by an environmental education program to reinforce both the importance of water reuse and the need to internalize the environmental externalities of the water cycle.

13.4 Economic Incentives to Promote Water Reuse

It is known that sanitation and water reclamation services often do not receive sufficient funding to cover operating and maintenance costs, plus capital costs. Subsidies are required to adequately cover these services at least temporarily. The existence of these subsidies does not mean that the sector should depend on this form of financing without taking advantage of market conditions or possible incentives to improve sustainability. Given the huge importance of the reuse and recovery of resources in WWTPs, the application of commercial and financial innovations should be promoted to guarantee the sustainability of circular economy models. Certainly, the treatment and reuse of water is an activity that requires a significant initial contribution of resources. The cost of investments is high, and this represents an important barrier to the implementation of this type of project. Furthermore, in many geographical areas, they are not considered as priority activities. It is evident that the recovery of resources in the treatment processes could contribute to the financial sustainability of these projects, achieving a paradigm shift in the sector. In this way, dependence on traditional public financing could be replaced by innovative financing and a more market-based business model. The potential of these projects can be exploited by improving their profitability and reducing their dependence on public rates. This implies the identification and development of new markets for reclaimed water, biogas, and biosolids. The design of business plans is essential to implement water reuse projects.

On the other hand, the challenge of achieving the Sustainable Development Goals will require the involvement of the private sector with an important role in investing in new technologies and in the sustainability of the water reuse projects. In fact, this type of project represents a great opportunity to achieve public–private financing agreements (Rodriguez et al. 2020). It would be a question of implementing a mixed financing scheme that would include a combination of subsidies or concessions together with private capital and debt. Revenue would come from user fees and the sale of treated wastewater and recovered by-products. Public funding for these wastewater treatment projects would be justified by their benefits to public health and the environment. Furthermore, in many countries, water rates are lower than those required to guarantee the principle of full-cost recovery in these projects. Sometimes, political and social criteria determine the amount of these fees. It is important to highlight that the amount of the subsidies over time must be contemplated in the project's own financial plan. It would only be an incentive to start the project and would never be indiscriminately applied, which usually goes against the efficiency and competitiveness of the business. The design of a tool that facilitates the preparation of these plans in a practical and simple way would be especially useful for the promotion of these circular economy projects.

These subsidies could be especially useful in the early stages of development of a water reuse project. It is assumed that initially both the treated water and the recovered resources should be offered at a price lower than the cost of obtaining it. The aim is to facilitate the users' access to these resources as much as possible, avoiding potential economic or trust barriers. Once the use of these resources is normalized, always with the maximum guarantees, cost recovery can be considered, avoiding excessive dependence on subsidies. The increased demand for these resources will contribute not only to the sustainability of the project itself but also to avoid pressure on conventional water resources along with a firm commitment to the circular economy in the water sector. The increasing scarcity of water and the influence of climate change will undoubtedly contribute to this development.

There is a growing number of studies in the literature demonstrating the feasibility of water reuse projects not only when it comes to covering costs but also generating significant benefits (Lazarova et al. 2013; World Bank 2019). For this reason, the role of subsidies would help start the project and then should decrease with the progress of the activity. Both the initial amount and the expected evolution of the required subsidy would be included in the corresponding business plan. As with any innovative initiative or proposal, this type of project should have an income-and-expense forecast plan that guarantees its financial viability and, therefore, its sustainability in the market. Table 13.2 summarizes the various cost-recovery options in wastewater treatment plants, focused on the energy saving, and sale of both biosolids and nutrients and water. All of these cost-recovery options allow to promote reclaimed water since wastewater treatment plants became a source of raw materials.

The various streams and savings included in Table 13.2 represent the successful implementation of circular economy projects. The analysis of these experiences makes it possible to verify the enormous market potential that the products recovered from wastewater have, both in terms of treated water, energy and biosolids. With a good business plan, these three types of products have proven their ability to generate profits. In terms of the potential of the project, it is important to consider the type of purpose and the existence or absence of alternative resources. For example, for industrial use, some experiences (Indian Institutes of Technology 2011; Lazarova et al. 2013; Rodriguez et al. 2020) demonstrate that cost coverage is perfectly feasible, especially in areas with water scarcity and high water rates.

Considering that the irrigation and aquifer recharge are the main uses of reclaimed water, the use of subsidies may be required since the treated water should be offered at attractive prices or even free, at least at the beginning of the project. In areas with increasing water scarcity, low-productivity land, or expensive fertilizers, farmers will be interested in using treated water with nutrients to increase their crop yields. In addition, in an area where there are social, environmental, and even health benefits due to the use of treated water, payment mechanisms could be applied to users for the enjoyment of these externalities. In addition, the use of treated water could avoid

Table 13.2 Potential revenue streams and savings from resource recovery for wastewater-treatment plants (Modified based on Rodriguez et al. 2020)

Energy	Biosolids and Nutrients	Water
Revenue	*Revenue*	*Revenue*
Sale of biogas or electricity	Sale of phosphorus as fertilizer	Sale of treated wastewater, especially in water-scarce areas
Sale of carbon credits	Sale of biosolids as compost	
Tipping fees for the collection of organic matter (in co-digestion)		
Savings	*Savings*	*Savings*
Using self-generated electricity in the plant	If the biosolids are given away for free (for agriculture, to restore degraded land, etc.), the utility saves transport costs and landfill fees	Discharge fee/tax
Improving energy efficiency		

paying the discharge fee. All of this will contribute to cost recovery and sustainability of the project.

Regarding the use of biogas, the potential viability of the project will depend on the existing rates for gas in each territorial area. If prices are high, the biogas generated can be sold to the distribution companies themselves. It can also be used for self-consumption in the treatment plant and saving electricity costs. The size of the installation is very relevant to determining the profitability of the project due to the influence of economies of scale. There are a good number of examples (Indian Institutes of Technology 2011; Lazarova et al. 2013; Rodriguez et al. 2020) that demonstrate the viability of these projects.

13.5 Conclusions

The influence of climate change on water shortage and the overexploitation of water reservoirs lead to uncertainty for water resources management. In order to correct this situation, reclaimed water has become the best option to meet the demand. The use of reclaimed water requires reconsidering the traditional water management model to adopt a new strategy to promote water reuse. This new paradigm means moving towards a circular economy approach in which wastewater is no longer seen as waste, but as a valuable resource in the context of water scarcity. A combination of regulations, incentives, and the participation of all stakeholders will be required to transform the traditional criteria of water management to the circular economy approach. This chapter highlights the real market opportunity to promote water reuse, considering the positive environmental impact related to consolidating the wastewater treatment plants as an unconventional water source. In this sense, the role of the authorities is essential both to justify the existence of advantages and to support the adoption of this type of exchange agreement between farmers and local authorities and so promote water reuse. The relevance and potential of reclaimed water requires good management of wastewater, especially regarding the water quality achievement, optimization of treatment processes, and cost-recovery principle, while considering the implementation of new treatment technologies and innovations.

Despite the obvious benefits of water reuse, there are some economic variables, such as technology costs and the final price of reclaimed water, that act as barriers to reclaimed water implementation. To remedy this situation, a strict application of the cost-recovery principle is needed. Considering the circular economy approach, participation of all stakeholders (governments, public and private managers, and society) in water reuse projects is fully justified due to the generation of social and environmental positive externalities, such as increasing water availability and reducing the impact of effluent disposal. This chapter highlights the importance of reclaimed water to achieve suitable water management under the circular economy approach, addressing the information about environmental and economic issues related to waster reuse. Hence, decision-makers will be able to carry out innovative water reuse projects in those areas without overexploitation of conventional water sources.

References

Ancev T, Azad S, Hernández-Sancho F (2017) New directions in productivity measurement and efficiency analysis: counting the environment and natural resources. In: Ancev T, Samad MA, Hernández-Sancho F (eds) New directions in productivity measurement and efficiency analysis: counting the environment and natural resources. Edgard Elgar Publishing, pp 1–282

Bellver-Domingo A, Fuentes R, Hernández-Sancho F (2017) Shadow prices of emerging pollutants in wastewater treatment plants: quantification of environmental externalities. J Environ Manage 203(1):439–447. https://doi.org/10.1016/j.jenvman.2017.08.025

Bellver-Domingo A, Hernández-Sancho F (2018) Environmental Benefit of improving wastewater quality: a shadow prices approach for sensitive areas. Journal of Water Economics and Policy 4(2). https://doi.org/10.1142/S2382624X17500084

Bellver-Domingo A, Fuentes R, Hernández-Sancho F, Carmona E, Picó Y, Hernández-Chover V (2018) Monetary valuation of salicylic acid, methylparaben and THCOOH in a Mediterranean coastal wetland through the shadow prices methodology. Journal of Science of the Total Environment 627:869–879. https://doi.org/10.1016/j.scitotenv.2018.01.303

Bellver-Domingo A, Fuentes R, Hernández-Sancho F (2019) Reclaimed water and irrigation: a cost-benefit analysis for desalination of WWTPs effluents. J Desalination Water Treat Sci Eng 166:193–201

BIO by Deloitte (2015) Optimising water reuse in the EU—final report prepared for the European Commission (DG ENV), Part I. In: collaboration with ICF and Cranfield University

Dinar A (2000) Political economy of water pricing reforms. Dinar A (ed) The political economy of water pricing reforms. Oxford University Press, Oxford, pp 1–25

European Commission (2020) Regulation on minimum requirements for water reuse enters into force. European Union website. https://ec.europa.eu/environment/water/reuse.htm

European Union (2016) Guidelines on integrating water reuse into water planning and management in the context of the WFD. European Commission

GWI (2012) Paying the price for reuse in Spain. J Glob Water Intell 13(12)

Hanjra MA, Drechsel P, Mateo-Sagasta J, Otoo M, Hernández-Sancho F (2015) Assessing the finance and economics of resource recovery and reuse solutions across scales. In: Drechsel P, Qadir M, Wichelns D (eds) Wastewater: economic asset in an urbanizing world. Springer Dordrecht, pp 113–136. https://doi.org/10.1007/978-94-017-9545-6_7

Hernández-Sancho F, Uriaga A, De Las Fuentes K, Bis B, Chiru E, Balazs B, Wintgens T (2006) Feasibility studies for water reuse projects: an economical approach. Desalination 187(1–3):253–261. https://doi.org/10.1016/j.desal.2005.04.084

Hernández-Sancho F, Lamizana-Diallo B, Mateo-Sagasta J, Qadir M (2015a) Economic valuation of wastewater—the cost of action and the cost of no action. United Nations Environment Programme

Hernández-Sancho F, Molinos-Senante M, Sala-Garrido R (2015b) Pricing for reclaimed water in Valencia, Spain: externalities and cost recovery. In: Dinar A, Pochat V, Albiac-Murillo J (eds) Water pricing experiences and innovations. Springer International Publishing, pp 431–443. https://doi.org/10.1007/978-3-319-16465-6_22

Hernández-Sancho F, Molinos-Senante M (2015) Wastewater management and reuse. In: Dinar A, Schwabe K (eds) Handbook of water economics. Edward Elgar Publishing Limited, pp 239–261

Hernández-Sancho F, Lamizana-Diallo B, Ingram W (2017) Valuing environmental and health impacts from no action in wastewater management. In: Ancev T, Samad MA, Hernández-Sancho F (eds) New directions in productivity measurement and efficiency analysis: counting the environment and natural resources. Edward Elgar publishing, pp 217–232

Indian Institutes of Technology (2011) Review of wastewater reuse projects worldwide: collation of selected international case studies and experiences. GRB EMP: Ganga River Basin Environment Management Plan. Indian Institutes of Technology

Kirhenstein I, Cherrier V, Jarritt N, Farmer A, Paoli G, Delacamara G, Psomas A (2016) EU-level instruments on water reuse. Final Report to Support the Commission's Impact Assessment Luxembourg: European Commission. https://doi.org/10.2779/974903

Lazarova V, Asano T, Bahri A, Anderson J (2013) Milestones in water reuse. The best success stories. IWA Publishing, London

Menegaki AN, Mellon RC, Vrentzou A, Koumakis G, Tsagarakis KP (2009) What's in a name: framing treated wastewater as recycled water increases willingness to use and willingness to pay. J Econ Psychol 30(3):285–292

Molinos-SenanteM, Hernández-Sancho F, Sala-Garrido R (2013) Tariffs and cost recovery in water 610 reuse. J Water Resour Manag 27:1797–1808. https://doi.org/10.1007/s11269-012-0111-4

Molinos-Senante M, Hanley N, Hernández-Sancho F, Sala-Garrido R (2014) The principles of economic evaluation and cost benefit analysis implemented in sewage treatment plants. In: Stamatelatou K, Tsagarakis K (eds) Sewage treatment plants: economic evaluation of innovative technologies for energy efficiency. IWA Publishing, pp 15–32

Mudgal S, Long L, Saïdi N, Wisniewska L (2015) Optimising water reuse in the EU: Final report—Part I. Luxembourg: European Union

Raso J (2013) Updated report on wastewater reuse in the European Union. European Union; TYPSA Consulting Engineers & Architects

Rodriguez DJ, Serrano HA, Delgado A, Nolasco D, Saltiel G (2020) From waste to resource: shifting paradigms for smarter wastewater interventions in Latin America and the Caribbean. Washington, DC: World Bank. https://openknowledge.worldbank.org/handle/10986/33436

Smol M, Adam C, Preisner M (2020) Circular economy model framework in the European water and wastewater sector. J Mater Cycles Waste Manage 22:682–697. https://doi.org/10.1007/s10163-019-00960-z

Tsagarakis KP, Georgantzis N (2003) The role of information on farmers willingness to use recycled water for irrigation. Journal of Water Science and Technology: Water Supply 3(4):105–113. https://doi.org/10.2166/ws.2003.0051

World Bank (2019) From waste to resource: shifting paradigms for smarter wastewater interventions in Latin America and the Caribbean. Background paper V: financial incentives for the development of resource recovery projects in wastewater. Washington, DC. World Bank. https://openknowledge.worldbank.org/handle/10986/33384

Chapter 14
The Future of Unconventional Water Resources

Manzoor Qadir, Vladimir Smakhtin, Sasha Koo-Oshima, and Edeltraud Guenther

Abstract The water scarcity challenge continues to grow and intensify in arid and semi-arid areas. There is a need to build a diversified portfolio of water management strategies to face this challenge. With unconventional water resources as the common theme, the following strategies have the potential to help address global water scarcity: (1) promoting further research and practice on both technical and nontechnical aspects of unconventional water resources; (2) ensuring that unconventional waters provide benefits, not cost to the environment; (3) positioning unconventional waters as a reliable source of water in times of uncertainty; and (4) supporting complementary and multidimensional approaches such as addressing water scarcity and climate change together because most climate change impacts are expressed through water issues. Such a focus on unconventional water resources needs to continue and be supported by on-the-ground projects in water-critical areas to connect water experts, practitioners, young professionals, the private sector, the media, and policymakers to learn and exchange pertinent knowledge and practices.

Keywords Water augmentation · Water research · Water futures · Climate change · Sustainable development

---

M. Qadir (✉) · V. Smakhtin
United Nations University Institute for Water, Environment and Health (UNU-INWEH), 204-175 Longwood Road South, Hamilton, ON L8P 01, Canada
e-mail: Manzoor.Qadir@unu.edu

V. Smakhtin
e-mail: vladimir.smakhtin@unu.edu

S. Koo-Oshima
Land & Water Division, Food and Agriculture Organization (FAO), Viale delle Terme di Caracalla, 00153 Rome, Italy
e-mail: Sasha.Koo@fao.org

E. Guenther
United Nations University Institute for Integrated Management of Material Fluxes and of Resources (UNU-FLORES), Ammonstrasse 74., 01067 Dresden, Germany
e-mail: Guenther@unu.edu

M. Qadir et al. (eds.), *Unconventional Water Resources*,
https://doi.org/10.1007/978-3-030-90146-2_14

Water scarcity constitutes a major risk to the global economy as it presents diverse challenges in ensuring adequate supply of desirable quantity and quality of water, particularly in arid and semi-arid regions of the world (World Bank 2018). Of the water scarce population worldwide, about 90% live in developing countries (Middleton et al. 2011) where water quality deterioration is an associated challenge, which further compromises access to potable water amid a changing climate (Qadir et al. 2013).

Water scarcity is driven by a population increase, uneven distribution of water resources and population densities, industrialization, higher living standards, a dietary shift toward more animal products, and deteriorating water quality. As global water crisis looms, the achievement of Sustainable Development Goal (SDG) 6 and water-related targets embedded in other SDGs remains doubtful (United Nations 2018).

While water demand has increased over time and is expected to remain imperative in the future, conventional water resources are not sufficient to meet growing water demand in water scarce areas. Even though water-use efficiency techniques have been improved over time, they have their limits. Thus, water scarce areas must sustainably access and utilize every available quality water resource to minimize the growing pressure on scarce water resources. The good news is that recent research and practice have shown the potential of unconventional water resources to narrow the water demand–supply gap (UN-Water 2020). In addition, the recognition of unconventional water resources is growing in high-income and upper-middle-income countries. Such awareness needs to be replicated in low-income and lower-middle-income countries.

In the context of the conclusions and key recommendations stemming from the book chapters on different aspects of unconventional water resources, along with the synthesis of recent literature, there is a need to consider a range of water augmentation strategies, including but not limited to: (1) promoting further research and practice on both technical and nontechnical aspects of unconventional water resources; (2) ensuring that unconventional waters provide benefits, not cost to the environment; (3) positioning unconventional waters as a reliable source of water in times of uncertainty; and (4) supporting complementary approaches such as addressing water scarcity and climate change together because most climate change impacts are expressed through water issues.

Promoting research and practice on unconventional waters: Recent years have witnessed a surge in research and practice related to *technical* and *nontechnical* aspects of unconventional water resources. While such research and practice have demonstrated the potential that unconventional water resources can offer, some unconventional water resources need further research. For example, technical developments related to weather modification, while steadily improving, still reflect limitations in the detailed understanding of cloud dynamics and microphysics, precipitation patterns and formation, as well as limitations in accurate precipitation measurements and potential gains from cloud seeding. Further testing and evaluation of physical concepts and seeding strategies are critically important. The acceptance of weather modification can only be improved by increasing the number of well executed experiments and building the basis for scientific results that are consistent and positive (UN-Water 2020).

On another front of harvesting water from the air, there are several examples of functional systems of fog water collection, which demonstrate that such systems can increase water availability for nearby communities where this valuable resource is available. At the same time, there are also fog harvesting projects that have not been implemented successfully. The reasons reported for unsuccessful projects include a lack of community engagement and gender mainstreaming, limited funding and a lack of financial feasibility mechanism, inadequate local stakeholder management, and a limited capacity to maintain fog collection equipment. Some fog water collection projects were terminated at various implementation stages due to at least one of these factors. There is a need to address such challenges to ensure that productive and sustainable fog water collection systems are suitable for fog collection in specific areas.

There is a growing interest in recent years in (1) moving water physically via iceberg towing from Antarctica to water-scarce countries, and (2) offloading treated ballast water for potential uses based on its quality. Harnessing the potential of such water resources needs further research because there are challenges and trade-offs of long-distance iceberg towing that need to be addressed in the process of making iceberg towing a successful and sustainable strategy in the long run (Lewis 2015). In the case of ballast water, there is a need to analyze and refine ballast water treatment systems and associated infrastructure to facilitate the distribution of treated ballast water.

There are also *nontechnical challenges* beyond technology development and its implementation. Such nontechnical challenges are particularly relevant to developing countries, and they can be described as follows, but not exclusively limited to:

- Weak and fragmented institutional arrangements, limited human resources capable to tackle the complex issues arising in the process of development and maintenance of unconventional water resources (UN-Water 2020);
- Inadequate financial innovations and assessments for the comparative evaluation of the economics of 'no action' on water scarcity and 'action' in addressing water scarcity via managing unconventional waters effectively;
- Rigid and outdated policies and often insufficient and ineffective political support; and
- Lack of stakeholders' engagement while implementing pertinent projects.

The *institutional challenges* suggest that there is often considerable diversity at the federal-level and local-level institutions, as well as private sector partnerships for the development, management, and distribution of different types of water resources—in certain cases, there are multiple federal ministries or local institutions dealing with water resources management. Second, the collaboration across relevant ministries and local institutions is limited because of generally unclear and at times overlapping assignments and mandates—in certain cases, there are bureaucratic and administrative impediments in managing water resources at different scales (Wichelns and Qadir

2015). At the community level, local institutions, service providers, utilities, and associated communities are the key to implementing projects involving certain unconventional water resources to address local water shortages. For example, a major cause of underperforming or failed water harvesting, fog water collection, and greywater reuse projects is weak local institutions to support community-based collaboration (Qadir et al. 2018). To achieve successful implementation of community-led projects and their sustainability in the future, it is necessary to ensure support from the local institutions.

There is *a critical shortage of skilled human resources* to deal with the complexity of the diverse range of technological interventions and innovations in accessing and producing most unconventional water resources, such as municipal wastewater, agricultural drainage water, fog water, weather modification for cloud seeding, ballast water treatment and reuse, icebergs selection and transportation, desalinated water, and offshore deep groundwater. The assessment of critical capacity gaps is a crucial step to design and implement *need-specific capacity development* activities. In a global project addressing capacity development on the safe and productive use of wastewater in agriculture in 71 developing countries (Liebe et al. 2013), capacity gaps were identified in the following key areas: (1) economics of wastewater treatment and use of treated wastewater in agriculture; (2) environmental impact assessment of using untreated or inadequately treated wastewater; (3) health risks and their management; and (4) gender, social, and cultural aspects of wastewater management. These capacity gaps were addressed by a consortium of partners by organizing pertinent capacity development workshops. Given current technological developments and access to mobile networks, communication technologies can effectively support capacity building activities (Dodson 2014).

A major challenge in undertaking the *economic analysis* of unconventional water resources is the general opinion that their development is based on high technology costs. Such seemingly high costs without undertaking comprehensive economic analyses and innovative financing mechanisms restrict the development of certain unconventional water resources and scaling up their use (Hanjra et al. 2015). In fact, these types of economic analyses do not consider the alternate water supply options such as tankers or long-distance water transportation from wells, including the costs in the form of women's time and labor, girls missing school days, and poor health of associated communities, particularly women and girls (Qadir et al. 2018).

With the aim of recycling and reusing water resources, the *circular economy* is a path towards harnessing the potential of some unconventional water resources. For example, vast amounts of valuable energy and agricultural nutrients can be recovered from municipal wastewater during the treatment process (Qadir et al. 2020). Brine generated from desalination plants can be used as a source of valuable minerals and rare-earth elements such as lithium, strontium, thorium, and rubidium. These metals are used to fabricate critical components of numerous products, including airplanes, automobiles, smart phones, and biomedical devices. There is a growing realization that with the development and implementation of clean-energy technologies and sustainable products, processing and manufacturing industries will also require large amounts of rare metals and other valuable elements (International Desalination

Association 2019). In such a scenario, recovery of precious metals from brine would offer an additional economic opportunity, while ensuring that the post-recovery brine would be managed in environmentally acceptable protocols.

There is a need to undertake *comprehensive economic assessments* in given settings of the costs of ignoring needed investments and benefits by introducing innovative financial mechanisms to support and prioritize various types of unconventional water resources. This can be achieved by identifying the full range of potential benefits associated with specific unconventional waters by using approaches that should be credible. The valuation of the benefits of 'action' or, alternatively the valuation of the costs of 'no action', is necessary to make a case for the needed investments in harnessing the potential of unconventional water resources to address water scarcity (Hernández-Sancho et al. 2015).

Pertinent policies and governance structures for unconventional water resources may vary across regions and countries. For example, there are large differences between developed and developing countries about policy issues related to wastewater management. In developed countries, most wastewater is treated and used for irrigation in treated form. The guidelines on safely managed wastewater are in place. Policy issues apply largely to financial, economic, and environmental factors of wastewater treatment systems. Public officials and water management agencies motivate greater use of treated wastewater by providing financial incentives and increasing public awareness of the safety and benefits of using wastewater on farms, golf courses, and urban landscapes. Although the policy issues in developing countries do address financial and economic aspects, while considering investments in wastewater management, such investments are not enough and at times delayed in achieving targeted treatment of wastewater to meet the desired quality (Wichelns and Qadir 2015). Thus, treatment of wastewater remains limited in developing countries because investments in treatment facilities have not kept pace with persistent increases in population and the consequent increases in wastewater volumes. As a result, much of the wastewater is not treated, and untreated and inadequately treated wastewater is largely used for irrigation by small holders in informal settings with little ability to optimize the volume or quality of the wastewater they receive (Drechsel et al. 2015).

Despite the growing importance of fog water collection in dry areas, there is a *lack of national water policies, economic incentive mechanisms, and action plans* that consider fog water collection as a means of addressing local water shortages in areas where there is abundant fog. As water policies and action plans do not place atmospheric moisture harvesting on the public policy agenda, the uptake of the potential of fog water collection systems is hampered and likewise the associated benefits (Klemm et al. 2012).

Given the fact that certain unconventional water resources have only begun to be recognized as significant contributors to alleviating water scarcity, there is limited guidance on developing *governance aspects* for these resources. Such aspects should stem from the objective of providing water of desired quantity and quality for current needs, while minimizing the impact on the people, ecosystems, and transboundary areas, inclusive of incentive mechanisms. In addition, *flexible policy frameworks* are the key to ensure successful implementation of research-based technical and nontechnical interventions in support of unconventional water resources in the overall water resources management planning and implementation strategies.

The *involvement of relevant multi-stakeholders* has emerged as a major factor in the ability of governments to successfully address and overcome challenges associated with water management policies and projects (OECD 2015). In the case of unconventional water resources, greater acceptance, trust, and ownership of water augmentation projects are crucial in engaging relevant stakeholders effectively to ensure economic, environmental, health, and social benefits and their sustainability in the long run.

To achieve successful and sustainable implementation of community-based projects, it is necessary to integrate local institutions and associated communities as stakeholders to promote their involvement and commitment (UN-Water 2020). Based on a multi-stakeholder policy dialogue and a comprehensive analysis of 69 case studies in the water sector worldwide, OECD (2015) has outlined the following principles for creating the conditions required to ensure *participatory and adaptive stakeholder engagement* by (1) mapping all potentially beneficial and negatively affected stakeholders along with their expected motivations, skepticisms, and interactions; (2) defining the ultimate line of decision making, the objectives of stakeholder engagement, and the expected use of inputs; (3) allocating adequate financial and human resources and sharing needed information for result-oriented stakeholder engagement; (4) assessing regularly the process and outcomes of stakeholder engagement to learn, adjust, and improve accordingly; (5) embedding engagement processes in clear legal and policy frameworks, organizational principles, and with responsible authorities; and (6) customizing the type and level of engagement to the needs and keeping the process flexible to meet changing circumstances.

In the water-sector projects, engaging relevant stakeholders in the early stages of decision-making is critical to secure support for reforms, to raise awareness about water risks and costs, to increase water users' willingness to pay, and to address conflicts. In this regard, public institutions can play a key role in ensuring more bottom-up decision-making processes. Given the importance of unconventional water resources and the need for financial resources to harness their potential, involvement of the *private sector* needs to be encouraged to invest in projects involving unconventional water resources.

The integration of multiple scientific communities in addressing the research questions is critical to promote and make progress with projects addressing the various aspects of unconventional water resources. In this regard, environmental processes can inspire disruptive innovations (the natural sciences perspective), ecological theories can provide an impetus for innovative water quality protection and wastewater treatment processes (the engineering perspective), antimicrobial resistance and micro-pollutants can be addressed as global health challenges (the medical perspective), water-supply security may be addressed in terms of water quality and quantity, but also water-related risks and their effects on human health, ecosystems, and economic developments along with individual behavior in households and companies, but also in the political and societal arenas (the social and behavioral sciences perspective).

Ensuring unconventional waters provide service, not cost to the environment: There are *contrasting environmental trade-offs* related to the processes of developing unconventional water resources. While desalination provides a valuable and reliable source of water, it also generates hypersaline brine, which poses an environmental challenge if not managed adequately. Substantial efforts, innovation, and research are currently invested to: (1) reduce the volume of brine being produced by increasing the efficiency of the desalination process; and (2) treat and use the produced brine in economically viable and environmentally friendly ways (Jones et al. 2019). In recent years, desalination science has developed several brine-concentration and mineral-extraction technologies that enable the creation of commercially viable products (International Desalination Association 2019). Extracting minerals from seawater is a more environmentally friendly enterprise than terrestrial mining. Moreover, seawater extraction does not require freshwater for processing. There are developments in new brine-concentration technologies that may result in significant reductions of brine discharge into the sea (UN-Water 2020). Over the past five years, many countries with large desalination plants have initiated the implementation of comprehensive programs for green desalination, aiming at reducing both the amount and types of chemicals used in the production of desalinated water. These initiatives aim to gradually convert all existing desalination facilities to low-input chemical plants by benefitting from the latest advances in desalination science and technology (International Desalination Association 2019).

In contrast to the production process for desalinated water, which needs safe disposal of brine to ensure environmental compliance, the production process of treated wastewater minimizes the discharge of untreated wastewater into the environment and contributes to reducing the pollution of other water bodies. In addition, wastewater treatment provides environmental benefits such as minimizing eutrophication—the phenomenon of excess nutrients in a body of water causing dense plant growth and aquatic animal deaths due to a lack of oxygen (Qadir et al. 2020). Depending upon the levels of contaminants present in wastewater, continued and uncontrolled irrigation with untreated or inadequately treated wastewater may result in groundwater contamination, through the movement of a wide range of chemical pollutants, such as nitrates and specific metals and metalloids, to groundwater and their gradual build-up (Ensink et al. 2002). Such accumulation in wastewater-irrigated soils may lead to potentially harmful metals and metalloids reaching phytotoxic levels and entering the food chain, affecting human and animal health.

As large-scale offshore freshwater development has not taken place at any field site yet, long-term pumping of offshore freshwater may result in reversals in groundwater flow directions offshore that could impact the health of the benthic community. There could be possible onshore land subsidence and seawater intrusion from the extraction of offshore groundwater. Thus, it is crucial to assess social and economic considerations in making decisions related to developing offshore freshwater (Yu and

Michael 2019). Besides, developing offshore freshwater may impact the marine environment by damaging flora and fauna in the surrounding seabed and water column (Martin-Nagle 2020). Despite the risks during times of drought, utilization of offshore freshwater by coastal megacities may represent an important water source for coastal residents.

Positioning unconventional waters as a reliable source in times of uncertainty: Despite increasing water scarcity and deteriorating water quality in water-scarce areas, the water sector is making gradual progress towards cost-effective and sustainable water management solutions that are expected to transform water resources management beyond conventional sources, while tapping into unconventional water resources (UN-Water 2020). Another notable trend seen is water professionals considering the critical role of unconventional water resources in building a possible water-secure future in which such waters are recognized as a precious and reliable resource in times of uncertainty.

Because the COVID-19 pandemic has potentially delayed the achievement of the 2030 Sustainable Development Agenda, the achievement of SDG 6 in ensuring water and sanitation for all has suffered similarly, although the world was already off track with SDG 6 achievement before the pandemic. With all its detrimental impacts, however, the COVID-19 pandemic has created a unique opportunity to rethink global development challenges and risks, including the water crisis. As the emergence of new infectious diseases is likely to increase in the coming years and possibly decades due to disruptive factors, such as uncontrolled human action impacting ecosystems, there would be a need to ensure an adequate and reliable supply of water as the world gets used to living with pandemics. Unconventional water resources can play a role in such situations. For example, several studies have demonstrated COVID-19 detection in wastewater, which can capture the rise and fall of novel coronavirus cases in a region. Such detection could be used as an early warning system to take timely actions against the virus outbreak (Randazzo et al. 2020; Larsen and Wigginton 2020). As half of global wastewater produced is released to the environment untreated (Jones et al. 2021), there is limited scope to detect COVID-19-like viruses in such large volumes of wastewater, which are released mostly in developing countries. Thus, there is a need for rethinking and implementing a paradigm shift to promote safely managed wastewater where (1) wastewater is considered as a valuable resource and its potential is harnessed rather than constituting only a waste stream with the challenges and obligations on the disposal of such waste streams; and (2) wastewater is considered as an early warning system and a vehicle for COVID-19 detection.

Other sources of unconventional waters, such as desalinated water, can provide reliable potable water supplies required for handwashing facilities to minimize pandemic risks while supporting the achievement of other SDGs such as SDG 3 on ensuring healthy lives and promoting human well-being. Community-based unconventional water resources such as fog water and harvested rainwater can also contribute to water resources supplies in critical times, particularly in water-scarce areas where the alternate sources of water supply are tankers, which may face substantial delays or provide even no service in times of restricted mobility during pandemic-like situations.

Supporting complementary and multidimensional approaches: The world at large is responding to the global water crisis and climate change risks in the 2030 Sustainable Development Agenda with two specific goals—SDG 6 on ensuring water and sanitation for all and SDG 13 on taking urgent actions to combat climate change. As certain climate-change impacts are expressed through water, climate change and water are closely interconnected. Climate change increases the likelihood of extreme droughts in arid and semi-arid areas, which are often home to several transboundary river basins (UN-Water 2020). In addition, there is increased run-off in certain areas with more intense precipitation leading to higher levels of pollution washed into waterways. Despite such interconnectivity, the water crisis and climate change are at times addressed in silos, which turn into a major roadblock in the journey to achieve water-related sustainable development amid changing climate in the SDG era and beyond.

Harnessing the potential of unconventional water resources and integrating such potential into water resources management strategies and plans at the transboundary, national, and local levels can go beyond narrowing the water demand-supply gap by (1) diversifying water supply resources, and (2) developing the resilience of water-scarce communities against climate change (UN-Water 2020). The examples from both developed and developing countries reveal that, given the supporting policies and political will, certain types of unconventional water resources can effectively be used for aquifer recharge as well as ecosystem services for environmental protection, water quality improvement, sustainable development, and human well-being amid challenges triggered by climate change and deteriorating water quality. For example, treated wastewater stored through aquifer recharge can provide a reliable supply of water during drought periods and times of interseasonal and interannual water shortages, reverse falling groundwater levels, and reduce water losses associated with leakage and evaporation. Ecosystem services generated by treated wastewater can support directly or indirectly human well-being by avoiding or minimizing water pollution, restricting overexploitation of groundwater, and recycling and reusing essential nutrients and water (Drechsel et al. 2015). Desalination of seawater or highly brackish water is an important water augmentation opportunity, extending water supplies beyond what is available from the hydrological cycle, providing a climate-independent and steady supply of high-quality water even during times of extreme droughts (Jones et al. 2019).

References

Dodson LL (2014) A foggy desert: equitable information flow for a fog water system in southwest Morocco. Ph.D. dissertation. University of Colorado Boulder

Drechsel P, Qadir M, Wichelns D (2015) Wastewater: economic asset in an urbanizing world. Springer Science+Business Media Dordrecht, The Netherlands, p 282

Ensink JH, Van Der Hoek W, Matsuno Y, Munir S, Aslam MR (2002) Use of untreated wastewater in peri-urban agriculture in Pakistan: risks and opportunities. Research report 64. Colombo, Sri Lanka: International Water Management Institute

International Desalination Association (2019) Sustainable management of desalination plant concentrate—desalination industry position paper—energy and environment committee of the IDA. In: Proceedings of IDA world congress, Dubai, October 2019, ref: IDAWC19-EEC. https://idadesal.org/wp-content/uploads/2019/10/EEC-White-Paper-10-13.pdf

Hanjra MA, Drechsel P, Wichelns D, Qadir M (2015) Transforming urban wastewater into an economic asset: opportunities and challenges. In: Drechsel P, Qadir M, Wichelns D (eds) Wastewater: economic asset in an urbanizing world. Springer Science+Business Media Dordrecht, The Netherland, p 271–278

Hernández-Sancho F, Lamizana-Diallo B, Mateo-Sagasta J, Qadir M (2015) Economic valuation of wastewater: the cost of action and the cost of no action. Division of Environmental Policy Implementation, United Nations Environment Programme, Nairobi, Kenya, p 68

Jones E, Qadir M, van Vliet MTH, Smakhtin V, Kang S (2019) The state of desalination and brine production: a global outlook. Sci of the Total Environ 657:1343–1356. https://doi.org/10.1016/j.scitotenv.2018.12.076

Jones ER, van Vliet MTH, Qadir M, Bierkens MFP (2021) Country-level and gridded estimates of global wastewater production, collection, treatment, and re-use. Earth Syst Sci Data 13:237–254

Klemm O, Schemenauer RS, Lummerich A, Cereceda P, Marzol V, Corell D, Fessehaye GM (2012) Fog as a fresh-water resource: overview and perspectives. Ambio 41:221–234. https://doi.org/10.1007/s13280-012-0247-8

Larsen DA, Wigginton KR (2020) Tracking COVID-19 with wastewater. Nat Biotechnol 38:1151–1153. https://doi.org/10.1038/s41587-020-0690-1

Liebe J, Ardakanian R, Mateo-Sagasta J, Medlicott K, Qadir M, Raschid-Sally L, Drechsel P (2013) Proceedings of the UN-water project on the safe use of wastewater in agriculture. UN-water decade programme on capacity development (UNW-DPC), Bonn, Germany. http://www.ais.unwater.org/ais/pluginfile.php/62/course/section/29/proceedings-no-11_WEB.pdf

Lewis CJ (2015) Iceberg harvesting: suggesting a federal regulatory regime for a new freshwater source. Boston CollE Environ Aff Law Rev 42:439–472. http://lawdigitalcommons.bc.edu/ealr/vol42/iss2/6

Martin-Nagle R (2020) Governance of offshore freshwater resources. Brill|Nijhoff. https://doi.org/10.1163/9789004421042

Middleton N, Stringer LC, Goudie A, Thomas DSG (2011) The forgotten billion: MDG achievement in the drylands. United Nations Development Programme (UNDP) and United Nations Convention to Combat Desertification (UNCCD). https://catalogue.unccd.int/39_Forgotten_Billion.pdf

OECD (2015) Stakeholder engagement for inclusive water governance, OECD studies on water. OECD Publishing, Paris. https://doi.org/10.1787/9789264231122-en

Qadir M, Noble AD, Chartres C (2013) Adapting to climate change through improving water productivity of soils in dry areas. Land Degrad Dev 24:12–21

Qadir MGC, Jiménez RL, Farnum LLD, Smakhtin V (2018) Fog water collection: challenges beyond technology. Water 10(4):372

Qadir M, Drechsel P, Cisneros BJ, Kim Y, Pramanik A, Mehta P, Olaniyan O (2020) The global and regional potential of wastewater as a water, nutrient, and energy source. Nat Res Forum 44(1):40–51. https://doi.org/10.1111/1477-8947.12187

Randazzo W, Cuevas-Ferrando E, Sanjuán R, Domingo-Calap P, Sánchez G (2020) Metropolitan wastewater analysis for COVID-19 epidemiological surveillance. Int J Hyg Environ Health 230:113621. https://doi.org/10.1016/j.ijheh.2020.113621

United Nations (2018) Sustainable development goal 6 synthesis report 2018 on water and sanitation. New York, UN. https://www.unwater.org/publications/sdg-6-synthesis-report-2018-on-water-and-sanitation/

UN-Water (2020) UN-water analytical brief on unconventional water resources. Geneva, Switzerland. https://www.unwater.org/publications/un-water-analytical-brief-on-unconventional-water-resources/#:~:text=At%20the%20same%20time%20water,change%20adaption%2C%20and%20sustainable%20development

Wichelns D, Qadir M (2015) Achieving sustainable irrigation requires effective management of salts, soil salinity, and shallow groundwater. Agric Water Manag 157:31–38

World Bank (2018) Beyond scarcity: water security in the middle east and North Africa. MENA Development Report. The World Bank: Washington, DC. https://openknowledge.worldbank.org/handle/10986/27659

Yu X, Michael HA (2019) Offshore pumping impacts onshore groundwater resources and land subsidence. Geophys Res Lett 46(5):2553–2562. https://doi.org/10.1029/2019gl081910

The manufacturer's authorised representative in the EU is Springer Nature Customer Service Centre GmbH, Europaplatz 3, 69115 Heidelberg, Germany. If you have any concerns regarding our products, please contact ProductSafety@springernature.com

Printed and bound by CPI Group (UK) Ltd, Croydon, CR0 4YY
30/09/2026
02225491-0008